Reinsurance: Actuarial and Statistical Aspects

WILEY SERIES IN PROBABILITY AND STATISTICS

Established by *Walter A. Shewhart and Samuel S. Wilks*

Editors: *David J. Balding, Noel A. C. Cressie, Garrett M. Fitzmaurice, Geof H. Givens, Harvey Goldstein, Geert Molenberghs, David W. Scott, Adrian F. M. Smith, Ruey S. Tsay*

Editors Emeriti: *J. Stuart Hunter, Iain M. Johnstone, Joseph B. Kadane, Jozef L. Teugels*

The ***Wiley Series in Probability and Statistics*** is well established and authoritative. It covers many topics of current research interest in both pure and applied statistics and probability theory. Written by leading statisticians and institutions, the titles span both state-of-the-art developments in the field and classical methods.

Reflecting the wide range of current research in statistics, the series encompasses applied, methodological and theoretical statistics, ranging from applications and new techniques made possible by advances in computerized practice to rigorous treatment of theoretical approaches. This series provides essential and invaluable reading for all statisticians, whether in academia, industry, government, or research.

A complete list of titles in this series can be found at
`http://www.wiley.com/go/wsps`

Reinsurance: Actuarial and Statistical Aspects

Hansjörg Albrecher
University of Lausanne
Switzerland

Jan Beirlant
KU Leuven, Belgium and University of the Free State,
South Africa

Jozef L. Teugels
KU Leuven
Belgium

WILEY

This edition first published 2017

Registered Offices
John Wiley & Sons, Inc., 111 River Street, Hoboken, NJ 07030, USA
John Wiley & Sons Ltd, The Atrium, Southern Gate, Chichester, West Sussex, PO19 8SQ, UK

Editorial Office
9600 Garsington Road, Oxford, OX4 2DQ, UK

For details of our global editorial offices, customer services, and more information about Wiley products visit us at www.wiley.com.

Wiley also publishes its books in a variety of electronic formats and by print-on-demand. Some content that appears in standard print versions of this book may not be available in other formats.

Library of Congress Cataloging-in-Publication Data

Names: Albrecher, Hansjörg, author. | Beirlant, Jan, author. | Teugels, Jozef L., author.
Title: Reinsurance : actuarial and statistical aspects / by Hansjörg Albrecher, University of Lausanne, Switzerland; Jan Beirlant, Katholieke Universiteit Leuven, BE, University of the Free State, South Africa; Jozef L. Teugels, Katholieke Universiteit Leuven, BE.
Description: Hoboken, NJ : John Wiley & Sons, [2017] | Series: Statistics in practice | Includes bibliographical references and index. |
Identifiers: LCCN 2017017184 (print) | LCCN 2017033182 (ebook) | ISBN 9781119419938 (pdf) | ISBN 9781119419945 (epub) | ISBN 9780470772683 (cloth)
Subjects: LCSH: Reinsurance. | Actuarial science.
Classification: LCC HG8083 (ebook) | LCC HG8083 .A43 2017 (print) | DDC 368/.0122–dc23
LC record available at https://lccn.loc.gov/2017017184

Cover image: Comparison of reinsured claim amounts for some combinations of claim sizes *Xi* and corresponding insured sums *Qi* for QS reinsurance with *L* xs *M* reinsurance
Cover design by Wiley

Set in 10/12pt Warnock by SPi Global, Pondicherry, India

Printed in the UK

Contents

Preface

Reinsurance is a fascinating field. Several of the challenges of classical insurance are amplified for reinsurance, particularly when it comes to dealing with extreme situations like large claims and rare events. This poses particular challenges for the modelling of claims and their occurrence, which often needs to be based on only few data points. In addition, in terms of better diversification of usual-scale risk on the local and global level as well as in terms of the development of innovative and sustainable techniques to deal with risks of an unusual kind, reinsurers play a crucial role in the insurance process.

This also reflects on practitioners and researchers involved in such topics, as they have to rethink classical models in order to cope successfully with the respective challenges. Over the years, there has been enormous research activity on problems connected to reinsurance. Close to 40% of the references in our literature list have appeared over the last 10 years, with a steep upward gradient over the last 3–4 years. While there exist some excellent classical textbooks on reinsurance either from the academic or the practitioner's side, our impression was that there was no modern reference book available that gave an overview of the academic research landscape in this field and also puts it in perspective with the practical viewpoint. The main reason for writing this book was to try to address this gap, at least for actuarial and statistical matters. As all the authors are from academia, there naturally remains a bias towards the academic angle. However, numerous and enlightening discussions with insurance and reinsurance practitioners over the last few years have motivated us to produce the current account, hoping and trying to further bridge the two worlds. The focus of the book is on modelling together with the statistical challenges that go along with it. We illustrate the discussed statistical approaches alongside six case studies of insurance loss data sets, ranging from MTPL over fire to storm and flood loss data. Some of the presented material also contains new results that have not yet been published in the research literature. We hope that the material presented can trigger new research questions and foster the communication between (re)insurance practitioners and academics working in these fields. One of our main goals was to give an up-to-date overview of the relevant research literature and to frame it to questions that matter in reinsurance practice. Since this a vast topic, we naturally had to take various compromises and we apologize for possible omissions on either side.

The book is written for researchers with an interest in reinsurance problems, for graduate students with a basic knowledge of probability and statistics as well as for practitioners in the field.

We start with a general introduction to the field in Chapter 1, presenting some basic facts and motivations for reinsurance activities. We also introduce the six real-life case studies that will accompany the considerations throughout the book. In Chapter 2, we discuss the most common reinsurance forms and their properties, together with some practical aspects of their implementation. Chapter 3 is dedicated to motivating and developing models for claim size distributions that are commonly used. Here we emphasize those aspects from actuarial mathematics that are relevant for reinsurance. Reinsurance is often invoked in the presence of large claims, therefore we need a thorough discussion of models capable of catching the essentials of what actuaries would call *large*. Chapter 4 contains detailed guidelines on how to proceed in the model choice when actually facing data. Throughout the text, we illustrate the presented procedures for our case studies. Chapter 5 proceeds with models for claim numbers, both from a conceptual and a practical viewpoint. We also provide guidelines for a statistical analysis of data sets in this context. The two ingredients (claim numbers and claim sizes) are then used in Chapter 6 for the aggregation of the claims. Emphasis is put on the aggregation of independent risks, and we describe both numerical and asymptotic methods in detail. The case of dependence in the aggregation process is also discussed briefly, although not in detail, as the results typically are very sensitive to the particular dependence structure used in the modeling process, and often the number of data points does not allow one such model to be decisively favored over another. It is beyond the scope of this book to discuss all such approaches. Chapter 7 treats important actuarial aspects of reinsurance pricing, once a distribution for the individual (or aggregate) risk is available (or, rather, decided upon). In Chapter 8 we discuss some guidelines on possible criteria for the choice of reinsurance forms and the respective consequences on the optimal choice of contracts. The identification of optimal reinsurance forms has been a very active research field recently and it is impossible to reflect all these contributions in one book chapter in an exhaustive way. We instead provide an overview of some of the main approaches and contributions alongside a structure in terms of decision criteria, with an emphasis on the intuition behind the results. Since stochastic simulation is an essential tool in many models relevant in reinsurance, we cover this topic in Chapter 9 and discuss some variance reduction techniques that can help to considerably speed up calculations. Chapter 10 then examines some further topics. We first provide more information on large claim analysis, and continue with an overview of alternative risk transfer products, which can serve as a complement to traditional reinsurance. We also highlight the role of finance in reinsurance and finish with a section on catastrophe insurance. Within the chapters and in particular at the end of the chapters we provide links for further reading.

Many of the topics dealt with in the book apply to both non-life and life insurance. Even when there is a clear emphasis on non-life insurance throughout, we hope that some of our attempts may help to also be of service to life insurance. As the title suggests, this book is about (traditional) actuarial as well as statistical aspects arising in reinsurance. As is outlined in Chapter 1, reinsurance also serves financial and management purposes in practice. Correspondingly, the role of capital is nowadays an important ingredient in managing and steering reinsurance companies, and financial pricing techniques for reinsurance contracts as well as general capital management tools eventually have to complement the actuarial approach. While we do consider such aspects when discussing the pricing and the possible choice of contracts in

Chapters 7, 8 and 10, it is beyond the scope of this book to treat and reflect the merging of actuarial and financial principles in the amount of detail this may deserve from a general perspective.

The idea for writing this book was born in the legendary and productive environment of EURANDOM, Eindhoven. We would like to thank this institution for its continuing support over the years as well as the University of Lausanne and KU Leuven for generous support for extended research visits that enabled the book to progress. We also thank Sophie Ladoucette, MunichRe, and the Versicherungsverband Österreich for providing data for our case studies.

We would like to thank all the people with whom we had interesting discussions about the topic over the recent years, including the participants of the Summer School of the Swiss Actuarial Association in Lausanne in 2015, as well as short course participants in Paris, Johannesburg, Lisbon, Lyon, Luminy, Yerevan, Warsaw, and Hong Kong.

Particular thanks for stimulating discussions or advice in earlier and later stages of the book writing go to Jose Carlos Araujo Acuna, Katrien Antonio, Peiman Asadi, Alexandru Asimit, Anastasios Bardoutsos, Arian Cani, Michel Dacorogna, Dalit Daily-Amir, Michel Denuit, François Dufresne, John Einmahl, Karl-Theodor Eisele, Michael Fackler, Damir Filipović, Hans U. Gerber, Alois Gisler, William Guevara-Alarcon, Jürgen Hartinger, Christian Hipp, Frans Koning, Yuriy Krvavych, Sandra Kurmann, Sophie Ladoucette, Stéphane Loisel, Franz Prettenthaler, Christian Y. Robert, Robert Schall, Matthias Scherer, Thorsten Schmidt, Wim Schoutens, Johan Segers, Wim Senden, Stefan Thonhauser, Joël Wagner, Roel Verbelen, Robert Verlaak, Leonard Vincent, Jean-François Walhin, Gord Willmot, and Mario Wüthrich. Special thanks go to Tom Reynkens for his tremendous effort writing an R package with this book and producing the plots linked with the statistical procedures. Further thanks go to William Guevara-Alarcon and Dominik Kortschak for help with the R codes underlying the illustrations in Chapter 9, and to Roel Verbelen and Tom Reynkens for their significant contribution to the splicing methods. We will maintain a webpage connected to the book at

`http://www.hec.unil.ch/halbrech_files/reinsurance.html`

where we also intend to keep a list of misprints and remarks. We are grateful to receive relevant material sent to us by email. The R package ReIns can be found at the CRAN page

`cran.r-project.org/package=ReIns`

Hansjörg Albrecher, Jan Beirlant, and Jozef L. Teugels
Lausanne and Leuven,
December 2016

1
Introduction

1.1 What is Reinsurance?

A *reinsurance contract* is an agreement in which one party (the *reinsurer*) agrees to indemnify another party (the *reinsured*, the *first-line insurer* or also the *ceding company, cedent*) for specified parts of its underwritten insurance risk. In turn, the cedent pays to the reinsurer a *reinsurance premium* for this service. That is, in reinsurance the principle of insurance is lifted up one level, so an insurance company seeks itself the possibility of replacing parts of its future loss by a fixed premium payment (much like a policyholder does when entering an insurance contract). There are many reasons why such a risk transfer from the insurer to the reinsurer can be desirable for both parties, as well as for the economy in general, and we will outline a number of them in Section 1.2.

While reinsurance can be seen as a particular form of insurance, and naturally shares various common features with it, reinsurance is also quite distinct from primary insurance in a number of aspects. These include the type and magnitude of risks under consideration, the type of data available for the risk analysis, the diversification possibilities, demand/supply peculiarities of contracts quite different from the primary insurance market, and also the fact that reinsurance is a form of risk sharing among two "professional" insurance entities, so that the necessary guidelines for regulation can be quite different.

(Non-life) reinsurance contracts are typically written for one year, and one distinguishes between *obligatory treaties*, where a binding agreement is specified that applies to all risks of a specified risk class, and *facultative* arrangements, which are negotiated on each individual risk. A *facultative treaty* is then a contract where the cedent has the option to cede and the reinsurer has the option to decline or accept classified risks of a particular business line. In practice many contracts actually involve several reinsurers (e.g., the contract is negotiated with a primary reinsurer, and other reinsurers then participate proportionally in the reinsurance coverage, or a second reinsurance contract with another reinsurer is written for parts of the remaining risk of the cedent after a first contract). The relationship between insurer and reinsurer is often of a long-term nature, which also has an effect on the way reinsurance premiums are negotiated. Finally, there is no relation between a reinsurer and the individual policyholders of the

Reinsurance: Actuarial and Statistical Aspects, First Edition.
Hansjörg Albrecher, Jan Beirlant and Jozef L. Teugels.

Table 1.1 Global premium volume 2015 (in US$ billions).

	Primary insurance	Reinsurance
Life and health	2500	65
Non-life	2000	170

Source: SwissRe.

underlying risks. A reinsurer may itself enter a reinsurance contract with another insurance company on parts of the reinsured risk, and such a procedure is called *retrocession*.

Table 1.1 gives a feeling for the size of the global reinsurance market in comparison to the primary insurance market. One sees that in terms of premium volume, reinsurance is only employed for a small fraction of the primary insurance risk. However, typically the reinsured risk is the one that is complicated to assess and handle (this is one of the main reasons why it is reinsured!), which makes this type of risk particularly challenging for actuaries, statisticians, and other risk professionals. Worldwide, there are about 200 reinsurance companies today, and many of these are also acting as primary insurers in the market.

1.2 Why Reinsurance?

Let us look at why an insurance company is interested in buying reinsurance. The main function of insurance companies is to take risk. This is similar to the business model of other financial organizations, and both types leverage the capital provided by shareholders through raising debt. However, insurers raise debt by selling policies to insureds, which makes the debt very risky (due to uncertainty around the timing and severity of claims), whereas financial debt would typically rather have pre-determined expiry and face value (severity). This leveraging activity is a competitive advantage, but also makes the companies vulnerable to distress and insolvency, creating the demand for risk management. Among the available risk management tools, *risk transfer* through reinsurance then plays an important role in improving the company's overall risk profile. Let us look at some of the main motivations for the insurer to buy reinsurance as a means of risk transfer (several of which are not independent of each other):

- **Reducing the probability to suffer losses that are hard to digest**
  This is a rather general statement and many of the items below are in fact refinements of this criterion. It should be kept in mind that for an insurance company buying reinsurance means passing on some of its insurance business (i.e., its core activity), and hence typically the goal is to keep the reinsured part small. However, reduction of risk exposure can be desirable or necessary for the reasons outlined below.
- **Stabilizing business results**
  Entering a reinsurance contract reduces the volatility of the cedent's financial result, as random losses are replaced by a (typically deterministic) premium payment. That is, reinsurance can be a means to steer the volatility of an insurance company towards a

desired level, and the latter can have particular advantages (e.g., with respect to taxes, capital requirements and market expectations).

- **Reducing required capital**
  Reducing the aggregate risk will reduce the required capital to bear such risks, and in view of capital costs this may be desirable. Concretely, if the reinsurance premium (together with the administration costs) is smaller than the gain resulting from the corresponding reduction of capital, the reinsurance contract is desirable. In fact, due to the ongoing shift towards risk-based regulation, the notion of capital and its management becomes a central issue for insurance companies, and reinsurance then should be understood as a tool in this context. This corresponds to an important *finance function* of reinsurance as a substitute for capital, freeing up capacity.
- **Increasing underwriting capacity**
  In the presence of a reinsurance contract, only a certain part of the risk is assumed by the insurer, and hence under otherwise identical conditions an insurance company can afford to underwrite more and larger policies (see Chapter 2 for details), which may be desirable for various reasons, including market share targets, testing and entering of new markets, gaining (data) experience in certain business lines or regions etc. It also can lead to enhanced liquidity.
- **Accessing benefits from larger diversification pools**
  Often the primary insurers' business model is restricted to a local area, in which case attempts to look on their own for diversification possibilities outside of that market for the more dangerous part of the risks would be very costly and inefficient. Reinsurers, on the other hand, typically act on an international level and therefore have more possibilities for diversifying such risks. Consequently the amount of capital needed to safeguard these risks in the portfolio can be considerably lower for a reinsurer and so the risk transfer produces economic gain through attractive reinsurance premiums.

We mention a few further motivations:

- **Reducing tax payments**
  Equalization reserves (i.e., reserves for volatility of claims and their arrivals over longer time periods, which is, for instance, particularly important for catastrophe risks) of insurance companies are taxed in most legislations. If such reserves are paid to a reinsurance company in the form of a reinsurance premium (or, alternatively, into a respectively created captive structure, cf. Section 10.2), then the taxation pattern becomes more favorable, as for reinsurers and captives (often located in tax-favorable countries) different tax rules may apply.
- **Other legal issues**
  Reinsurance can be a helpful tool to resolve legal constraints such as regulatory compliance. For instance, if an insurance company does not have a formal license to write business in a certain country, a solution can be to find a local insurer with such a license and act as a reinsurer for this local company.
- **Financial solutions**
  The reinsurer can serve as a facilitator for financial solutions. Examples include reducing (expected) financial distress costs by providing run-off solutions (cf. Section 10.2) and portfolio transfers to other companies or the capital markets as well as setting up securitization transactions like issuing bonds.

- **Protection against model risk**
  Insurance activities are designed on the basis of stochastic models for the underlying risks. For the aggregate performance, both the understanding of the marginal risks as well as of the dependence between them is important.[1] However, every model is an imperfect description of reality, and the less experience and data one has, the higher the uncertainty about whether the model underlying the business plan is appropriate. Reinsurance is a way to mitigate model inadequacy (e.g., concerning the tails of the risks or their dependence).
- **Support in risk assessment, pricing, and management**
  In certain situations an insurance company does not have enough data points or manpower available to analyze the risks (in particular their tails), and passing on those risks to an entity with respective experience is a natural procedure, which is often much cheaper than dealing with such risks by other means. This also includes business expansions to new regions or business lines, in which the reinsurer may already have experience from earlier activities. In fact, reinsurance contracts often have a certain consultancy component, as the reinsurer may share its expertise and data on the respective risks with the cedent.

On the society level, reinsurance allows insurers to write more business, which makes insurance more broadly available and affordable. This can foster economic growth and increase stability at large. Reinsurance enables risks to be insured that otherwise would not be insurable, and assigning premiums to (i.e., quantifying) risks can also provide incentives for more risk-adequate behavior and possibly risk prevention.

For all these reasons, reinsurance serves as a tool to increase the efficiency of the marketplace. When designing reinsurance contracts, all these aspects will play some role. The goal of this book is to focus on the actuarial elements involved in the process as well as the statistical challenges that appear in this context.

1.3 Reinsurance Data

As for primary insurance, in the reinsurance business one will be interested in the statistical analysis of claim information for different types of business lines (such as car liability insurance or fire insurance), where one can obtain claim information on the individual claim level. Due to the nature of the reinsurance contract, there are, however, additional challenges with respect to the type of claim data.

Consider, for instance, the case of non-proportional reinsurance where the reinsurer will pay (parts of) the excesses over some threshold, say M. The ceding company then does not need to provide all claim information to the reinsurer. For example, information may be provided on those events only for which the incurred claim amount I (i.e., the estimate of the amount of outstanding liabilities) is larger than a certain percentage of M. Then, as long as I stays below that reporting threshold during the development process, the claim will not be known to the reinsurer and hence the

---

1 Here, dependence can be causal (e.g., the occurrence of a claim triggers another claim) or due to common risk drivers. An appropriate modelling of dependence can be a considerable challenge, particularly when only few data points are available and the number of dependent risks is high.

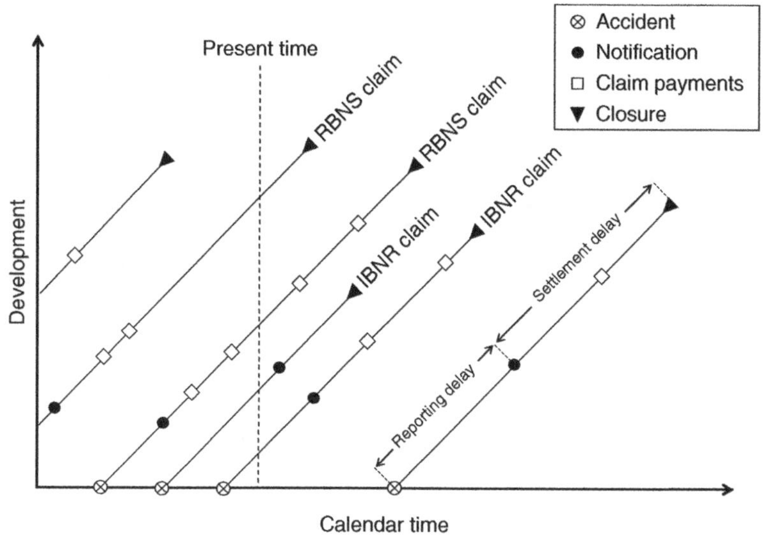

Figure 1.1 Claim development scheme.

incurred value is *left truncated* in such a case. For some lines of business, development times can be quite large (up to several decades) so that, at the time of evaluation, the cumulative payments are still a lower bound for the ultimate claim amount. The data then are *censored.* In practice, companies use claim development methods to forecast the ultimate claim amounts. Of course these also yield uncertain information, which hampers the statistical analysis. Hence, in reinsurance we face incomplete information, due to incurred but not reported (IBNR) and reported but not settled (RBNS) claims (the latter are also frequently called *open case estimates*). This is illustrated in Figure 1.1. The development of claims progresses with calendar time, and when the notification does not arrive before the present evaluation time (e.g., because the incurred value is too low), the data are left truncated (IBNR). If the claim is notified to the reinsurer but not completely settled before the evaluation time, the information is censored (RBNS).

Throughout the development of the book we will make use of the real data examples described in the following sections to illustrate the practical statistical side of implementing reinsurance treaties.

1.3.1 Case Study I: Motor Liability Data

We here present a data set on motor third-party liability (MTPL) data, gathering information about two direct insurance companies operating in the EU, named A and B hereafter. The data come from an observation period between 1991 and 2010, with evaluation date at January 1, 2011. All amounts are corrected in order to reflect costs in calendar year 2011, with inflation and super-inflation taken into account. For every claim, the payments in a given year were aggregated in a single observation. For Company A 16 years and for Company B 20 years of data are available. In the subsequent chapters we will analyze the two data sets separately: the statistical analysis of the losses will show different characteristics for these companies. For Company A, the exact

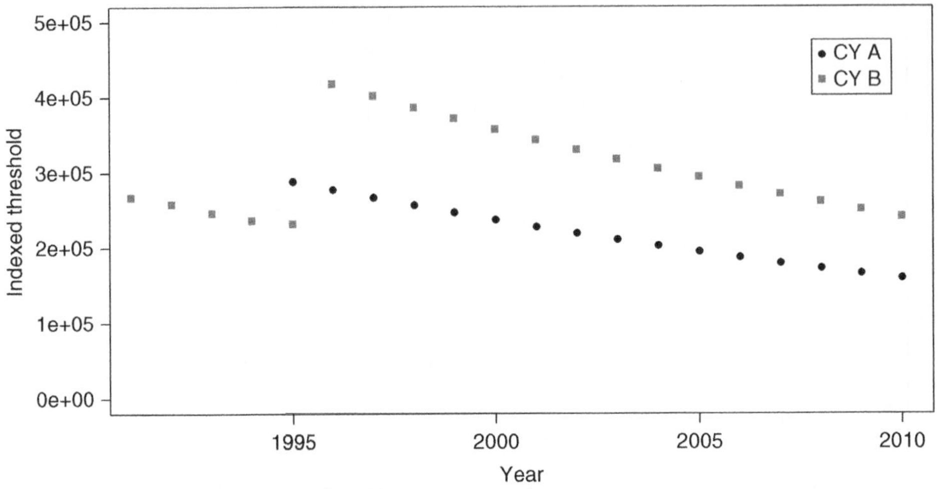

Figure 1.2 Indexed reporting thresholds of Companies A and B.

occurrence dates of the claims are also available, so the analysis of the counting process can be performed more accurately for that claim data set.

Per development year and per claim the *aggregate payment* and *incurred loss* are given. The incurred loss at a given moment in time is the sum of the already paid amount and a reserve for further payments, proposed by company experts at that moment. A claim enters the database from the moment the incurred value exceeds the reporting threshold as given in Figure 1.2. Once a claim has been reported, it stays in the data set even if the associated incurred loss falls below the reporting threshold at some point later. When estimating the loss experienced by the reinsurer, one needs to model the *reporting delay* between the accident time and the year where the claim was first reported to the reinsurer, that is, when the incurred loss I_1 first exceeds the reporting threshold. Indeed, claims that have occurred close to the evaluation time at the end of 2010 can still be IBNR to the reinsurer. In Figure 1.3 the histogram of the reporting delays is given. Given that the accident dates were only reported for Company A, we restrict the plot to this data set. The delay time is then obtained from the difference between the reporting year and accident date, rounded off in years, using the reporting threshold of the particular accident year (see Figure 1.2).

For Company A one has 849 claims of which 340 are completely developed, while the sample size for Company B is 560 of which 225 are fully developed. In Figure 1.4 we show the development of four selected claims. The cumulative payments (aggregated on a yearly basis) are indicated by a full line, while the incurred values are given by dashed lines. When payments and incurred meet, the claim is closed. The characteristics of the four depicted claims are given in Table 1.2.

Note that the information concerning the loss values and development periods is right censored since for the claims which are not fully developed at the end of 2010, the loss as well as the development time at the end of 2010 are only lower bounds for their final value. In Table 1.3 the observed numbers of claims per accident year and per

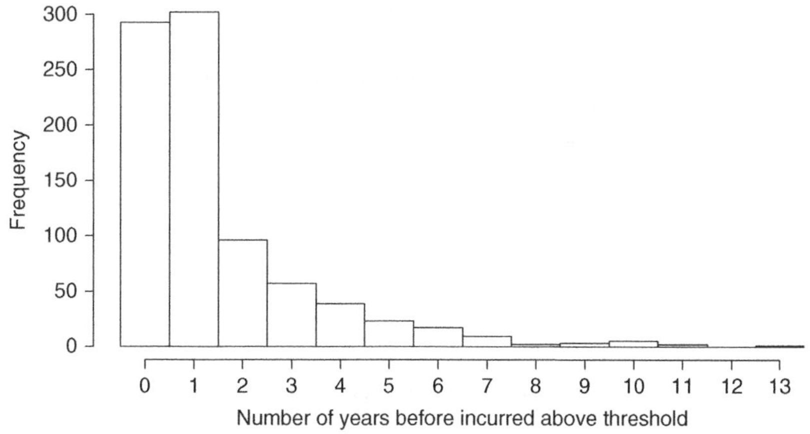

Figure 1.3 Company A: histogram of reporting delays.

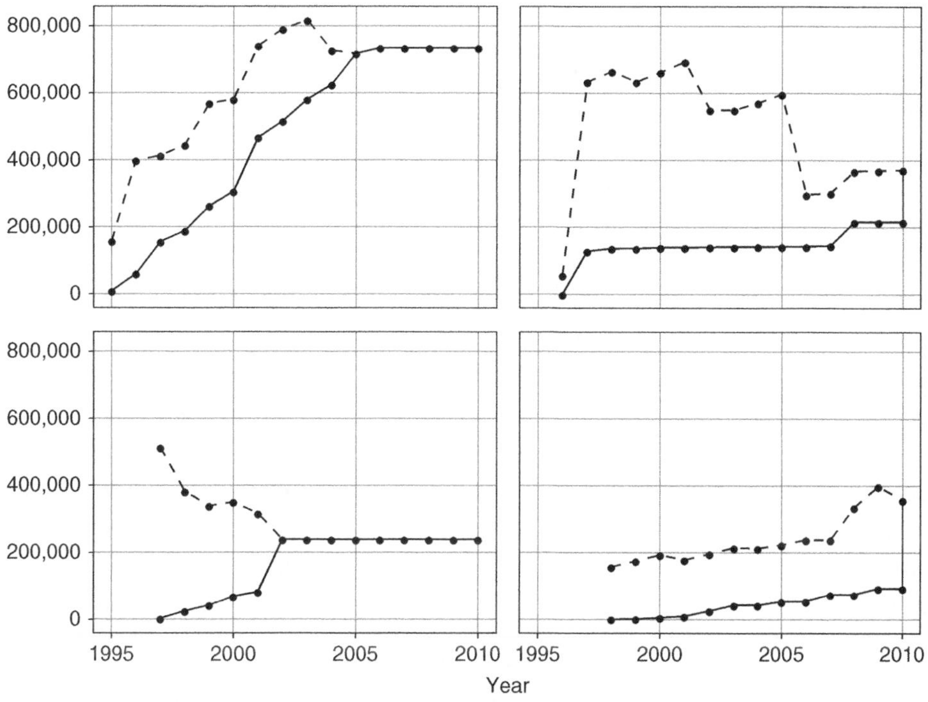

Figure 1.4 MTPL data: development pattern of four particular claims.

development time up to 2010 (in years, DY_{2010}) are given for Company A. Clearly the amount of censoring increases with increasing accident year.

In Figures 1.5 and 1.6, time plots of the incurred loss data of Company A and Company B, respectively, are given as a function of accident year.

Table 1.2 Company A: characteristics of the claims from Figure 1.4

Claim	**Reporting year**	**Closing year**	**Development time (years)**
Top left	1995	2005	10
Top right	1996	-	≥ 15
Bottom left	1997	2002	5
Bottom right	1998	-	≥ 13

Table 1.3 Company A: observed number of claims per accident year and per number of development years in 2010 (DY_{2010})

								DY_{2010}									Nr.		Prop.
	1	2	3	4	5	6	7	8	9	10	11	12	13	14	15	16	censored	Total	non-censored
1995	0	0	2	3	4	8	1	1	2	2	2	2	1	1	2	2	11	44	0.75
1996	0	1	0	0	4	1	6	2	3	4	3	4	2	3	0		14	47	0.7
1997	0	1	3	3	1	3	6	6	4	3	2	2	5	1			10	50	0.8
1998	0	0	0	6	3	8	8	7	5	4	4	4	0				21	70	0.7
1999	0	0	0	3	2	4	3	2	1	4	4	3					17	43	0.6
2000	0	1	1	1	4	6	8	6	2	7	2						19	57	0.67
2001	0	0	1	2	5	4	6	4	9	3							23	57	0.6
2002	0	0	1	2	5	1	3	2	1								27	42	0.36
2003	0	2	2	5	7	6	5	1									38	66	0.42
2004	0	0	1	6	8	5	4										41	65	0.37
2005	0	0	1	2	6	1											46	56	0.18
2006	0	0	2	2	2												43	49	0.12
2007	0	0	0	4													65	69	0.06
2008	0	0	0														69	69	0.00
2009	0	0															55	55	0.00
2010	0																10	10	0.00
Censored	10	55	69	65	43	46	41	38	27	23	19	17	21	10	14	11	509		
Total	10	60	83	104	94	93	91	69	54	50	36	32	29	15	16	13		849	0.60

The classical statistical procedure to estimate the distribution of right censored random variables is given by the Kaplan–Meier estimator of the distribution function. This estimator is discussed in more detail in Chapter 4. Note from these plots that about half of the claims are expected to demand a development period of at least 10 years.

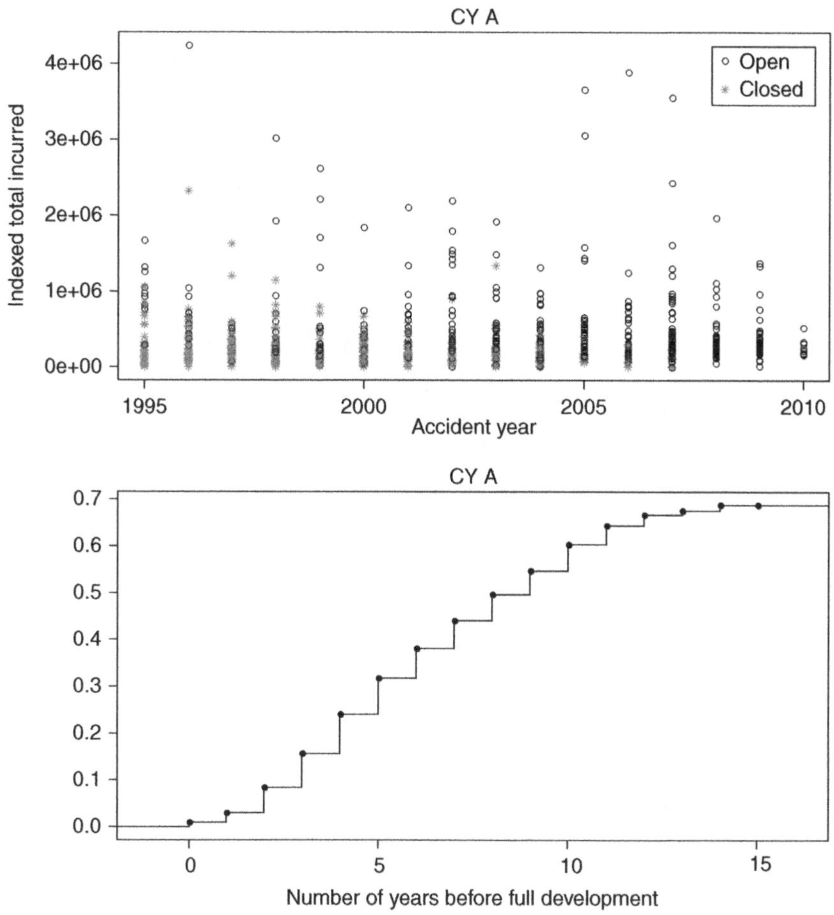

Figure 1.5 Company A: incurred losses (top); Kaplan–Meier estimator for the distribution function of the number of development years (bottom).

Alongside the aggregate payment and the incurred loss, when analysing the risk many companies compute *ultimate loss amounts* for claims that are still under development. These ultimates are statistical estimates of the final loss. The ultimate value of course equals the final aggregate payment in case the claim is closed during the period of study. In practice, the ultimate estimates for non-closed claims are often primarily based on chain ladder development factors based on paid and incurred loss triangles (e.g., see Wüthrich and Merz [797] and Radtke et al. [638]), but then applied on the individual loss data, see also Drieskens et al. [308]. In Figure 1.7 scatterplots of the ultimate against the incurred losses for the data of the two companies are given. Note that the regression fits on these scatterplots for the claims that are still open at the end of the observation period indicate a linear relation between ultimate and incurred values with a negligible intercept: ultimate $= a \times$ incurred, for some $a > 1$.

Finally, in Figure 1.8 we plot the daily cluster sizes for the claims of Company A. Up to three claims per day were observed.

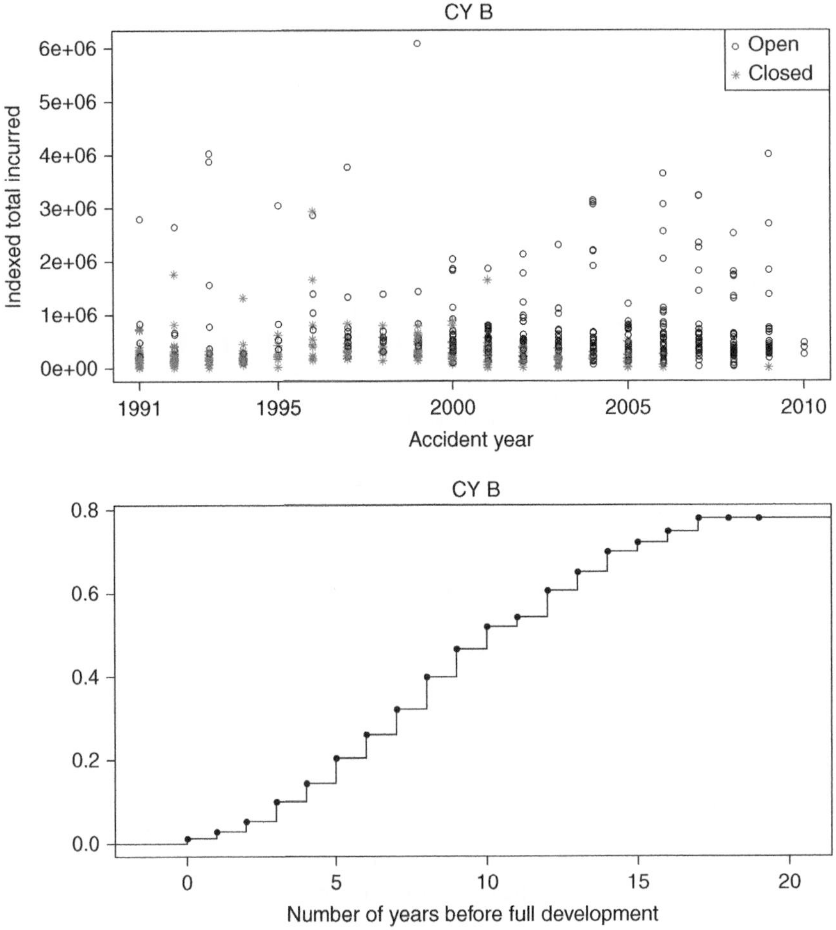

Figure 1.6 Company B: incurred losses (top); Kaplan–Meier estimator for the distribution function of the number of development years (bottom).

1.3.2 Case Study II: Dutch Fire Insurance Data

We will use claim data from the Dutch fire insurance market between 2000 and 2015, provided by a reinsurance company. The date for every fire is known, together with the type of building and regional information. Here the development times are short. Figure 1.9 depicts the logarithm of the claim sizes as a function of time as well as the daily cluster sizes (one sees up to five claims per day). The loss data are indexed to 2015. The reporting threshold equals a value equivalent to 2 million Dutch guilders up to 2002, after which 1 million Euros is used.

1.3.3 Case Study III: Austrian Storm Claim Data

Sometimes individual claim data are not available, and instead claims aggregated over time or regions have to be used. As an illustration, we will use data from historical storm losses of residential buildings in Austria in the period 1998–2009, aggregated over two-digit postcode regions. This data set contains 36 storm events and was provided by the

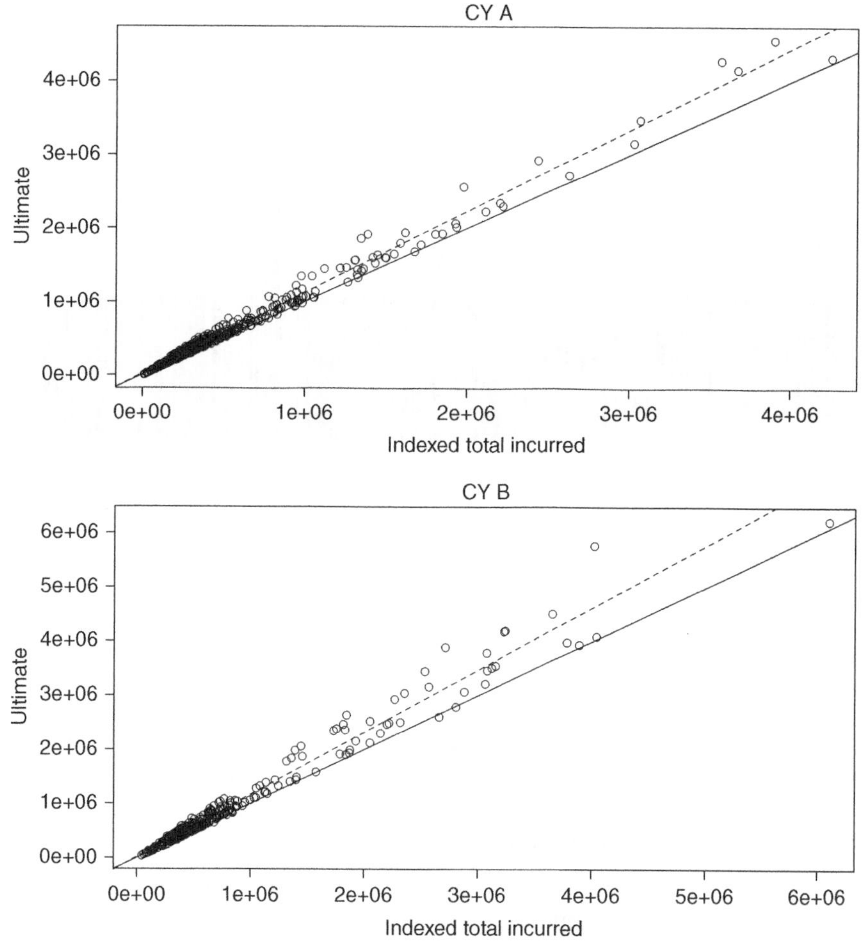

Figure 1.7 Ultimate versus incurred losses with least squares regression fit for the open claims of Company A (top) and Company B (bottom).

Austrian Association of Insurance Companies (VVÖ). The data are indexed according to the building value index and normalized with respect to the overall building stock value in the respective year. Using actual wind fields of each storm on a fine grid, Prettenthaler et al. [633] formulated a building-stock-value-weighted wind index W for each region and storm, and then developed a stochastic model relating wind speed and actual losses (expressed per million of the building stock value). Figure 1.10 depicts the losses of the 36 storms in the data set as a function of this wind index W for Vienna and the province of Upper Austria. Here one studies the distribution of the loss data as a function of W in a regression setting.

1.3.4 Case Study IV: European Flood Risk Data

Floods rank amongst the most wide-reaching natural hazards. Losses from floods show an increasing trend which (to a considerable extent) is attributable to socio-economic factors, including population growth, economic development and construction

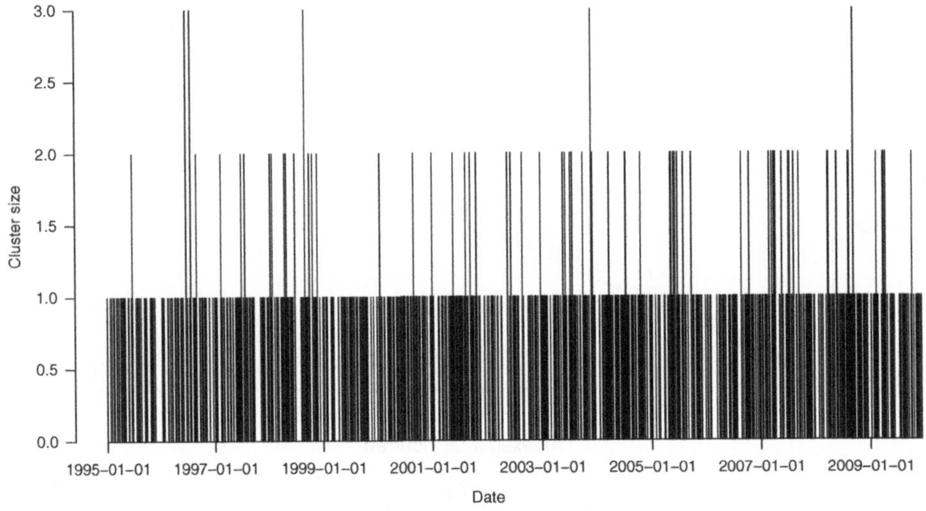

Figure 1.8 Company A: observed cluster sizes of the claim number process.

activities in vulnerable areas. In Prettenthaler et al. [632] (indexed) flood loss data across Europe (provided by Munich Re NatCatSERVICE, 2014) were transformed into losses expressed as a percentage of building stock value, and then used to determine loss quantiles as required for flood risk management. Figure 1.11 depicts the respective aggregate annual losses for the period 1980–2013 for Germany and the UK.

1.3.5 Case Study V: Groningen Earthquakes

Next to loss amount data, reinsurers also need to analyze the physical phenomena causing damage. A classical example is earthquake risk. We discuss the Groningen earthquakes caused by gas extraction. The Groningen field is the largest gas field of Western Europe, with 2800 billion cubic metres available and 800 billion cubic metres left. The pressure inside the gas layers decreases due to the extraction, and the layers on top collapse. This collapse does not happen homogeneously, which causes the earthquakes. Hundreds of earthquakes have been detected since 1986 with magnitudes between 2 and 3 on the Richter scale, and 14 larger than 3 (Figure 1.12). The damage to houses and public buildings was substantial, with many buildings needing reinforcements. The largest observed magnitude was 3.6 (Huizinge, August 2012). In this context, the estimation of the maximum possible magnitude is the main goal. Depending on the research team, maximum magnitudes between 3.9 and 5 were predicted, see for instance Bourne et al. [157].

1.3.6 Case Study VI: Danish Fire Insurance Data

It is quite common to combine reinsurance forms across various lines of business (LoB), so modelling the dependence of the different LoB is important. To illustrate the appropriate multivariate models and statistical methodology, we will use the Danish fire insurance data set containing information on 2167 fire losses over the period

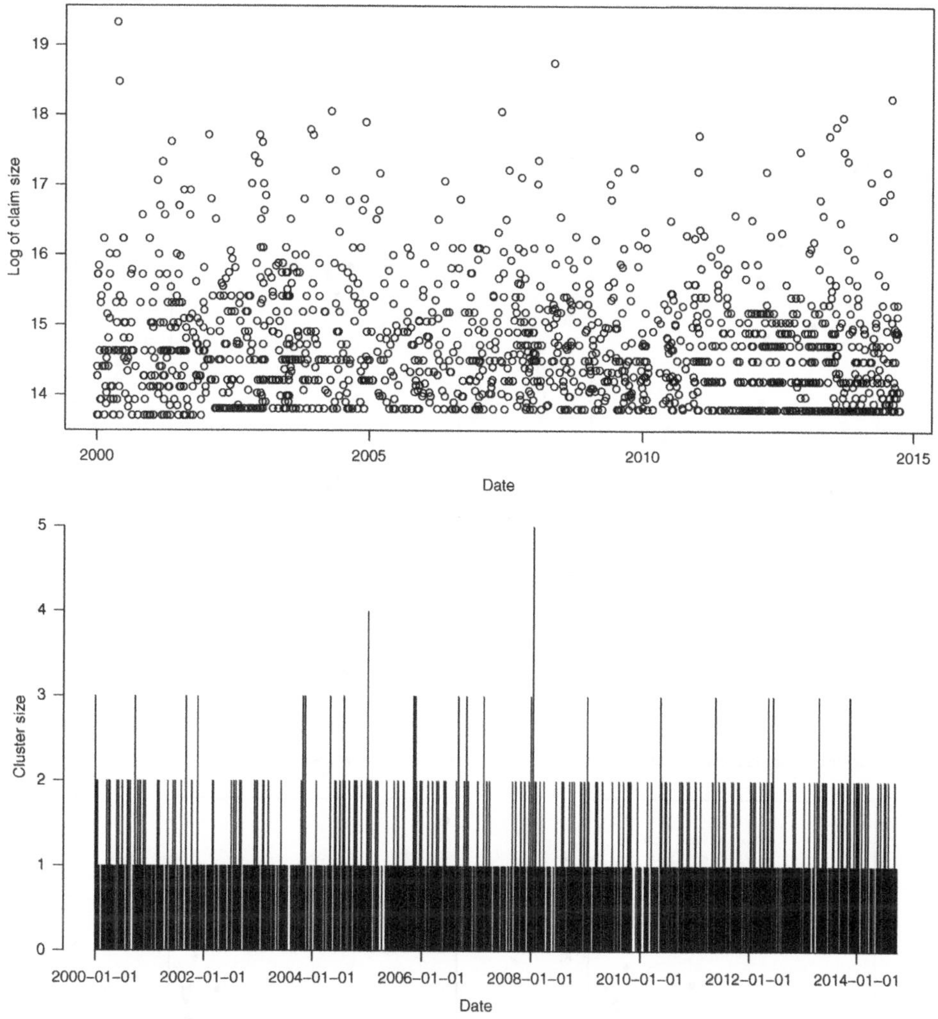

Figure 1.9 Dutch fire insurance claims: log-claims as a function of time for Dutch fire insurance (top); observed cluster sizes of the claim number process (bottom).

1980–1990. The data have been adjusted for inflation to reflect 1985 values and are expressed in millions of Danish kroner. The total loss amount X_i of the ith claim is subdivided into damage to building ($X_{i,1}$), damage to content ($X_{i,2}$) (e.g., furniture and personal property) and loss of profits ($X_{i,3}$). A claim is only registered if the total loss exceeds 1 million kroner, that is, $X_{i,1} + X_{i,2} + X_{i,3} \geq 1$. This data set was collected at the Copenhagen Reinsurance Company and can nowadays be seen as a folklore example as it has been studied extensively over the years in the academic literature (e.g., see Embrechts et al. [329]). In Figure 1.13 a scatterplot matrix is given for the log-transformed data. On the diagonal, histograms of the logarithm of the marginal losses are given. Note that several claims exhibit losses in only one or two of the components (for only 517 claims there is a loss in each of the three components).

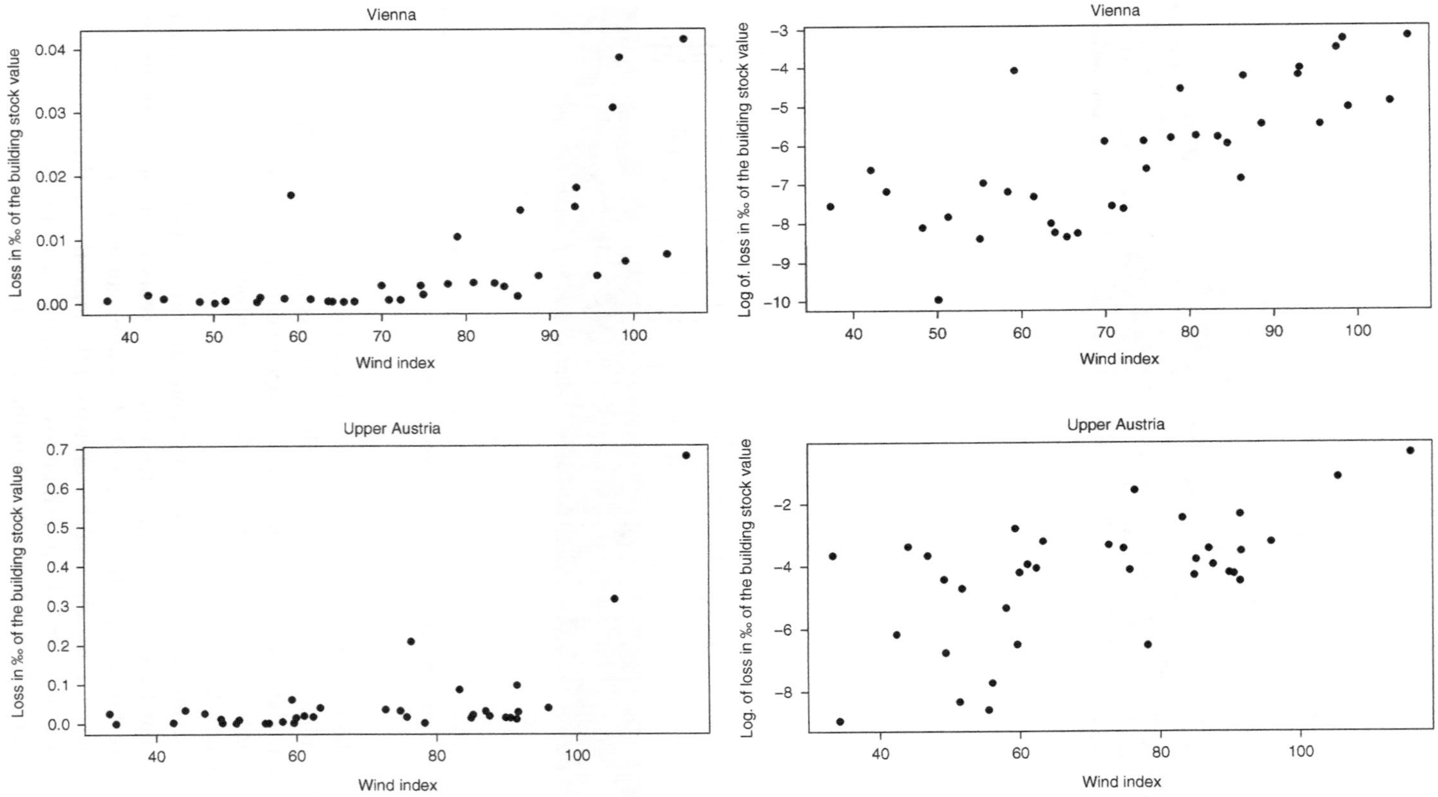

Figure 1.10 Normalized loss data against wind index W for Vienna (top) and Upper Austria (bottom); original scale (left) and log-scale (right).

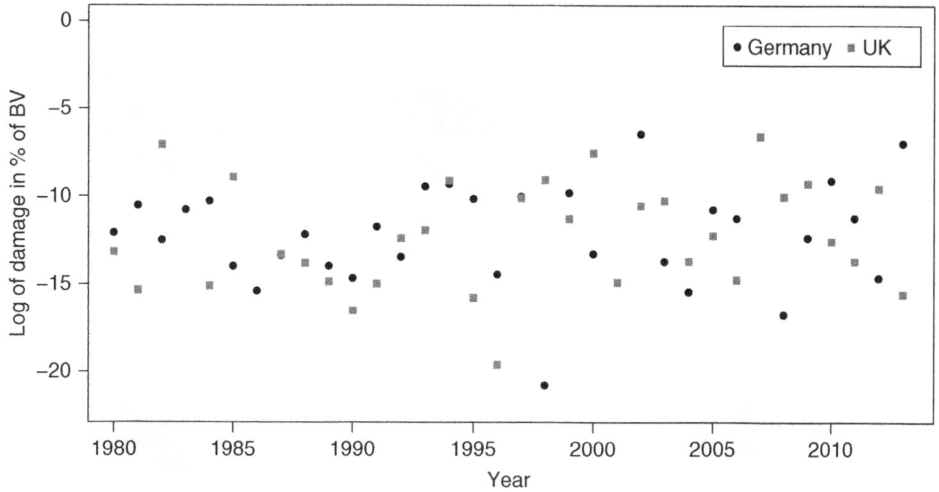

Figure 1.11 Flood risk: aggregate annual losses (in log scale) by percentage of building value for Germany and the UK.

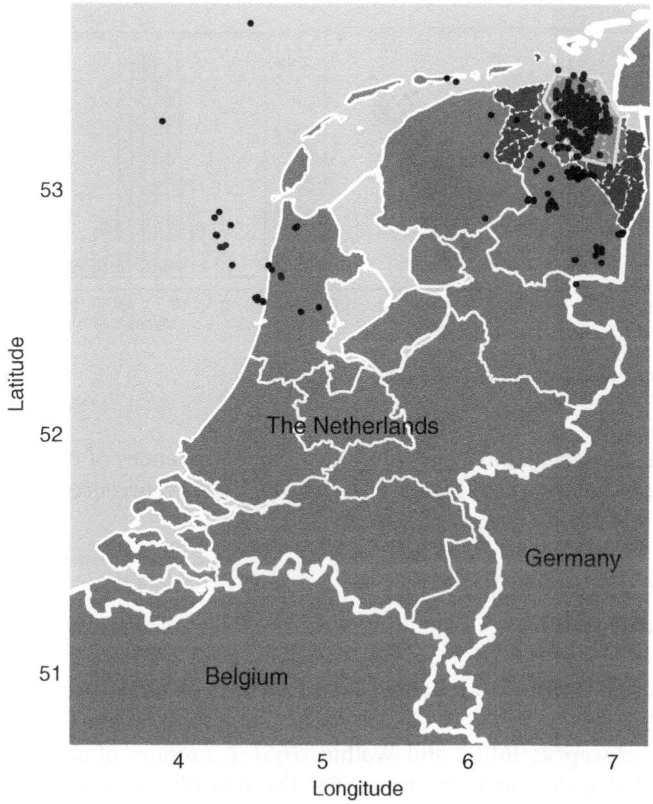

Figure 1.12 Induced (dark points) earthquakes in the northern part of the Netherlands with magnitudes larger than 1.5.

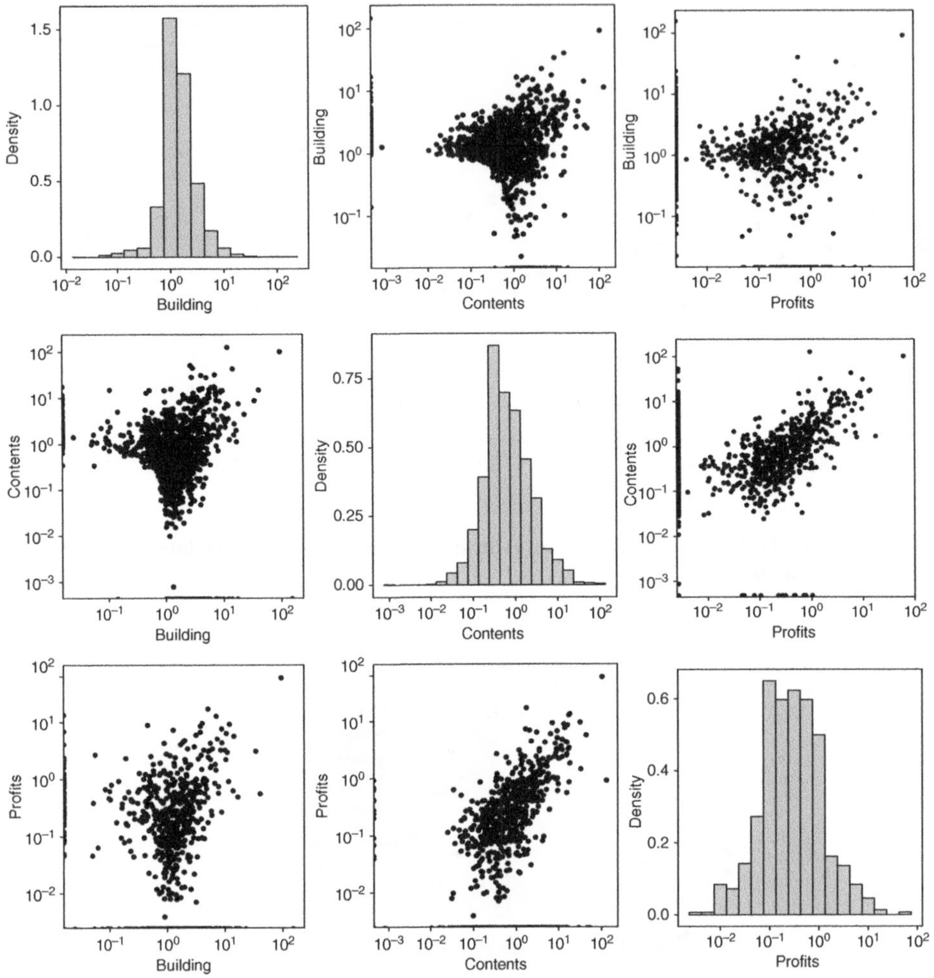

Figure 1.13 Danish fire insurance data: scatterplot matrix on the log-scale.

The occurrence dates are also given and hence simultaneous occurrences of claims for the three components can be observed. Figure 1.14 illustrates the occurrences and the cluster sizes when all portfolio components were affected.

1.4 Notes and Bibliography

There are a number of classical textbooks available which provide a general introduction to reinsurance, for example Carter [185], Gerathewohl [382], Grossmann [409], Strain [710], Gastel [375], Schwepcke [686], and Walhin [765]. A number of articles in Teugels and Sundt [741] also deal with the topic. For the role of reinsurance in risk management, see D'Outreville [598]. More recent and shorter overviews can be found in Liebwein [543], Albrecher [13], Outreville [599], Bernard [120], and

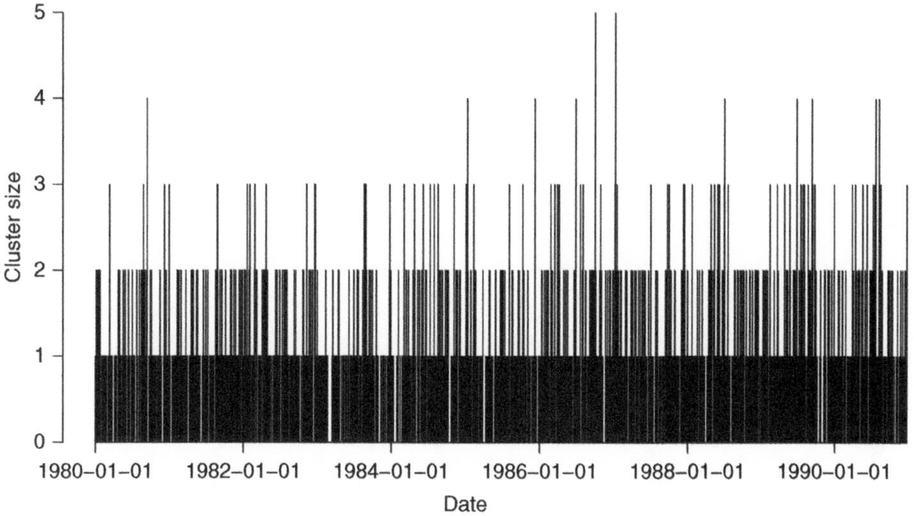

Figure 1.14 Danish fire insurance data: cluster arrivals, selecting the claim dates where each component is addressed.

Deelstra et al. [267]. Furthermore, a number of basic textbooks on risk theory contain sections on reinsurance. Examples include (in alphabetic order) Beard et al. [93], Beekman [94], Borch [149], Bowers et al. [157], Bühlmann [165], Cramér [234], Daykin et al. [254], De Vylder [263], Gerber [383], Goovaerts et al. [402], Heilmann [435], Kaas et al. [475], Klugman et al. [491], Lundberg [551], Mack [555], Rolski et al. [652], Schmidt [677], Seal [690], Straub [711], and Sundt [716]. For a discussion on the challenges and opportunities of reinsurance as an international business, see Göbel [392]. A recent overview from a practical perspective can be found in Swiss Re [725]. The increasing role of the notion of capital and capital management in running insurance and reinsurance companies, which can be seen as an ongoing change of paradigm in the insurance industry, is highlighted in Dacorogna [243], see also [689] and Krvavych [510, 512].

The number of 200 reinsurance companies can be compared with the more than 10,000 primary insurance companies in the market today (using economic arguments, Powers and Shubik [627] in fact claim that the "optimal" number of reinsurers in the market is connected to the number of primary insurers by a square-root rule).

Historically, the first documented reinsurance contract dates back to 1370, when the cargo of a ship sailing from Genoa to Sluis (near Bruges in Flanders) was reinsured by the direct insurer for the more dangerous part of the journey from Cadiz to Sluis (interestingly, the contract did not state the premium, which most likely was done to avoid usury discussions). The first reinsurance company was founded much later, in 1846, in Cologne after the big fire of Hamburg in 1842, and the first retrocession contract seems to date back to 1854, involving Le Globe Compagnie d'Assurance contre L'incendie. Soon the (nowadays) major European reinsurance companies were founded, and the American Life Reinsurers followed in the early 20th century. For a detailed account of the history of reinsurance, see Kopf [496], Holland [451], and Borscheid et al. [154].

2

Reinsurance Forms and their Properties

Let $\{X_i; i \in \mathbb{N}\}$ be random variables denoting the claim sizes that the first-line insurer experiences and let $\{N(t); t \geq 0\}$ be a counting process, where $N(t)$ represents the number of claims up to time $t > 0$ (measured in years). Then the *total* or *aggregate claim amount* at time t for the first-line insurer is given by

$$S(t) = \begin{cases} \sum_{i=1}^{N(t)} X_i & \text{if } N(t) > 0, \\ 0 & \text{if } N(t) = 0. \end{cases}$$

Recall that most (non-life) contracts are written for the duration of one year, so the static random variable $S(1)$ will be of prime interest in many applications.

In a reinsurance contract, this aggregate claim size is now sub-divided into

$$S(t) = D(t) + R(t),$$

where $D(t)$ is the deductible (retained) amount that stays with the first-line insurer after reinsurance and $R(t)$ is the amount paid by the reinsurer. For many reinsurance contracts the splitting will be defined on the individual risks X_i, and in this case we write $X_i = D_i + R_i$ (or just $X = D + R$ for short, in case they all follow the same distribution).

We will now discuss the most common obligatory reinsurance forms and their properties. We start with proportional (also called *pro-rata*) treaties.

2.1 Quota-share Reinsurance

The simplest possible reinsurance form is *quota-share* (QS) *reinsurance*, which is a fully proportional sharing of the risk, that is,

$$R = a \cdot X \text{ and } R(t) = a \cdot S(t)$$

for a *proportionality factor* $0 < a < 1$.

This form of reinsurance is popular in almost all insurance branches, particularly due to its conceptual and administrative simplicity. In general the first-line insurer will also

Reinsurance: Actuarial and Statistical Aspects, First Edition.
Hansjörg Albrecher, Jan Beirlant and Jozef L. Teugels.

cede to the reinsurer a similarly determined proportion of the premiums (see Chapter 7 for details). If the distribution of X is available, one immediately has

$$\mathbb{P}(R \le x) = F_X\left(\frac{x}{a}\right), \quad \mathbb{P}(D \le x) = F_X\left(\frac{x}{1-a}\right)$$

expressed in terms of the cumulative distribution function (c.d.f.) $F_X(x) = \mathbb{P}(X \le x)$. For the aggregate risk, correspondingly

$$\mathbb{P}(R(t) \le x) = \mathbb{P}(S(t) \le x/a), \quad \mathbb{P}(D(t) \le x) = \mathbb{P}(S(t) \le x/(1-a))$$

and for the moment-generating function

$$\mathbb{E}(e^{sR(t)}) = \mathbb{E}(e^{(as)S(t)}), \quad \mathbb{E}(e^{sD(t)}) = \mathbb{E}(e^{((1-a)s)S(t)}), \tag{2.1.1}$$

so one only needs to evaluate the moment-generating function of $S(t)$ at a different argument. As a result, the rth moments ($r \in \mathbb{N}$) are given by

$$\mathbb{E}(R^r) = a^r\,\mathbb{E}(X^r), \quad \mathbb{E}(D^r) = (1-a)^r\mathbb{E}(X^r).$$

Note that both the coefficient of variation and the skewness coefficient ν do not change under a QS treaty:

$$\mathrm{CoV}(R(t)) = \frac{\sqrt{\mathrm{Var}\,(R(t))}}{\mathbb{E}(R(t))} = \mathrm{CoV}(D(t)) = \mathrm{CoV}(S(t)),$$

$$\nu_{R(t)} = \mathbb{E}\left(\frac{R(t) - \mathbb{E}(R(t))}{\sqrt{\mathrm{Var}\,(R(t))}}\right)^3 = \nu_{D(t)} = \nu_{S(t)}.$$

2.1.1 Some Practical Considerations

QS reinsurance can be understood as (virtually) increasing the available solvency capital. To see that in a simple example, consider the probability

$$\mathbb{P}(\nu + P(t) - S(t) > 0)$$

that at some time $t > 0$ the capital ν together with the received premiums $P(t)$ suffices to cover the claims $S(t)$. Then, after entering a QS treaty and assuming that premiums are shared with the same proportion, this probability changes to

$$\mathbb{P}(\nu + (1-a)P(t) - (1-a)S(t) > 0) = \mathbb{P}\left(\frac{\nu}{1-a} + P(t) - S(t) > 0\right).$$

In practice, a further positive effect of QS reinsurance is to improve the premium-to-surplus ratio: according to statutory accounting principles implied by the regulator, an insurer typically has to immediately include in the balance sheet all the expenses connected to issuing a policy, but the respective premium can only be entered gradually over the duration of the policy; the correspondingly needed *unearned premium reserve* considerably reduces the surplus and a QS arrangement will improve this situation, as it reduces that reserve and the expenses simultaneously (see, for example, [585]). QS contracts are often used at the initiation of smaller companies to broaden their chances for underwriting policies and to gain experience in a new market with a limited amount of risk. For reinsurers, in turn, a QS arrangement can also have the advantage of gaining claim experience in that particular market, which may be useful in other related portfolios.

QS arrangements are easy to combine, that is, an insurer can have simultaneous QS contracts on the same portfolio with different reinsurers. Also, due to the proportional share that is left with the insurer, the risk of some forms of *moral hazard* (like sloppy claim settlement procedures) is avoided.

One of the main shortcomings of QS reinsurance is that, due to its form, *all* claims are partly reinsured, not just the largest of them. This is often not ideal, as claims from small policies could have easily been borne by the insurer alone (and passing on those parts of the portfolio is a non-attractive loss of insurance business).

2.2 Surplus Reinsurance

A reinsurance form that improves on the disadvantages of QS treaties, but keeps its main advantages, is *surplus reinsurance*, which is a proportional reinsurance form for which the proportionality factor depends on the coverage limit in the underlying policy (sum insured). Let Q_i be the sum insured (policy limit) of claim X_i. For a fixed *retention line* M the reinsured amount is then given by

$$R_i = \left(1 - \frac{M}{Q_i}\right) X_i \cdot 1_{\{Q_i > M\}}, \quad D_i = X_i 1_{\{Q_i \leq M\}} + M \frac{X_i}{Q_i} 1_{\{Q_i > M\}}, \tag{2.2.2}$$

where $1_{\{A\}}$ denotes the indicator function of event A. Altogether,

$$R(t) = \sum_{i=1}^{N(t)} R_i, \quad D(t) = \sum_{i=1}^{N(t)} D_i.$$

The ratio $V_i := X_i/Q_i$ is called the *loss degree* of claim X_i. With a surplus reinsurance each claim with an insured sum below M is fully kept by the insurer, and otherwise the relative participation of the reinsurer in the claim payment is larger the larger the underlying sum insured is (see Figure 2.1). Consequently, this reinsurance form retains the advantages of the proportionality for each claim payment, but only reinsures claims from larger policies. Due to the proportionality feature, the determination of premiums is again rather simple. In some cases Q_i is alternatively the probable maximum loss (PML) of claim X_i (see Chapter 7). From the definition, it becomes clear that the maximum retained size of each claim is M ("the line"). The surplus reinsurance contract homogenizes the portfolio of the first-line insurer, as illustrated in the following simple example.

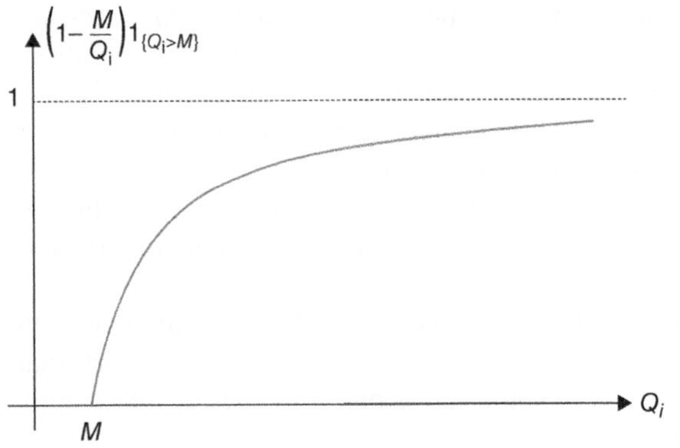

Figure 2.1 Proportionality factor of the reinsurer as a function of insured sum.

Example 2.1 Assume there are 100 independent policies in an insurance portfolio. For each policy, a claim occurs with probability 0.01 within the next year. For 70 policies, the claim size is Q_1 and for 30 policies the claim size is $Q_2 > Q_1$, given a claim occurs (i.e., for simplicity here the claim size is always equal to the policy limit). Then the insurer expects one claim during this year and an aggregate claim payment of $E(S(1)) = 0.7Q_1 + 0.3Q_2$. If the insurer charges this amount as the overall premium, this will be sufficient to cover this one expected claim if it is one with insured sum Q_1, but not if it comes from a policy with insured sum Q_2. However, if the insurer buys surplus reinsurance with $M = Q_1$ for a pure premium of $\mathbb{E}(R(1)) = 0.3(Q_2 - Q_1)$, then the remaining amount $\mathbb{E}(D(1)) = Q_1$ is sufficient to cover the retained amount of that expected claim, no matter from which type of policy it comes.

In order to determine the distributional properties of the retained and reinsured amount under surplus reinsurance, it is helpful to consider the insured amount of a claim as a random variable with (c.d.f) F_Q (based on frequencies of the sums insured specified in the policies of the portfolio and the respective claim occurrence probabilities, for example in Example 2.1 Q would have a two-point distribution with $\mathbb{P}(Q = Q_1) = 1 - \mathbb{P}(Q = Q_2) = 0.7$).[1] The distributions of the quantities R and D are then given by

$$\mathbb{P}(D \le x) = \int_0^\infty \mathbb{P}\left(X \le x \max\{1, y/M\} | Q = y\right) \, dF_Q(y), \tag{2.2.3}$$

$$\mathbb{P}(R \le x) = \int_M^\infty \mathbb{P}\left(X \le \frac{x}{1 - \frac{M}{y}} \middle| Q = y\right) \, dF_Q(y) \, . \tag{2.2.4}$$

1 Such an approach is akin to the philosophy of the collective risk model, where a heterogeneous portfolio is treated as a homogeneous one, but equipped with a mixture distribution for the claim size to take into account the heterogeneity.

For the moments of D, we have

$$
\begin{aligned}
\mathbb{E}(D^r) &= \int_0^\infty \mathbb{E}(D^r|Q=y)\,dF_Q(y)\\
&= \int_0^M \mathbb{E}(X^r|Q=y)\,dF_Q(y) + \int_M^\infty M^r\mathbb{E}\left(\left(\frac{X}{Q}\right)^r|Q=y\right)\,dF_Q(y)\\
&= \int_0^\infty \min\{y^r, M^r\}\mathbb{E}(V^r|Q=y)\,dF_Q(y). \qquad (2.2.5)
\end{aligned}
$$

In practice it may often be reasonable to assume that the loss degree is independent of the sum insured (particularly if the sums insured do not vary too much across policies). In that case, (2.2.5) simplifies to

$$
\mathbb{E}(D^r) = \mathbb{E}(V^r)\int_0^\infty \min\{y^r, M^r\}\,dF_Q(y).
$$

For the reinsured amount, the respective expression is slightly more involved, but for the first moment one easily gets

$$
\mathbb{E}(R) = \int_0^\infty \max\{y-M,0\}\mathbb{E}(V|Q=y)\,dF_Q(y)
$$

and under independence of V and Q

$$
\mathbb{E}(R) = E(V)\int_0^\infty \max\{y-M,0\}\,dF_Q(y).
$$

Surplus reinsurance is very popular, particularly in fire insurance, as well as property, accident, engineering and marine insurance. Typically, there is an upper limit $Q_i \le (k+1)M$ ("k lines") in the treaty, that is, the ceded share is capped by

$$
R_i = \min\left\{1-\frac{M}{Q_i}, 1-\frac{1}{k+1}\right\}X_i \cdot 1_{\{Q_i>M\}},
$$

and the remaining part for the policies with larger sums insured is then negotiated on a facultative basis. Also, for certain policies the insurer may decide to retain several, say $m < k$, lines and only reinsure the remaining $k-m$ lines (e.g., see [585]). In general, it is not uncommon to apply a *table of lines*, that is, different retention lines to various groups of similar risks. The retention line is then often chosen in a way to aim for the same maximum loss (*method of inverse claim probability*) or average loss (*method of inverse rate*) for each policy (cf. [526]).

2.3 Excess-of-loss Reinsurance

We now move on to non-proportional reinsurance forms. The simplest case is the (per risk) *excess-of-loss* (XL) *reinsurance* defined by

$$R(t) = \sum_{i=1}^{N(t)} (X_i - M)_+, \qquad D(t) = \sum_{i=1}^{N(t)} \min(X_i, M), \tag{2.3.6}$$

for some pre-defined *retention* M,[2] that is, the reinsurer agrees to pay for each claim the excess over the retention M. Typically, this will only be agreed upon up to a certain limit L, leading to

$$R(t) = \sum_{i=1}^{N(t)} \min\{(X_i - M)_+, L\},$$

$$D(t) = \sum_{i=1}^{N(t)} \left(\min\{X_i, M\} 1_{\{X_i \le M+L\}} + (X_i - L) 1_{\{X_i > M+L\}}\right),$$

and one refers to such a treaty as L xs M, characterized by the *layer* $[M, M+L]$ (layer size L). Note that $(X-u)_+ := \max\{X-u, 0\}$. The ratio $(M+L)/M$ is referred to as the *relative layer length*.

This reinsurance form is very popular in casualty and fire insurance, as it reduces the exposure of the ceding company in an effective way and has an intuitive and simple form. The premium calculation is, however, considerably more involved than for proportional reinsurance forms (see Chapter 7). Figure 2.2 schematically depicts the reinsured claim amounts under a QS, surplus and XL treaty.

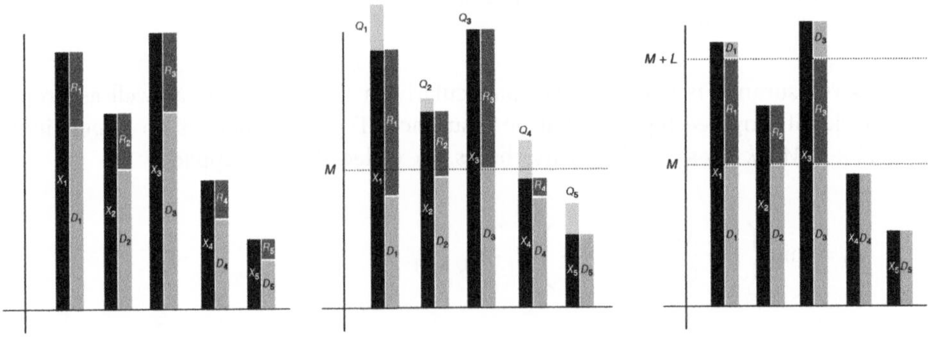

Figure 2.2 Comparison of reinsured claim amounts for some combinations of claim sizes X_i and corresponding insured sums Q_i for QS reinsurance with $a = 0.3$ (left), surplus reinsurance (middle), and L xs M reinsurance (right).

---

2 Depending on the region, the terms *deductible*, *priority* (continental Europe) and *attachment point* (US) are also used for M.

2.3.1 Moment Calculations

Consider the random variable

$$\tilde{X} := \min\{(X-u)_+,\ v\} = \min\{X, u+v\} - \min\{X, u\} \tag{2.3.7}$$

for any $u, v \geq 0$. If $u = M$, $\tilde{X}$ refers to the reinsured amount R of a single risk in a v xs M treaty. On the other hand, for $u = 0$ and $v = M$, $\tilde{X}$ refers to $D = \min(X, M)$ in an ∞ xs M contract. That is, studying distributional properties of $\tilde{X}$ will be relevant for both parties involved in the XL contract.

If F_X denotes the distribution function of X, then one gets

$$F_{\tilde{X}}(z) := \mathbb{P}(\tilde{X} \leq z) = \begin{cases} F_X(u+z) & \text{if } 0 \leq z < v, \\ 1 & \text{if } v \leq z. \end{cases}$$

and for the Laplace transform of $\tilde{X}$

$$\widehat{F}_{\tilde{X}}(s) := \mathbb{E}(e^{-s\tilde{X}}) = F(u) + \int_{u+}^{u+v-} e^{-s(z-u)}\, dF_X(z) + e^{-sv}(1 - F_X(u+v)).$$

For the kth moment we get

$$\tilde{\mu}_k := \mathbb{E}(\tilde{X}^k) = k \int_0^{\infty} (1 - F_{\tilde{X}}(z))\, z^{k-1}\, dz = k \int_0^{v} (1 - F_X(u+z))\, z^{k-1}\, dz. \tag{2.3.8}$$

From this expression one can read off the first moments

$$\mathbb{E}(D) = \int_0^M (1 - F_X(z))\, dz, \qquad \mathbb{E}(R) = \int_M^{\infty} (1 - F_X(z))\, dz \tag{2.3.9}$$

for the retained and reinsured single claim amount in an ∞ xs M treaty, which has the appealing optical interpretation of sub-dividing the area between $F_X(z)$ and the constant line 1 (see Figure 2.3).

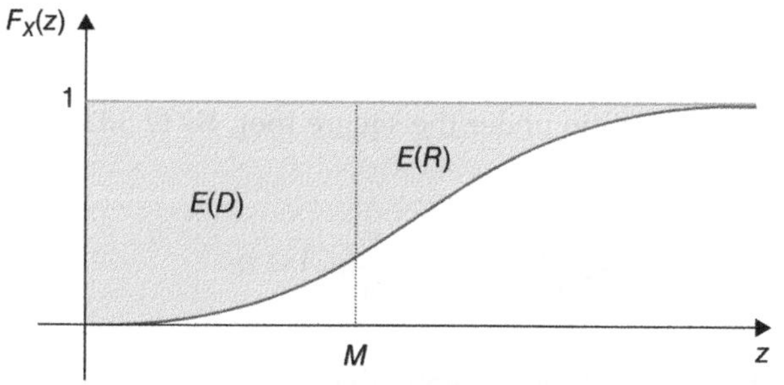

Figure 2.3 Graphical interpretation for the splitting of the expected claim size between insurer and reinsurer.

For the case L xs M one analogously obtains

$$\mathbb{E}(R) = \int_M^{M+L} (1 - F_X(z))\, dz. \tag{2.3.10}$$

These simple expressions will play a crucial role in the pricing of XL treaties in Chapter 7. From (2.3.8) one also gets the inequalities

$$(1 - F_X(u + v))v^k \le \tilde{\mu}_k \le (1 - F_X(u))v^k.$$

On the other hand, $E(\tilde{X}^k) \le v\, E(\tilde{X}^{k-1})$ so that

$$\tilde{\mu}_k \le v\, \tilde{\mu}_{k-1}. \tag{2.3.11}$$

For the variance, one has

$$\mathrm{Var}(\tilde{X}) = 2 \int_0^v (1 - F_X(u + z))z\, dz - \left(\int_0^v (1 - F_X(u + z))\, dz \right)^2.$$

Since the partial derivative with respect to v is $2(1 - F_X(u + v)) \int_0^v F_X(u + z)\, dz \ge 0$, the variance is non-decreasing in v and we get the bound

$$\mathrm{Var}(\tilde{X}) \le 2 \int_u^\infty (z - u)(1 - F_X(z))\, dz - \left(\int_u^\infty (1 - F_X(z))\, dz \right)^2.$$

The quantity on the right is non-increasing in u and therefore smaller than the same expression where we put $u = 0$, but the latter then refers to the variance of the original risk X, so that we get for any $u, v \ge 0$

$$\mathrm{Var}(\tilde{X}) \le \mathrm{Var}(X).$$

The coefficient of variation can be written as

$$\mathrm{CoV}(\tilde{X}) = \left(\frac{\tilde{\mu}_2}{\tilde{\mu}_1^2} - 1 \right)^{1/2}$$

and hence depends monotonically on the ratio under the square root. By (2.3.11) the derivative of that ratio is

$$\frac{\partial}{\partial v} \frac{\tilde{\mu}_2}{\tilde{\mu}_1^2} = \frac{1}{\tilde{\mu}_1^2} \left(\frac{\partial}{\partial v}\tilde{\mu}_2 - \frac{2\tilde{\mu}_2}{\tilde{\mu}_1} \frac{\partial}{\partial v}\tilde{\mu}_1 \right) \ge \frac{1}{\tilde{\mu}_1^2} \left(\frac{\partial}{\partial v}\tilde{\mu}_2 - 2v \frac{\partial}{\partial v}\tilde{\mu}_1 \right) = 0.$$

If we write $\mathrm{CoV}(u, v) := \mathrm{CoV}(\tilde{X})$ for any u, v, then it follows that

$$\mathrm{CoV}(u, v) \le \mathrm{CoV}(u, \infty).$$

The dependence on u is more intricate. Let us introduce the *retention distribution*

$$G_{u,v}(x) := \begin{cases} \frac{F_X(u+x)-F_X(u)}{F_X(u+v)-F_X(u)} & \text{if } 0 \le x \le v, \\ 1 & \text{if } x \ge v, \end{cases} \tag{2.3.12}$$

with moments $v_k := k\int_0^\infty (1 - G_{u,v}(x))\, x^{k-1}dx$. If we abbreviate $r := F_X(u+v) - F_X(u)$ then it is easy to show that

$$r v_k = \tilde{\mu}_k - v^k(1 - F_X(u+v)). \tag{2.3.13}$$

The latter relation is handy in rewriting the partial derivative of $\mathrm{CoV}(u,v)$ with respect to u. Indeed, it easily follows that

$$\frac{\partial}{\partial u}\tilde{\mu}_1 = -r \qquad \text{and} \qquad \frac{\partial}{\partial u}\tilde{\mu}_2 = -2rv_1,$$

but then

$$\tilde{\mu}_1^3 \frac{\partial}{\partial u}\frac{\tilde{\mu}_2}{\tilde{\mu}_1^2} = \left(\tilde{\mu}_1 \frac{\partial}{\partial u}\tilde{\mu}_2 - 2\tilde{\mu}_2 \frac{\partial}{\partial u}\tilde{\mu}_1\right) = -\tilde{\mu}_1 2rv_1 + 2\tilde{\mu}_2 r = 2r\left(\tilde{\mu}_2 - \tilde{\mu}_1\, v_1\right).$$

Replacing in the last expression the moments $\tilde{\mu}_k$ by their analogues v_k from (2.3.13), we get

$$\tilde{\mu}_1^3 \frac{\partial}{\partial u}\frac{\tilde{\mu}_2}{\tilde{\mu}_1^2} = 2r^2\left(v_2 - v_1^2\right) + 2rv\,(v - v_1)(1 - F_X(u+v)),$$

which is positive. Indeed, the quantity $v_2 - v_1^2$ is the variance of the distribution $G_{u,v}$ while by definition $v_1 \le v$. This then shows that the requested partial derivative is non-negative and hence that $\mathrm{CoV}(u,v)$ is also increasing in u. In particular,

$$\mathrm{CoV}(0,v) \le \mathrm{CoV}(u,v) \le \mathrm{CoV}(u,\infty)$$

comparing the risk of the different layers. Applying the left inequality for $v = \infty$ and the right one for $u = 0$ we also get

$$\mathrm{CoV}(0,v) \le \mathrm{CoV}(0,\infty) = \mathrm{CoV}(X) \le \mathrm{CoV}(u,\infty),$$

where the quantity in the middle is the coefficient of variation for the original claim size. For moment calculations for the aggregate claims of each party under an XL treaty see Chapter 6.

2.3.2 Reinstatements

Many L xs M contracts in practice have in addition an *(annual) aggregate deductible* (*AAD*) and an *(annual) aggregate limit* (*global layer*) (*AAL*) (often a multiple of L), so that

$$R(t) = \min \left\{ \left(\sum_{i=1}^{N(t)} \min\{(X_i - M)_+, L\} - AAD \right)_+ , AAL \right\}. \tag{2.3.14}$$

In the following we assume $AAD = 0$ for simplicity. A very common variant (particularly in property and casualty insurance) is such a contract with k *reinstatements*, that is, at the beginning only an initial premium P_0 for the coverage of a first "layer" (or "liability") $\min\{R(t), L\}$ is paid. When a claim occurs, that layer is (partially) used up, and it can be refilled with later premium payments (*reinstatement premiums*) (see Example 2.2). Altogether then $AAL = (k+1)L$. More details on the respective premium schemes are discussed in Section 7.4.3.

Example 2.2 Figure 2.4 illustrates a 100 xs 100 treaty for an initial premium P_0 with one reinstatement (for an additional premium P_1). The light-grey bar depicts the current reinsurance coverage of the layer throughout the claim history. For the first claim, $X_1 = 150$, the reinsurer pays the entire part in excess of the retention $M = 100$, which uses up half of the coverage of the layer. This half is now reinstated (using the first half of the overall available reinstatement, here illustrated by a radially contured area), so that again the entire layer is covered. At this point in time the additional reinsurance premium payment $P_1/2$ takes place. For the second claim, $X_2 = 175$, the reinsurer will hence pay $R_2 = 75$, which puts down the remaining coverage for a next claim to 25. There is, however, still the second half of the reinstatement available (for another premium payment $P_1/2$), raising the coverage again to 75. For the third claim, $X_3 = 225$, the reinsurer consequently pays $R_3 = 75$. Since the contract contains only one reinstatement, there is no further reinsurance coverage, and the last claim, $X_4 = 150$, is paid entirely by the cedent.

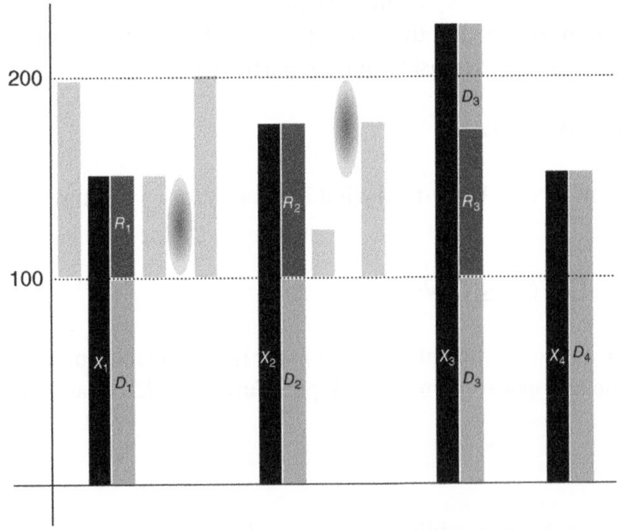

Figure 2.4 100 xs 100 treaty with one reinstatement and claims $X_1 = 150, X_2 = 175, X_3 = 225$, and $X_4 = 150$.

Note that with such a reinstatement clause, the premium payment is no longer deterministic, but depends on the loss history during the contract. Such reinstatements can be negotiated to be automatic or optional for the first-line insurer. Clearly, this variant of XL contracts is attractive for the cedent, as some premiums only have to be paid if more coverage is needed, and there is less financial burden for the cedent in the beginning. With such clauses, it is easier to agree on premiums, and one can interpret the setup as a loss participation scheme of the cedent, which also reduces moral hazard. The actuarial analysis of such contracts is, however, more involved. Early discussions of XL contracts with reinstatements can be found in Sundt [717, 718] and Rytgaard [663]. The resulting distribution of aggregate claim sizes is studied in Walhin and Paris [767] and Hürlimann [455], and for corresponding ruin probabilities see Walhin and Paris [768] and Albrecher and Haas [417]. For the pricing of such contracts see Section 7.4.3.

2.3.3 Further Practical Considerations

There are many issues to be taken care of in the concrete implementation of an XL treaty. One is that the way in which inflation influences the retention may not be the same as for the claim sizes, and corrections of inflation effects (and implementing respective clauses in the contracts) are an integral part of XL treaty analysis (see also Chapter 7). Walhin et al. [766] provide an overview of financial, economic, and commercial aspects in the practical pricing of XL treaties. On the data side, particularly for the lines of business relevant for XL treaties (such as liability), some claims take several or even many years until settlement, so respective reserving techniques on an individual claims level have to be set up and one faces statistical challenges (see Chapter 4).

It is not uncommon to combine XL treaties with adjacent layers. In practice, one often refers to the *rate on line* (ROL)

$$\text{ROL} = \frac{\text{premium for layer}}{\text{size of layer}}$$

and the *payback period* 1/ROL, which reflects the average number of years it takes to collect premiums for this layer so that one payment of the full layer is reimbursed (one distinguishes *working layers* (small ROL), *middle layers* (medium ROL), and *CAT layers* (high ROL); this terminology is mainly used for casualty risk). Finally, layers far in the tail are sometimes referred to as *capacity layers*. For property losses the respective expressions are *frequency layer, estimated normal loss layer, PML layer, PML protection layer*, and *CAT layer*.

Whereas the usual XL treaty (2.3.6) is defined *per risk*, this may not be an effective reinsurance form if one faces many (but possibly not excessively large) claims (*frequency risk*). One way to deal with this problem is a *cumulative XL* (*per-event XL, cat XL*) contract, where first all claims that can be attributed to the same event are aggregated to $X_j^c = \sum_i X_i$ and then the reinsurer pays the excess of that aggregate claim over a specified priority M_c, that is,

$$R(t) = \sum_{\text{events}} \left(X_j^c - M_c\right)_+ .$$

Mathematically, this reduces again to an XL-type contract, with the number of claims now replaced by the number of events and the individual claims replaced by the aggregate claim per event. That is, if there are enough data available to model the event process and the respective aggregate claim distribution for an event, one can again use the same techniques as for the per risk XL. Cumulative XL treaties are popular in property, marine hull, motor hull, personal accident, and natural catastrophe insurance.

It is evident that in XL reinsurance there is an *adverse selection*, namely that an insurer seeks protection particularly for risks that are hard to model or have heavy tails, but often with very limited past claim experience. Some aspects of this phenomenon will be dealt with later in the book. Another issue in practice can be moral hazard in its various forms. In addition to what was mentioned above, another example is that if a claim is already larger than the retention, there may not be the same incentives for the cedent to be very careful with the exact settlement of the claim (as this is costly). A finite layer size L is already a partial remedy to this problem, another one is that the reinsurer only pays a (pre-specified) fraction of the original reinsured amount in the XL treaty (such contracts are sometimes referred to as *change-loss contracts*).

Variants of per-risk XL are *adverse development covers* (dealing with run-offs) and *industry-loss warranties* (cf. Section 10.2).

2.4 Stop-loss Reinsurance

A *stop-loss* (SL) reinsurance treaty is defined by

$$R(t) = \left(\sum_{i=1}^{N(t)} X_i - C \right)_+ ,$$

that is, the aggregate loss that the cedent faces over the time interval is capped at the *priority* C, and the reinsurer takes care of the excess. Although SL is a natural alternative to XL reinsurance that relieves the cedent from tail risk completely, some of the problems of XL (like the moral hazard issue) are even amplified for this reinsurance form. In almost all cases implemented in practice there is also an upper limit U specified in the contract:

$$R(t) = \min \left\{ \left(\sum_{i=1}^{N(t)} X_i - C \right)_+ , U \right\}. \tag{2.4.15}$$

To avoid some of the problems, such a reinsurance form is typically combined with a proportional and/or XL cover *inuring* to the SL treaty (i.e., the SL feature only applies to the risk remaining after the other treaties) (cf. Section 2.6). It is also common to include an additional *unconditional deductible* in (2.4.15) to lower the administration costs connected to very small losses.

A SL contract is particularly useful if the assignment of claims to a particular event for a per-event XL is difficult (such as hail, agriculture or frostiness of waterpipes). However, on the administrative side such a contract can be quite tedious, as all claims have to be considered (and potentially agreed upon) by both parties and modelling for the aggregate tail can be a challenge. In addition, all small claims contribute to the reinsured amount, a feature that makes this reinsurance form less effective. As a consequence, such contracts are typically expensive and not so frequently applied (an exception being, for example, German windstorm reinsurance). A SL cover is, however, a popular reinsurance form between connected institutions. The distributional properties of a SL contract can in principle be deducted from the corresponding ones of single risks in an XL cover (cf. Section 2.3.1), but due to the clauses in the aggregation structure the analysis can in concrete cases still look quite different.

2.5 Large Claim Reinsurance

Consider the ordering of the claims

$$X_{1,N(t)} \leq X_{2,N(t)} \leq \cdots \leq X_{N(t)-1,N(t)} \leq X_{N(t),N(t)}.$$

In a *large claim reinsurance* contract the reinsurer agrees to cover the largest r claims, where $r \geq 1$ is a fixed number, that is,

$$R(t) = \sum_{i=1}^{r} X_{N(t)-i+1,N(t)}.$$

This is an intuitive treaty against the risk of large claims for the cedent, and it leads to challenging mathematical questions for the distribution of $R(t)$ and $D(t)$, which we will discuss in some detail in Section 10.1, particularly also with respect to asymptotic properties. A variant is drop-down XL with

$$R(t) = \sum_{i=1}^{N(t)} \min\{L_i, (X_{N(t)-i+1,N(t)} - M_i)_+\}, \tag{2.5.16}$$

where different retentions and layer sizes are applied to the respective order statistics (e.g., see Ladoucette and Teugels [522, 524]). While the implementation of such treaties seems quite reasonable, they are nevertheless not very popular in modern reinsurance practice, probably due to the intricate mathematical details and the considerable model risk, as particularly for the largest claims in the portfolio one only has limited means to justify on a statistical basis assumptions on the claim distribution. A variant of (2.5.16) that has some popularity is called *second event retention (or DD)*, which is an XL cover, where the retention and the size of the layer also vary for different claims over time, but are then ordered by chronological appearance rather than size.

A further variant of large claim reinsurance that was implemented for some time in certain countries is called ECOMOR (Excédent du Coût Moyen Relatif). Introduced by Thépaut [742], it was defined by

$$R(t) := \sum_{i=1}^{r} X_{N(t)-i+1,N(t)} - rX_{N(t)-r,N(t)} = \sum_{i=1}^{N(t)} \left(X_i - X_{N(t)-r,N(t)}\right)_+,$$

that is, it is an XL-type treaty, where the retention is given by the $(r+1)$th largest claim. The respective mathematical analysis is slightly more complicated than for large claim reinsurance contracts, and we will give some details in Section 10.1.3. One motivation for the introduction of such an (a priori) random retention was the advantage that it is (by definition) prone to the same inflation forces as the claims themselves. At the same time, there is an intuitive drawback of the ECOMOR construction: if during the contract period an additional claim appears that raises the applied retention, it can happen that the reinsured amount decreases, although the overall burden for the cedent has increased by the arrival of this additional claim, a feature that is not desirable and hence moral hazard can be a problem. Up to now ECOMOR treaties have enjoyed some academic popularity due to their mathematical challenges, but these are also to some extent the reason why this reinsurance form is not used in current practice. There are, however, some reinsurance treaties in force that to some extent mimick ECOMOR features, such as an SL cover on all claims that are larger than some threshold.

2.6 Combinations of Reinsurance Forms and Global Protections

Even for a relatively simple property portfolio a combination of different types of reinsurance protections is quite common. Clearly, this complicates the analysis of the contracts and makes it even more important to clearly understand the implications of each type of contract on the shape of the retained and reinsured risk. One of the main reasons for implementing such combinations is heterogeneity. Heterogeneity is induced by the difference of sums insured per policy, but also differences in coverage (theft, third party protection, etc.) and in type (simple, commercial or industrial risks, etc.).

Combinations of proportional and non-proportional reinsurance protections are quite frequent. If one combines non-proportional reinsurance types, a logical order needs to be respected: first a per-risk XL, second a per-event XL, and finally a SL. A QS can be ceded in any order. When applied, a surplus treaty (possibly preceded by a QS) should be the first in the series. A surplus after a per-event XL is logically impossible, while a surplus after a per-risk XL will imply that the risks with higher sum insured, which are systematically ceded to the surplus, also have a higher potential to lead to XL payments. So one should recover parts of the XL premium on a pro rata basis from the surplus reinsurers, which entails some challenges in practice.

There seems to be an increasing trend that ceding companies ask for *global protections* of their portfolios. One example is the protection by the reinsurer of two (or more) lines of business, for instance of fire and MTPL. This allows better advantage to be taken of the diversification possibilities. In a *multiline XL* coverage, layers from different lines of business are combined in one treaty with a *global* (or *multiline*) annual aggregate deductible AAD which is taken to be rather large (to keep the reinsurer's payments and the respective reinsurance premium in a reasonable range).

As an example consider a classical fire treaty with three layers (2000 xs 1000; 3000 xs 3000; 1000 xs 6000) with reinstatements, and a MTPL treaty with three layers (3000 xs 2000; 5000 xs 5000; ∞ xs 10000) and unlimited free reinstatements. An alternative coverage could consist of a global multiline treaty combining a fire layer 2500 xs 500 with a 4000 xs 1000 MTPL layer with unlimited free reinstatements, and a multiline retention of 1000. These two layers then form one treaty with one global premium, which is complemented by the two remaining fire layers and the two remaining MTPL layers. In order to price such a multiline coverage one needs to model the possible dependence between the different components or lines of business. Analysis of the Danish fire insurance data will be used to illustrate the multivariate joint modelling of different components.

2.7 Facultative Contracts

Currently about 15% of reinsurance contracts (in terms of premium volume) are not of an obligatory treaty form, but are contracts negotiated on individual risks. This includes in particular coverage for non-standard risks such as reinsurance for skyscrapers, powerplants etc. As there are virtually no a priori rules for such contracts and the agreed risk-sharing mechanisms can vary widely, they cannot be treated systematically. Due to their individual nature the premiums are typically considerably higher than for standardized obligatory treaties. Often, the insurer tries to avoid including risks in a treaty that are particularly prone to large losses as this may have a bad effect on premiums in the following year, and so the insurer will aim for a facultative reinsurance for those risks.

Further examples of non-standard reinsurance forms are discussed in Section 8.8.

2.8 Notes and Bibliography

In addition to the general references already given at the end of Chapter 1, it should be noted that some reinsurance companies regularly publish online information on practical details and reinsurance market trends, see, for example, the webpages `http://www.swissre.com/sigma/` and `http://www.munichre.com`.

3

Models for Claim Sizes

In this chapter we deal with models that are useful for describing individual and collective claim size data and physical measurement data of natural hazards such as earthquakes and floods. Because we want to give special attention to the modelling of large claims we provide some general background on this topic. In what follows we will try to distinguish between claims that are considered small and claims that are large. It is intuitively clear that reinsurance contracts will depend heavily on whether or not the individual claims should be considered large.

3.1 Tails of Distributions

One of the main reasons for taking reinsurance is the possible appearance of large claims. While this sounds like an obvious statement, a useful and acceptable definition of what is meant by a large claim is far from obvious. Among the possible examples of distributions, some are better suited to model large claims than others. In view of future applications to actuarial topics one definitely needs a way of making a difference between average claims and large claims. An acceptable guideline is to compare claim distributions with the exponential distribution. In some sense the exponential distribution acts as a *splitting distribution* between small and large. A first and rather vague criterion would be to check whether the claim distribution under consideration has a *fatter tail* than the exponential distribution or not. If it does not, then we could call the distribution *super-exponential*; in the alternative case we might call the distribution *sub-exponential* but unfortunately this term has already been standardized in the probabilistic literature.

We will call a distribution F *super-exponential* if $1 - F(x)$ is bounded above by a decreasing exponential. A more quantifiable definition can be given in terms of $\widehat{F}$, the Laplace transform of F, in that $\widehat{F}$ has a strictly negative abscissa of convergence σ_F. Following Taylor [731] we will indeed conclude that the super-exponential class of distributions is a reliable family of light-tailed claim size distributions. The exponential distribution itself also satisfies this criterion of super-exponentiality. The aggregate claim distribution (but also other risk quantities such as ruin probabilities) often exhibit exact or approximate exponential behavior when the underlying claim size distribution is super-exponential.

Reinsurance: Actuarial and Statistical Aspects, First Edition.
Hansjörg Albrecher, Jan Beirlant and Jozef L. Teugels.

For a distribution F with a fatter tail than any exponential the abscissa of convergence σ_F of $\hat{F}$ will be zero. Unfortunately this property is not sufficiently specific to be useful as a definition. Instead one uses some classes of distributions with $\sigma_F = 0$ that have a bit more structure. In particular we will deal later with sub-exponential distributions, Pareto-type distributions, extreme value distributions, etc. If we take the claim size distribution from such a class, the corresponding aggregate risk and ruin quantities will show no trace of exponential behavior.

3.2 Large Claims

Before we offer candidates for claim size distributions, we need to remind the reader that one of our main objectives is to provide adequate models for large claims. This section contains a few thoughts on what we might righteously call a large claim and how one can perhaps distinguish it from others. For an early general discussion on the role of large claims see the summary report by Albrecht [32]. For an attempt to define the even more difficult concept of a *catastrophic claim*, see Ajne et al. [10].

Consider all claims $\{X_1, X_2, \dots, X_n\}$ related to a specific portfolio. Let $S_n = \sum_{i=1}^{n} X_i$ be the total claim amount and consider the maximum value $X_{n,n}$. Under which conditions should we consider this largest claim to be actually *large*? More generally, which of the extreme order statistics could be considered to be large? For an attempt to define large, see Teugels [738]. Beirlant et al. [105] put large claims into a statistical and actuarial context. Here are some interpretations, the first two theoretically inspired.

(i) A claim could be called large when the total claim amount is predominantly determined by it. A possible interpretation would be to assume that S_n is large because $X_{n,n}$ is so. This might be interpreted by the condition

$$1 - F^{*n}(x) := \mathbb{P}(S_n > x) \sim \mathbb{P}(X_{n,n} > x)\ ,\ x \uparrow \infty.$$

For $n = 2$ this is equivalent to

$$1 - F^{*2}(x) \sim 2\,(1 - F(x))\ ,\ x \uparrow \infty. \tag{3.2.1}$$

The last equivalence follows from

$$\mathbb{P}\{X_{2,2} > x\} = 1 - F^2(x) = (1 - F(x))(1 + F(x)) \sim 2(1 - F(x)),\ x \uparrow \infty.$$

The relation (3.2.1) is precisely the definition that F belongs to the class $\mathcal{S}$ of *sub-exponential distributions*. A property of $\mathcal{S}$, proved by Chistyakov [211], is that if $F \in \mathcal{S}$ then for any non-negative integer n, as $x \to \infty$,

$$\lim_{x\to\infty} \frac{1 - F^{*n}(x)}{1 - F(x)} = n. \tag{3.2.2}$$

In the sub-exponential case, the tail $1 - F^{*n}$ is (up to the quantity n) just as heavy as that of $1 - F$.

Members of $\mathcal{S}$ are automatically members of the class of *long-tailed* distributions[1] denoted by $\mathcal{L}$, which means that for all $y \in \mathbb{R}$

$$\lim_{x\to\infty} \frac{1-F(x+y)}{1-F(x)} = 1 .$$

The class $\mathcal{S}$ and its subclasses have been constantly used as candidates for claim size distributions with a heavy tail but also in other probabilistic contexts like branching processes, queueing theory, etc. The major drawback of $\mathcal{S}$ is that it is only defined in terms of a limiting property, which is hard to verify in practice, and that it is defined by a non-parametric condition. Up to now the sub-exponential class has defied a representation. Actually, it remains a challenging problem to decide whether a set of actuarial data comes from a sub-exponential distribution or not. Further, it is known that $\mathcal{S}$ is not closed under convolution or under convex combinations. Cline and Samorodnitsky [195] have shown that nevertheless large subclasses of $\mathcal{S}$ are closed under product operations (see also the work by Rosinski et al. [657]). A variety of sufficient conditions for membership of $\mathcal{S}$ can be found in the literature, for example see Pitman [622], Teugels [737], Klüppelberg [493], Pinelis [619], Smith [704], and Foss et al. [356] for a recent survey. Refinements are available in Chover et al. [212] and Willekens [779].

Henceforth, practitioners avoid the class as such, going for distributions that are sub-exponential but that at the same time contain enough parameters. Only then will they be able to use data. Among the many parametrized examples in $\mathcal{S}$ we mention the log-normal distribution, the Pareto distribution and Pareto-type distributions to be defined next, as well as non-normal stable distributions.

(ii) Another way of defining large claims can start from the behavior of the maximum claim, which is then also closely related to the concept of the PML (cf. Chapter 2). This mathematical problem was considered in the work by Fisher and Tippett [353] and Gnedenko [391], and was further streamlined by de Haan [257]. The main question is the search for distributions of X for which there exist sequences $a_n > 0$ and b_n $(n = 1, 2, \dots)$ such that for all real values x (at which the limit is continuous)

$$\lim_{n\to\infty} \mathbb{P}\left(\frac{X_{n,n} - b_n}{a_n} \le x\right) = G(x) \tag{3.2.3}$$

for some non-degenerate distribution G. It was then shown that the extreme value distributions

$$G_\gamma(x) = \begin{cases} \exp\left(-(1+\gamma x)^{-1/\gamma}\right), & \text{if } \gamma \neq 0, \\ \exp\left(-e^{-x}\right), & \text{if } \gamma = 0, \end{cases} \tag{3.2.4}$$

are the only possible limits in (3.2.3), with $\gamma \in \mathbb{R}$ called the *extreme value index* (EVI). It has to be understood that G_γ has to be a proper distribution. This means

1 Here the concept of long-tailedness should not be confounded with long-tailed insurance portfolios, which refers to claims with long development times.

in particular that the range of G_γ extends over the interval $(-\frac{1}{\gamma}, +\infty)$ if $\gamma > 0$ (the *Fréchet–Pareto case*), over $(-\infty, -\frac{1}{\gamma})$ if $\gamma < 0$ (the *extremal Weibull case*), or over $(-\infty, +\infty)$ if $\gamma = 0$ (the *Gumbel case*).

For any specific γ, the *max-domain of attraction*, containing the distributions F for which there exist sequences $a_n > 0$ and b_n such that (3.2.3) holds, have also been described. The class of distributions in the max-domain of attraction can be defined in terms of the tail quantile function. Given that F is a distribution, its *quantile function* is defined by the inverse function

$$Q(p) = \inf\{x|F(x) \geq p\}, \quad p \in (0,1).$$

The *tail quantile function* is defined and denoted by

$$U(t) = Q(1 - 1/t), \quad t > 1.$$

The following condition is a necessary and sufficient condition for the existence of normalizing and centering constants for the weak convergence (3.2.3) of the maximum of a sample from the distribution F.

Definition 3.1 Let F be a distribution with tail *quantile function* $U(t)$. The distribution F belongs to the extremal class $C_\gamma(a)$ if there exists an EVI γ and an ultimately positive function a such that

$$\{U(xu) - U(x)\}/a(x) \to \int_1^u v^{\gamma-1}\, dv \tag{3.2.5}$$

for all $u \geq 1$ as $x \to \infty$.

Here the restriction to $u \geq 1$ can be broadened to $u > 0$. It follows from the general theory of regularly varying functions that a is automatically *regularly varying at infinity with index* γ: $a(x) = x^\gamma \ell(x)$ where ℓ is *slowly varying*, that is, a measurable and ultimately positive function that satisfies

$$\lim_{x\uparrow\infty} \frac{\ell(tx)}{\ell(x)} = 1,$$

for all $t > 0$. For more information see Bingham et al. [135].

It has been shown by de Haan in [257] that the above definition can be equivalently stated in terms of the original distribution. The alternative condition is

$$\frac{1 - F(t + uh(t))}{1 - F(t)} \to (1 + \gamma u)^{-\frac{1}{\gamma}}, \text{ as } t \to x_+, \tag{3.2.6}$$

for all u such that $1+\gamma u > 0$ and where $h \circ U = a$. For $\gamma = 0$ we read $(1+\gamma u)^{-\frac{1}{\gamma}} = e^{-u}$. The above relation holds locally uniformly in u. Furthermore if $\gamma > 0$, then $a(x)/U(x) \to \gamma$ as $x \to \infty$. However, if $\gamma < 0$, then F has a finite upper limit

$x_+ = \inf\{y : F(y) = 1\}$ and $a(x)/(x_+ - U(x)) \to -\gamma$. The limit distribution in (3.2.6), that is,

$$F(u) = 1 - (1 + \gamma u)^{-\frac{1}{\gamma}},$$

is called the *generalized Pareto distribution (GPD).*

In case $\gamma > 0$, the class C_γ equals the class of *Pareto-type distributions* defined by

$$1 - F(x) \sim x^{-\alpha}\ell(x), \ x \uparrow \infty, \tag{3.2.7}$$

where $\alpha = 1/\gamma > 0$ and ℓ is slowly varying. Note that $1 - F$ in (3.2.7) is regularly varying with index $-\alpha$.

When $\gamma \leq 0$, the underlying X has a tail that is lighter than Pareto-type distributions. In case $\gamma = 0$ the tail of the distribution of X can have a finite endpoint or infinite support.

Seemingly the first attempt to model large claims with a parameterized distribution is due to Benckert et al. [109]. Here the authors assume that the claim size distribution starts out as a Pareto distribution, this means that for large x, $1-F(x) \sim cx^{-\alpha}$ for some positive α. The distribution is then "cut off" at the point corresponding to the sum insured, in which the remaining mass of the Pareto distribution is concentrated. This then yields a model with negative γ. A bit later, Benktander [110] pointed out how the Pareto distribution itself (or its variants) could be used to model large claims. In particular he considered the Pareto class as a dividing class between claim size distributions for which all moments are finite and those for which most moments diverge. Pareto-type distributions have always been popular, for instance when modelling fire, storm and liability data, as will be illustrated in Chapter 4. For a survey of extreme value theory and relevant references, see Embrechts et al. [329], Beirlant et al. [100], and de Haan and Ferreira [258].

(iii) For statistical purposes one can make a distinction between tails that tend to 0 faster or slower than the exponential distribution near infinity. When for any $\lambda > 0$

$$\frac{\mathbb{P}(X > x)}{e^{-\lambda x}} \to 0 \text{ as } x \to \infty$$

the tail of X is termed *lighter than exponential* (LTE), while for

$$\frac{\mathbb{P}(X > x)}{e^{-\lambda x}} \to \infty \text{ as } x \to \infty$$

the tail of X is termed *heavier than exponential* (HTE).

(iv) When an actuary looks at claim data, he might suspect that large claims are likely. For example, when he tries to estimate the mean and/or the variance of the claim size distribution in a sequential way, he notices that the sample estimates do not converge. Even when he uses re-sampling techniques, the estimates fail to average out to a limiting value. One possible reason might be that the mean or the variance of the underlying claim size distribution actually does not exist because the tail of the distribution is too heavy. This could mean that the law of large numbers does not apply, preventing ultimate stabilization. A possible parameterized form to cope with such a phenomenon is to assume that F is a Pareto-type distribution with a small index α, as considered above.

(v) One could call a distribution heavy tailed if the ratio $S_n/X_{n,n}$ converges in distribution to a non-degenerate limit law. This approach leads to the class of Pareto-type distributions with $0 < \alpha < 1$, as shown by Breiman [163]. If the distribution F has a finite expected value μ then the analogous condition that $(S_n - n\mu)/X_{n,n}$ converges to a non-degenerate limit leads to a similar outcome but with $1 < \alpha < 2$.

(vi) Another possibility is that a number of the largest claims consume a fair portion of the total claim volume. In a formula, for some $p \in (0, 1)$, $X_{n,n} > pS_n$. Of course, the value of p should be large enough since otherwise all order statistics could be termed large. In a more general phrasing, one could look at

$$\frac{1}{S_n}\left(X_{n,n} + X_{n-1,n} + \ldots + X_{[n\alpha],n}\right)$$

for a value of α close to 1. For a quantification of this approach in terms of Lorenz curves, see Aebi et al. [9].

(vii) In a similar fashion we might interpret largeness by the condition that for some finite constant c when $n \uparrow \infty$,

$$\mathbb{E}\left\{\frac{S_n - n\nu}{X_{n,n}}\right\} \to c$$

where ν is the mean μ if it is finite while otherwise ν is 0. As shown in Bingham et al. [136] this condition again leads to the Pareto-type distributions with $\alpha \in (0, 1) \cup (1, 2)$.

Before continuing we need to stress the difference between a large claim and an *outlier*. While the first is a genuine member of the sample of claim sizes, an outlier is considered an extraneous value. Next to clear misprints, events can occur which are completely unexpected in view of all data before such an event. Using methods from extreme value analysis (EVA) one can estimate how unlikely certain events are in view of all prior information. When events with an extremely low likelihood do occur, however, one has to be ready to change the statistical models.

3.3 Common Claim Size Distributions

In this section we state the traditional examples of claim size distributions that are commonly considered in the actuarial literature. Some of these examples are simple while others are more elaborated variations. For other surveys of common claim size distributions, see Kupper [517], Ammeter [40] and Klugman et al. [491].

In many cases, distributions can be derived from a simple original by a transformation. Among the most popular are the following:

- replacement of the random variable X by a *Box–Cox transformation*, that is, $Y_\lambda := (X^\lambda - 1)/\lambda$, which for the limit $\lambda = 0$ gives $Y_0 = \log X$

- replacement of X by $\exp(X)$, where the resulting distribution is called a *log-distribution*
- replacement of X by a *normalized* version $aX + b$ for constants $a > 0$ and $b \in \mathbb{R}$
- replacement of X by $\frac{X}{a+X}$ with $a > 0$, a so-called *homeographic transformation*.

Note that such transformations may dramatically change the tail behavior of the distribution.

Because of the importance of extreme values in reinsurance, the *extreme value distribution* G_γ, $\gamma \in \mathbb{R}$, and the generalized Pareto distribution are important candidates for modelling purposes in view of the limit results (3.2.3) and (3.2.6). The sets of extreme value distributions and generalized Pareto distributions are one-parameter families of distributions ranging from light tails with a finite endpoint (with $\gamma \leq 0$), up to Pareto-type tails (when $\gamma > 0$). Applying a normalization to these families we obtain the location-scale versions with $\mu \in \mathbb{R}$ and $\sigma > 0$:

$$G_{\gamma,\mu,\sigma}(x) = \exp\left\{ -\left(1 + \gamma\frac{x-\mu}{\sigma}\right)^{-\frac{1}{\gamma}} \right\}$$

and

$$F_{\gamma,\mu,\sigma}(x) = 1 + \log G_{\gamma,\mu,\sigma}(x) = 1 - \left(1 + \gamma\frac{x-\mu}{\sigma}\right)^{-1/\gamma}, \tag{3.3.8}$$

where $1 + \gamma\frac{x-\mu}{\sigma} > 0$. The latter distribution has been used to model aggregate claim distributions in McNeil [567]. Condition (3.2.6) leads to the popular *peaks-over-threshold (POT)* approach in EVA, as discussed in Chapter 4.

We now list a number of examples of models with tails that are exponentially bounded and then turn to tails heavier than exponential.

3.3.1 Light-tailed Models

3.3.1.1 With EVI $\gamma < 0$

A classical example of a light-tailed distribution with finite endpount x_+ is given by the *beta* distribution with $x_+ = 1$ and distribution function

$$F(x) = \frac{1}{B(p,q)} \int_0^x u^{p-1}(1-u)^{q-1}du,$$

with extreme value index $\gamma = -1/q$ (here $B(p,q) = \Gamma(p)\Gamma(q)/\Gamma(p+q)$ denotes the beta function). The uniform distribution on $(0,1)$ is of course a special case with $p = q = 1$. This is then a possible model for loss degree data. Beta distributions can be constructed starting from a Pareto-type random variable Y (cf. (3.2.7)) through the transformation $X = x_+ - 1/Y$ leading to an extreme value index $\gamma = -1/\alpha$:

$$1 - F(x) = \mathbb{P}\left(Y > \frac{1}{x_+ - x}\right) = (x_+ - x)^\alpha \ell((x_+ - x)^{-1}),\ x < x_+.$$

Another way to produce a light tail with finite endpoint from a heavy-tailed distribution W is by conditioning on $W < T$ for some value T:

$$X =_d W|W < T.$$

Such an operation is called here *upper-truncation*.[2] A first reference in this respect is Benckert et al. [109]. See Clark [217] for a reference in enterprise risk management.

With T fixed, one can show that X is then light tailed with EVI $\gamma = -1$. When modelling large claims it appears appropriate to consider T sufficiently large, possibly with the meaning of a sum insured. Another example is found in the Gutenberg–Richter model for earthquake magnitudes, as will be discussed when treating earthquake data in Chapter 4.

3.3.1.2 With EVI $\gamma = 0$: the Gumbel Domain

1) Our first set of examples starts from the *exponential distribution* given by

$$F(x) = 1 - \exp(-\lambda x), \; x > 0, \lambda > 0.$$

The exponential distribution plays a central role in tail modelling, not least because of its *memoryless property*, that is, for all $s, t > 0$

$$\mathbb{P}(X > s + t|X > t) = \mathbb{P}(X > s).$$

Since the tail quantile function of the exponential distribution equals $U(t) = (\log\, t)/\lambda$, the exponential distribution belongs to $C_0(a)$, where $a(t) = 1/\lambda$. From the exponential distribution we get the following:

(i) A first Box–Cox transformation of the exponential distribution is known as the *Weibull distribution* and is defined by

$$F(x) = 1 - \exp(-\lambda\, x^{\tau}), \; x > 0.$$

For $\tau > 1$ the distribution is still super-exponential. Note the special case of the *Rayleigh distribution* which is obtained by putting $\tau = 2$. The Weibull distribution has $U(t) = \left(\frac{1}{\lambda}\log\, t\right)^{1/\tau}$ as the tail quantile function so that this distribution belongs to $C_0(a)$, where $a(t) \sim \frac{1}{\tau}\lambda^{-\frac{1}{\tau}}(\log\, t)^{-1+\frac{1}{\tau}}$.

(ii) An exponential change followed by a normalization yields the *logistic distribution* with explicit form

$$F(x) = 1 - \{1 + e^{\frac{x-\mu}{\sigma}}\}^{-1}, \; x \in \mathbb{R},$$

where μ is a real parameter and $\sigma > 0$. Note that this distribution is taken on the entire real line. One-sided versions are of course possible. The most common choice of the latter is the *one-sided logistic distribution*

$$F(x) = 1 - 2(1 + e^{x})^{-1}, \; x > 0.$$

---

2 This notion of truncation should not be confused with other truncation schemes, referring to missing data, such as in the case of non-reported claims because they are below a reporting level (cf. Chapter 4).

2) The *gamma distribution* also has a long tradition in claim size modelling. Its explicit form is

$$F(x) = \frac{1}{\Gamma(\alpha)} \int_0^{\lambda x} e^{-u} u^{\alpha-1} du, \ x > 0,$$

where $\alpha, \lambda > 0$. Further $\widehat{F}(s) = (1 + \frac{s}{\lambda})^{-\alpha}$ so that $\sigma_F = -\lambda$ and hence, the gamma distribution is super-exponential. For integer values of α the gamma distribution can be characterized as a sum of independent exponential random variables (and is then referred to as the *Erlang* distribution). For $\alpha = n/2$ and n an integer we find a *chi-squared distribution.*

Many other special Box–Cox forms are available. We mention here the *transformed gamma distribution,* obtained from the gamma distribution via a power transformation. We find

$$F(x) = \frac{1}{\Gamma(\alpha)} \int_0^{\lambda x^{\tau}} e^{-u} u^{\alpha-1} du, \ x > 0,$$

a distribution with three parameters.

When there is good reason to believe that a claim comes from one of several different risk classes and for each of these classes one has a good idea about the claim size distribution, then a mixing distribution will be a natural model. In this context, mixtures of Erlang distributions are very popular in claims modelling, for example see Willmot and Woo [790]. Such mixed Erlang distributions are used in Chapter 4 to produce global fits in combination with separate tail fits. A popular, tractable and more general class of super-exponential type in such a probabilistic construction context are *phase-type distributions* (see Bladt and Nielsen [139], [57, Ch. IX] and Asmussen et al. [65] for the statistical perspective). For a recent variant of infinite-dimensional phase-type distributions with finitely many parameters leading to a heavy-tailed distribution, see Bladt et al. [140].

3) When modelling claim size distributions, the *normal distribution* can hardly be advocated as a valuable model because claim sizes are non-negative. Nevertheless the distribution is still popular as an approximation. Some distributions derived from it have also found their way into the actuarial literature:

   (i) The *one-sided normal distribution* is a candidate, suggested already in Benktander et al. [115]. It has the density function

$$f(x) = \sqrt{\frac{2}{\pi}} e^{-x^2}, \ x > 0.$$

   (ii) The *inverse Gaussian distribution* is defined by the density function

$$f(x) = \sqrt{\frac{\beta\mu^2}{2\pi}}\, x^{-\frac{3}{2}}\, e^{-\frac{\beta}{2x}(x-\mu)^2}, \ x > 0,$$

where the two parameters β and μ are positive. Its Laplace transform is given by

$$\hat{F}(s) = \exp\left(-\beta\mu\left(\sqrt{1+\frac{2s}{\beta}}-1\right)\right),$$

from which it is easy to see that the density of the n-fold convolution is

$$f^{*n}(x) = \sqrt{\frac{\beta n^2\mu^2}{2\pi}}\, x^{-\frac{3}{2}}\, e^{-\frac{\beta}{2x}(x-n\mu)^2},\ x > 0.$$

Hence $\sigma_F = -\beta/2$ and the distribution is super-exponential. For further properties see Embrechts [323]. The closedness under convolution makes this distribution an interesting candidate for claim size modelling, probably Seal [692] was the first to consider it for this purpose. Later applications can be found in Gendron et al. [380], ter Berg [736] and Mack [555].

3.3.2 Heavy-tailed Models

3.3.2.1 With EVI $\gamma = 0$: the Gumbel Domain

1) Our first examples can be derived from the exponential distribution.

   (i) The *Weibull distribution*

   $$F(x) = 1 - \exp(-\lambda\, x^{\tau}),\ x > 0,$$

   is sub-exponential for $0 < \tau < 1$.

   (ii) The *second Benktander distribution* is defined by the expression

   $$F(x) = 1 - c\, a\, x^{b-1} \exp\left(-\frac{a}{b}x^b\right),\ x > 0,$$

   where a and c are positive constants and $0 < b < 1$.

2) Also the normal distribution can give rise to heavy tails after transformation.

   (i) The most popular such distribution is the *log-normal distribution,* defined as a two-parameter distribution of the form

   $$F(x) = 1 - \frac{1}{\sigma\sqrt{2\pi}}\int_x^{\infty} \exp\left\{-\frac{(\log\, u-\mu)^2}{2\sigma^2}\right\}\frac{du}{u} = \Phi\left(\frac{\log\, x-\mu}{\sigma}\right).$$

   Here $\mu \in \mathbb{R}$ while $\sigma > 0$. This important distribution belongs to $\mathcal{S}$, as shown in Embrechts et al. [326], and asymptotically has a tail heavier than the Weibull distribution, namely

   $$\overline{F}(x) \sim \frac{\sigma}{\log x\sqrt{2\pi}}\exp\left\{-\frac{1}{2}\left(\frac{\log x-\mu}{\sigma}\right)^2\right\},\ x \to \infty.$$

   In 1962, Benckert [108] suggested the use of the log-normal distribution for the modelling of industrial and non-industrial fire data. Ferrara [347] fitted a log-normal distribution to fire claim data. Further specific examples have been treated in the papers by Bennett et al. [116] and Dickmann [289], and Taylor [729] also illustrated the use of the log-normal. For an early application to windstorm and glass claims data see Ramlau-Hansen [640].

(ii) The *quasi-log-normal distribution* is defined by the following class containing three parameters

$$F(x) = 1 - b(x/x_0)^{\{-\alpha-\beta\log(x/x_0)\}},\ x > x_0,$$

where α, b and β are positive parameters. It captures the dominant component $\exp(-\beta\log^2 x)$ (for some $\beta > 0$) in the tail behavior of a log-normal distribution.

3.3.2.2 With EVI $\gamma > 0$

This class corresponds to the Pareto-type distributions as defined in (3.2.7).

1) By far the most popular distribution to generate heavy claims is the *Pareto distribution* with its transformed versions. It is the prime example of heavy-tailed distributions. The simplest possible definition is the *strict Pareto distribution* given for $\alpha > 0$ by

$$F(x) = 1 - (x/x_0)^{-\alpha},\ x > x_0 > 0.$$

The strict Pareto is sub-exponential for all values of α. It can also be seen as a log-distribution generated by an exponential random variable. The fact that this distribution is only defined from a positive value x_0 on, is often not considered a problem since it is mainly used to model large claims. More than that, this distribution is very popular in practice because of a certain type of lack-of-memory property: for any threshold $M > x_0$ one has

$$\mathbb{P}(X > x|X > M) = \begin{cases} \left(\frac{x}{M}\right)^{-\alpha}, & x > M, \\ 1, & \text{else}, \end{cases} \tag{3.3.9}$$

that is, the conditional excess is again Pareto-distributed, now with parameters (α, M), a property that is particularly attractive in XL reinsurance.

2) The *shifted Pareto distribution* is a two-parameter family defined by

$$F(x) = 1 - \beta^{\alpha}(\beta + x)^{-\alpha},\ x > 0.$$

It can be obtained from the strict Pareto by a simple shifting and rescaling or from the Pareto-type by a specialization of the slowly varying function. Note that its support is now the entire positive axis. For a treatment in an actuarial context see Seal [694]. It is also known as the *US–Pareto distribution* in actuarial circles. The GPD is actually the special case $\beta = 1/\alpha$.

3) Adding an additional power to the shifted Pareto distribution yields the versatile *Burr distribution* defined by

$$F(x) = 1 - \beta^{\alpha}(\beta + x^{\tau})^{-\alpha},\ x > 0.$$

This three-parameter distribution has received a lot of attention in the actuarial literature. The tail quantile function of the Burr distribution is

$$U(t) = \beta^{1/\tau}(t^{1/\alpha} - 1)^{1/\tau}.$$

The distribution therefore belongs to $C_{\frac{1}{\alpha\tau}}(a)$, where $a(t) \sim \frac{\beta^{1/\tau}}{\alpha\tau} t^{1/\alpha\tau}$. For a generalized Burr-gamma distribution, see Beirlant et al. [103]. For $\alpha = 1$ one finds the *log-logistic distribution.*

4) The *Fréchet distribution* defined by

$$F(x) = \exp\left(-x^{-1/\gamma}\right), \; x > 0,$$

is directly derived from the extreme value distributions (3.2.4), replacing X by $1+\gamma X$ when $\gamma > 0$, and is a popular model for heavy-tailed data on its own.

5) The gamma distribution also leads to heavy-tailed distributions after a transformation. We mention in particular the *log-gamma distribution,* obtained via a log-transformation from the gamma. We have

$$F(x) = \frac{1}{\Gamma(\alpha)} \int_0^{\lambda \log x} e^{-u} u^{\alpha-1} du, \; x > 0.$$

The log-gamma distribution is a Pareto-type distribution since it belongs to $C_{1/\lambda}(a)$ with $a(t) \sim \frac{1}{\lambda} U(t)$ $(t \to \infty)$. An illustration of the use of the log-gamma distribution as a claim size distribution for fire claims of dwellings is given in Ramlau-Hansen [640].

6) The *t-distribution* offers some possibilities to model heavy claims. Folding the two-sided t-distribution onto the positive half line gives a family of candidates for claim size distributions called the *one-sided t-distributions.* The density is given by

$$f(x) = \frac{2\Gamma\left(\frac{n+1}{2}\right)}{\sqrt{\pi n}\,\Gamma\left(\frac{n}{2}\right)} \left(1 + \frac{x^2}{n}\right)^{-\frac{n+1}{2}}, \; x > 0.$$

This distribution is of Pareto-type with $\alpha = n$.

Hogg and Klugman [449] have suggested the *log t-distribution* by applying first a logarithmic transformation to be followed by a normalization. There results a density with slightly more general parameters

$$f(x) = \frac{\lambda^\alpha \Gamma\left(\alpha + \frac{1}{2}\right)}{\sqrt{2\pi}\,\Gamma(\alpha)} \frac{1}{x} [\lambda + (\log x - \mu)^2]^{-\left(\alpha + \frac{1}{2}\right)}, \; x > 0.$$

As a special case one finds the *one-sided Cauchy distribution* with density

$$f(x) = \frac{1}{\pi} \frac{1}{1 + x^2}, \; x > 0.$$

7) Even the beta distribution leads to heavy tails after proper transformation. The *long-tailed beta distribution* is defined by the distribution

$$F(x) = \frac{1}{B(p,q)} \int_0^x u^{p-1}(1+u)^{-p-q} du, \ x > 0,$$

where $p, q > 0$. The long-tailed beta distribution is a member of $C_{1/q}(a)$, where $a(t) \sim \frac{1}{q} U(t)$.

8) A variation of the long-tailed beta distribution is obtained by a power transformation. This leads to a four-parameter family called *GB2* by Cummins et al. [238]. The distribution can be introduced by the explicit formula

$$F(x) = \frac{1}{B(p,q)} \int_0^{bx^\tau} u^{p-1}(1+u)^{-p-q} du, \ x > 0.$$

The distribution was introduced in an actuarial context by ter Berg in [735] under the name *power-ratio-gamma-distribution*, where statistical diagnostics are considered and references are given to theoretical properties. For an application of beta densities to loss data see, for example, Corro [228].

9) The *log-Pearson III distribution* is obtained from a gamma distributed random variable Y by the transformation $X = \exp(a + Y)$, where a is a constant. Flood distributions in the USA have been statistically modelled using this distribution.

10) The *Wakeby distribution* is another distribution that is used in connection with extremes in water studies. It is defined through the quantile function Q:

$$Q(p) = m + a(1 - (1-p)^b) - c(1 - (1-p)^{-d}), \ 0 < p < 1,$$

where the constants a, b and c are non-negative while d is strictly positive. The best way to look at this definition is through the eyes of the tail quantile function. For

$$U(ux) - U(x) = ax^{-b}(1 - u^{-b}) + cx^d(u^d - 1)$$

so that $F \in C_d(c\,d\,x^d)$.

11) The *first Benktander distribution* is a three-parameter distribution with an exponential, a power and a logarithmic component

$$F(x) = 1 - c\,x^{1-a} x^{-b \log x}(a + 2b \log x), \ x > 0,$$

where a, b and c are positive constants.

The list of distributions discussed above is summarized in Table 3.1, where the models are ordered from light to heavy classes, mentioning the sign of γ. For $\gamma = 0$ we also indicate if the tail is HTE or LTE.

Table 3.1

Sign of γ	Distribution	$1 - F(x)$	(x_-, x_+)
$\gamma < 0$	Beta	$\frac{1}{B(p,q)} \int_x^1 u^{p-1}(1-u)^{q-1} du$	$(0, 1)$
	Reversed Burr	$\beta^\alpha(\beta + (x_+ - x)^{-\tau})^{-\alpha}$	$(0, x_+)$
	Upper-truncated Pareto	$(x^{-\alpha} - T^{-\alpha})/(x_0^{-\alpha} - T^{-\alpha})$	(x_0, T)
$\gamma = 0$, LTE	Second Benktander	$cax^{b-1}e^{-\frac{a}{b}x^b}, b > 1$	$(0, \infty)$
	Weibull	$e^{-\lambda x^\tau}, \tau > 1$	$(0, \infty)$
$\gamma = 0$	Inverse Gaussian	$\sqrt{\frac{\beta\mu^2}{2\pi}} \int_x^\infty u^{-\frac{3}{2}} e^{-\frac{\beta}{2u}(u-\mu)^2} du$	$(0, \infty)$
	Gamma	$\frac{1}{\Gamma(\alpha)} \int_{\lambda x}^\infty e^{-u} u^{\alpha-1} du$	$(0, \infty)$
	Exponential	$e^{-\lambda x}$	$(0, \infty)$
$\gamma = 0$, HTE	Second Benktander	$cax^{b-1}e^{-\frac{a}{b}x^b}, 0 < b < 1$	$(0, \infty)$
	Weibull	$e^{-\lambda x^\tau}, \tau < 1$	$(0, \infty)$
	Log-normal	$1 - \Phi\left(\frac{\log x - \mu}{\sigma}\right)$	$(0, \infty)$
	Quasi-log-normal	$b(x/x_0)^{-\alpha-\beta\log(x/x_0)}$	(x_0, ∞)
$\gamma > 0$	Strict Pareto	$(x/x_0)^{-\alpha}$	(x_0, ∞)
	GPD	$\left(1 + \frac{\gamma}{\sigma}x\right)^{-1/\gamma}$	$(-\sigma/\gamma, \infty)$
	Burr	$\beta^\alpha(\beta + x^\tau)^{-\alpha}$	$(0, \infty)$
	Fréchet	$1 - \exp\left(-x^{-1/\gamma}\right)$	$(0, \infty)$

Table 3.1 (Continued)

Sign of γ	Distribution	$1 - F(x)$	(x_-, x_+)
	Log-gamma	$\frac{1}{\Gamma(\alpha)} \int_{\lambda \log x}^{\infty} e^{-u} u^{\alpha-1} du$	$(0, \infty)$
	One-sided t	$\frac{2\Gamma(\frac{n+1}{2})}{\sqrt{\pi n}\Gamma(\frac{n}{2})} \int_x^{\infty} (1 + \frac{u^2}{n})^{-\frac{n+1}{2}} du$	$(0, \infty)$
	GB2	$\frac{1}{B(p,q)} \int_{bx^{\tau}}^{\infty} u^{p-1}(1+u)^{-p-q} du$	$(0, \infty)$
	First Benktander	$c\, x^{1-a} x^{-b \log x}(a + 2b \log x)$	$(0, \infty)$

3.4 Mean Excess Analysis

Under an unlimited XL treaty with retention u, the expected amount to be paid by the reinsurer is given by $e(u)\bar{F}(u)$, where $e(u)$ is the *mean excess* amount

$$e(u) = \mathbb{E}(X - u|X > u).$$

Assuming $\mathbb{E}(X) < \infty$, the *mean excess function* or *mean residual life function* e is well defined, and its calculation for a random variable with tail function $\overline{F}$ starts from the formula

$$e(t) = \frac{\int_t^{x_+} \overline{F}(u)du}{\overline{F}(t)}. \tag{3.4.10}$$

On the other hand, the distribution function F can also be calculated from e if it exists:

$$1 - F(x) = (1 - F(0))\frac{e(0)}{e(x)} \exp\left(-\int_0^x \frac{du}{e(u)}\right). \tag{3.4.11}$$

In fact the first Benktander distribution was derived by applying (3.4.11) to the mean excess function $e(t) = t(a + 2b \log t)^{-1}$.

Thanks to its memoryloss property, the exponential distribution plays a central role when using e:

$$e(t) = \mathbb{E}(X) = \frac{1}{\lambda} \text{ for all } t \geq 0.$$

When the tail of the distribution of X is HTE, then we find that the mean excess function ultimately increases while for LTE tails e ultimately decreases. For example, for the Weibull distribution we obtain as $t \to \infty$

$$e(t) = \frac{t^{1-\tau}}{\lambda\tau}(1 + o(1))$$

yielding an ultimately decreasing (respectively increasing) e in case $\tau > 1$ (respectively $\tau < 1$). In the case of a Pareto-type distribution the function e ultimately has a linearly increasing behavior since when $\alpha > 1$

$$e(t) \sim \frac{t}{\alpha - 1} \text{ as } t \to \infty. \tag{3.4.12}$$

Distributions with a finite endpoint x_+ show a mean excess function that ultimately decreases and $e(x_+) = 0$.

Hence the mean excess function can play an important role in deciding for a HTE tail. This will be exploited in Chapter 4.

3.5 Full Models: Splicing

A good fit of the severity model over the entire range of loss sizes, from the many smaller to the few large ones, is essential in many practical situations. The traditional models listed above are often not able to capture the entire severity range. If one is restricted to the very large losses, the Pareto-like distributions frequently will be the best choice, but these heavy-tailed distributions rarely have the right shape to fit well below the tail area. One way to deal with this problem is by *splicing* a tail fit to the right of some large threshold t, with a model which fits the bulk of the data. The basic idea here is to stick pieces of two (or more) different models together. This fits in with *mixing models* where, as in a classical actuarial collective model, different processes $f_1, \dots, f_m$ act on different contracts with proportions $p_1, \dots, p_m$ ($\sum_{j=1}^m p_j = 1$) so that

$$f_X(x) = \sum_{j=1}^{m} p_j f_j(x).$$

Splicing concerns a specific kind of mixing reflecting that insurance data exhibit different statistical behavior over some subintervals of the outcome set of loss amounts due to different scrutinies. An *m-component spliced distribution* then has a density expressed as

$$f(x) = \begin{cases} \pi_1 \frac{f_1(x)}{F_1(c_1)-F_1(c_0)}, & c_0 < x \le c_1, \\ \pi_2 \frac{f_2(x)}{F_2(c_2)-F_2(c_1)}, & c_1 < x \le c_2, \\ \dots & \\ \pi_m \frac{f_m(x)}{F_m(c_m)-F_m(c_{m-1})}, & c_{m-1} < x < c_m, \end{cases} \tag{3.5.13}$$

with $\pi_j > 0$ and $\sum_{j=1}^m \pi_j = 1$, where f_j, respectively F_j, ($j = 1, \dots, m$) denote densities and distribution functions of random variables. Restrictions on the parameters can be imposed, requiring continuity, or even differentiability, of the density f at the junction points $c_1, \dots, c_{m-1}$.

Several splicing models using $m = 2$ components have recently been proposed. Motivated by the methods from EVA, Beirlant et al. [100, Sec. 6.2.4] proposed a composite exponential Pareto model for a motor insurance data set of the type

$$f(x) = \begin{cases} \left(1 - \frac{k}{n}\right) \frac{\lambda \exp(-\lambda(x-c_0))}{1-\exp(-\lambda(t-c_0))}, & c_0 < x \leq t, \\ \frac{k}{n\gamma} \left(\frac{x}{t}\right)^{-1/\gamma-1}, & x > t, \end{cases}$$

where k is the number of extremes referring to the number of exceedances above an appropriate threshold t.

An alternative version based on EVA developed in the next chapter consists of splicing a generalized Pareto distribution with a bulk model:

$$1 - F(x) = \begin{cases} 1 - F_1(x), & x \leq t, \\ (1 - F_1(t)) \left(1 + \frac{\gamma}{\sigma}(x - t)\right)^{-1/\gamma}, & x > t, \end{cases} \tag{3.5.14}$$

where F_1 is the distribution function of an appropriately chosen distribution for the modal part of a loss distribution. If F_1 is chosen to have a continuous density f_1, the density of (3.5.14) is given by

$$f(x) = \begin{cases} f_1(x), & x \leq t, \\ (1 - F_1(t))\frac{1}{\sigma} \left(1 + \frac{\gamma}{\sigma}(x - t)\right)^{-1-1/\gamma}, & x > t. \end{cases}$$

Lee et al. [531] considered a mixture of two exponentials

$$f_1(x) = p\lambda_1 \exp(-\lambda_1 x) + (1 - p)\lambda_2 \exp(-\lambda_2 x), \quad \lambda_1, \lambda_2 > 0.$$

Cooray and Ananda [226] proposed a composite log-normal Pareto model, which was suitably modified by Scollnik [687]. Scollnik and Sun [688] considered spliced Weibull–Pareto models, while Calderín-Ojeda and Kwok [180] also introduce splicing log-normal and Weibull models with a tail model. In Fackler [343] a classification of potential combinations for small and large losses is considered. Miljkovic and Grün [578] is another recent reference on this topic.

Following Scollnik [687], consider as an example splicing a log-normal distribution with density function

$$f_1(x) = \frac{1}{\sqrt{2\pi}x\sigma} \exp\left(-\frac{1}{2}\left(\frac{\log x - \mu}{\sigma}\right)^2\right), \quad x > 0$$

and a Pareto distribution with density

$$f_2(x) = \frac{\alpha t^\alpha}{x^{\alpha+1}}, \quad x > t.$$

The density of the composite model is then given by

$$f(x) = \begin{cases} \pi \frac{f_1(x)}{\Phi(\tau)}, & 0 < x \le t, \\ (1-\pi) f_2(x), & x > t, \end{cases}$$

with $\tau = (\log t - \mu)/\sigma$.

Some authors require smoothness at t. When splicing a log-normal and a Pareto distribution, imposing continuity at t leads to

$$\pi = \frac{\alpha\sigma\Phi(\tau)/\varphi(\tau)}{\alpha\sigma\Phi(\tau)/\varphi(\tau) + 1},$$

while differentiability at t leads to $\tau = \alpha\sigma$. Smoothness reduces parameters, which is appropriate in case data are scarce. On the other hand it links the geometries of the body and tail fits, reducing the flexibility the splicing is trying to offer.

Pigeon and Denuit [618] considered a mixed composite log-normal Pareto model, where one assumes that every observation X_i may have its own threshold c_{1i} ($i = 1, \ldots, n$), which are realizations of some non-negative random variable Θ. More specifically the case with Θ being gamma distributed was worked out in detail by these authors.

3.6 Multivariate Modelling of Large Claims

The max-domain of attraction in the multivariate case has been worked out in detail for marginal ordering: for d-dimensional vectors $\mathbf{x} = (x_1, \ldots, x_d)$ and $\mathbf{y} = (y_1, \ldots, y_d)$ the relation $\mathbf{x} \le \mathbf{y}$ is defined as $x_j \le y_j$, $j = 1, \ldots, d$. Moreover we use the notations $\mathbf{xy} = (x_1y_1, \ldots, x_dy_d)$, $\mathbf{x}^{-1} = (x_1^{-1}, \ldots, x_d^{-1})$ and $\mathbf{x}+\mathbf{y} = (x_1 + y_1, \ldots, x_d + y_d)$. Considering a sample of d-dimensional observations $\mathbf{X}_i = (X_{i,1}, \ldots, X_{i,d})$ ($i = 1, \ldots, n$), we denote the sample maximum by $\mathbf{M}_n$ with components

$$\mathbf{M}_{n,j} = \max_{i=1,\ldots,n} X_{i,j},\ j = 1, \ldots, d.$$

The distribution function of $\mathbf{M}_n$ of an independent sample $\mathbf{X}_1, \ldots, \mathbf{X}_n$ from a distribution function $F(\mathbf{x}) = \mathbb{P}(X_1 \le x_1, \ldots, X_d \le x_d)$ is given by

$$\mathbb{P}(\mathbf{M}_n \le \mathbf{x}) = \mathbb{P}(\mathbf{X}_1 \le \mathbf{x}, \ldots, \mathbf{X}_n \le \mathbf{x}) = F^n(\mathbf{x}),\ \mathbf{x} \in \mathbb{R}^d.$$

As in the univariate case, one needs to normalize $\mathbf{M}_n$ in order to obtain a non-trivial limit distribution as $n \to \infty$. The domain of attraction problem is then concerned with finding sequences $\mathbf{a}_n > \mathbf{0} = (0, \ldots, 0)$ and $\mathbf{b}_n$ such that there exists a d-variate distribution function G for which

$$F^n\left(\mathbf{a}_n\mathbf{x} + \mathbf{b}_n\right) \to G(\mathbf{x}),\ n \to \infty. \tag{3.6.15}$$

Again, as in the univariate case, we say that F is in the max-domain of attraction of G, and G is called a (multivariate) extreme value distribution.

Let F_j and G_j denote the jth marginal distribution functions of F and G, respectively. Then one easily derives from (3.6.15) that for $j = 1, \ldots, d$

$$F_j^n\left(a_{n,j}x_j + b_{n,j}\right) \to G_j(x_j),\ n \to \infty, \tag{3.6.16}$$

that is, G_j itself is a univariate extreme value distribution and F_j is in its domain of attraction. Below we will use the following general parametrization of G_j:

$$G_j(x_j) = \begin{cases} \exp\left(-\left(1+\gamma_j(x_j-\mu_j)/\alpha_j\right)^{-1/\gamma_j}\right), & \text{if } \gamma_j \neq 0, \\ \exp\left(-\exp\{-(x_j-\mu_j)/\alpha_j\}\right), & \text{if } \gamma_j = 0, \end{cases} \tag{3.6.17}$$

with γ_j the EVI for the jth margin.

Also, the notion of *max-stability* of G carries over from the univariate case, that is,

$$G^k(\boldsymbol{\alpha}_k\mathbf{x} + \boldsymbol{\beta}_k) = G(\mathbf{x}),$$

for any positive integer k and $\mathbf{x} \in \mathbb{R}^d$, with vectors $\boldsymbol{\alpha}_k > 0$ and $\boldsymbol{\beta}_k$ with $\mathbf{a}_n^{-1}\mathbf{a}_{nk} \to \boldsymbol{\alpha}_k$ and $\mathbf{a}_n^{-1}(\mathbf{b}_{nk} - \mathbf{b}_n) \to \boldsymbol{\beta}_k$ as $n \to \infty$.

An extreme value distribution function G can be reconstructed from its margins and its *stable tail dependence function* (STDF) l. This function is defined as

$$l(\mathbf{v}) = -\log G\left(Q_1(e^{-v_1}), \ldots, Q_d(e^{-v_d})\right),\ \mathbf{v} \in \mathbb{R}_+^d, \tag{3.6.18}$$

with Q_j the quantile function of the jth margin of G ($j = 1, \ldots, d$). One then gets

$$-\log G(\mathbf{x}) = l\left(-\log G_1(x_1), \ldots, -\log G_d(x_d)\right),\ \mathbf{x} \in \mathbb{R}^d. \tag{3.6.19}$$

The expression for an *extreme value copula*

$$C_G(u_1, \ldots, u_d) := G(Q_1(u_1), \ldots, Q_d(u_d))$$

then follows:

$$C_G(\mathbf{u}) = \exp\left(-l(-\log u_1, \ldots, -\log u_d)\right),\ \mathbf{u} \in [0,1]^d. \tag{3.6.20}$$

Note that the STDF describes the dependence between the components after transforming the margins to a standard exponential distribution, which is in contrast to the use of copulas where the margins are transformed to uniform $(0, 1)$ distributions.

A STDF l has the following properties:

(L1) homogeneity: $l(s\cdot) = s\,l(\cdot)$ for $s > 0$ (which follows from the max-stability)
(L2) $l(\mathbf{e}_j) = 1$ for $j = 1, \ldots, d$, where $\mathbf{e}_j$ is the jth unit vector in $\mathbb{R}^d$
(L3) $\max_{j=1,\ldots,d} v_j \leq l(\mathbf{v}) \leq v_1 + \ldots + v_d$ for $\mathbf{v} \in \mathbb{R}_+^d$
(L4) convexity: $l(\lambda\mathbf{v} + (1-\lambda)\mathbf{w}) \leq \lambda l(\mathbf{v}) + (1-\lambda)l(\mathbf{w})$ for $\lambda \in [0,1]$.

On the basis of (L1) it follows that an extreme value copula satisfies

$$C_G^s(\mathbf{u}) = C_G(u_1^s, \ldots, u_d^s),\ \mathbf{u} \in [0,1]^d. \tag{3.6.21}$$

The upper and lower bounds in (L3) are themselves STDFs: the lower bound corresponds to complete dependence $G(\mathbf{x}) = \min_{j=1,\ldots,d} G_j(x_j)$, whereas the upper bound corresponds to independence $G(\mathbf{x}) = G_1(x_1)\ldots G_d(x_d)$.

Finally we note that properties (L1) to (L4) do not characterize the class of STDFs, that is, a function l that satisfies (L1)–(L4) is not necessarily an STDF.

Classical examples of bivarate STDFs are

- the symmetric logistic model, with $l(x_1, x_2) = (x_1^{1/\tau} + x_2^{1/\tau})^\tau$, with $0 \le \tau \le 1$, where $\tau = 1$ corresponds to the independence case and $\tau = 0$ to the complete dependence case
- the Student(ν) distribution for which

$$l(x_1,x_2) = x_2 F_{\nu+1}\left(\sqrt{\nu+1}\frac{(x_2/x_1)^{1/\nu} - \theta}{\sqrt{1-\theta^2}}\right) + x_1 F_{\nu+1}\left(\sqrt{\nu+1}\frac{(x_1/x_2)^{1/\nu} - \theta}{\sqrt{1-\theta^2}}\right),$$

  where $F_{\nu+1}$ is the distribution function of the univariate $t_{\nu+1}$ distribution, and θ the Pearson correlation coefficient
- the Archimax model with mixed generator $l(x_1, x_2) = (x_1^2 + x_2^2 + x_1 x_2)/(x + y)$.

The copula of the distribution function F^n of the sample maximum $\mathbf{M}_n$ is

$$C_{F^n}(\mathbf{u}) = F^n\left(Q_1\left(u_1^{1/n}\right), \ldots, Q_d\left(u_d^{1/n}\right)\right) = C_F^n\left(u_1^{1/n}, \ldots, u_d^{1/n}\right).$$

If F is in the max-domain of attraction of G, then

$$\lim_{n\to\infty} C_F^n\left(u_1^{1/n}, \ldots, u_d^{1/n}\right) = C_G(\mathbf{u}),\ \mathbf{u} \in [0,1]^d,$$

or, as $n \to \infty$,

$$-n \log C_F\left(u_1^{1/n}, \ldots, u_d^{1/n}\right) \to -\log C_G(\mathbf{u}) = l\left(-\log u_1, \ldots, -\log u_d\right),$$

from which

$$n\left\{1 - F\left(Q_1\left(u_1^{1/n}\right), \ldots, Q_d\left(u_d^{1/n}\right)\right)\right\} \to_{n\to\infty} l\left(-\log u_1, \ldots, -\log u_d\right).$$

Setting now $-\log u_j = v_j$ ($j = 1, \ldots, d$) and approximating $u_j^{1/n}$ by $1 - v_j/n$, we find that a multivariate distribution function F is in the max-domain of attraction of an extreme value distribution with STDF l if the *tail dependence function* $1 - F(Q_1(1 - v_1), \ldots, Q_d(1 - v_d))$ converges in the following way to the STDF l of G:

$$\lim_{u\to\infty} u\left[1 - F(Q_1(1 - u^{-1}v_1), \ldots, Q_d(1 - u^{-1}v_d)\right] = l(\mathbf{v}),$$

which can be rewritten as

$$\lim_{u\to\infty} u\mathbb{P}\left(1 - F_1(X_1) \le u^{-1}v_1 \text{ or } \dots \text{ or } 1 - F_d(X_d) \le u^{-1}v_d\right) = l(\mathbf{v}), \tag{3.6.22}$$

or, when using the corresponding copula:

$$\lim_{u\to\infty} u\left[1 - C_F(1 - u^{-1}v_1, \dots, 1 - u^{-1}v_d)\right] = l(\mathbf{v}). \tag{3.6.23}$$

For more details concerning multivariate extreme value theory see Chapter 8 in Beirlant et al. [100].

Copulas and stable tail dependence functions which describe the dependence between the components are infinite-dimensional objects and therefore not always easy to handle. One can restrict to a parametric model, such as a logistic model, but alternatively one can summarize the main properties of the dependence structure in a number of well-chosen dependence coefficients. We restrict the list here to the bivariate case.

- The *extremal coefficient*

  $$\theta = l(1,1) \in [1,2],$$

  which equals $\theta = 2^\tau$ in the logistic model.
- The *coefficient of extremal dependence,* defined as

  $$\chi = \lim_{u\to 1} \mathbb{P}(U_2 > u | U_1 > u),$$

  where $U_j = F_j(X_j)$ $(j = 1,2)$. One calls (X_1, X_2) asymptotic independent if $\chi = 0$ and asymptotic dependent if $0 < \chi \le 1$. Approximations can be obtained for $u \to 1$ from

  $$\begin{aligned}\chi(u) &= 2 - \frac{\log C_F(u,u)}{\log u} \\ &= 2 - \frac{1 - C_F(u,u)}{1-u} + o(1) \\ &= \mathbb{P}(U_2 > u | U_1 > u) + o(1).\end{aligned}$$

  From (3.6.23) with $x_1 = x_2 = 1$ we obtain that for a bivariate distribution in the domain of attraction of an extreme value copula $\chi(u) \to_{u\to 1} 2-\theta \in [0,1]$. In particular, $\chi(u) = 2 - \theta$ is constant in u for a bivariate extreme value distribution. Hence in the logistic model one obtains $\chi = 2 - 2^\tau$.
- Transforming X_j to $Z_j = 1/(1-F_j(X_j))$ $(j = 1,2)$, which are standard Pareto distributed, $P(Z_j > z) = z^{-1}$, Ledford and Tawn [530] introduced a third dependence coefficient by assuming that the joint survivor function of (Z_1, Z_2) is regularly varying

  $$\mathbb{P}(Z_1 > z, Z_2 > z) = \mathbb{P}(\min(Z_1, Z_2) > z) = \mathcal{L}(z) z^{-1/\eta}, \; z > 0,$$

with $\mathcal{L}$ a slowly varying function. The extreme value index η of the random variable $\min(Z_1, Z_2)$ is termed the *coefficient of tail dependence.* Note that

$$\chi = \lim_{z\to\infty} \mathbb{P}(Z_2 > z|Z_1 > z) = \lim_{z\to\infty} \mathcal{L}(z)z^{1-1/\eta}.$$

We find that

- if $\eta = 1$ and $\lim_{z\to\infty} \mathcal{L}(z) = c \in (0,1]$, then $\chi = c \in (0,1]$
- if $\eta \in (0,1)$, or $\eta = 1$ with $\lim_{z\to\infty} \mathcal{L}(z) = 0$, then $\chi = 0$.

Hence η increases with stronger dependence within the class of asymptotic independence. If $1/2 < \eta < 1$ we have positive tail dependence. If $\eta = 1/2$ then extremes of Z_1 and Z_2 are nearly independent, and even exactly independent if $\mathcal{L}(z) = 1$. If $0 < \eta < 1/2$ then we have negative tail dependence.

As in the univariate case, the domain of attraction condition (3.6.15) can be cast in terms of exceedances over a high threshold. The event $\{\mathbf{X} \nleq \mathbf{t}\}$ is called an exceedance over the (multivariate) threshold $\mathbf{t}$. This means that there is at least one coordinate variable X_j that exceeds the corresponding threshold t_j, although the precise coordinate where this happens remains unspecified. We are then interested in the asymptotic distribution of the excess vector $\mathbf{a}_n^{-1}(\mathbf{X}-\mathbf{t})$ conditionally on $\mathbf{X} \nleq \mathbf{t}$, as $t_j \to \infty$, $j = 1, \ldots, d$. It was shown, for example Beirlant et al. [100] or Rootzén and Tajvidi [654], that if $F^n\left(\mathbf{a}_n\mathbf{x} + \mathbf{b}_n\right) \to G(\mathbf{x})$ as $n \to \infty$, and $0 < G(\mathbf{0}) < 1$,

$$\mathbb{P}\left(\mathbf{a}_n^{-1}(\mathbf{X} - \mathbf{b}_n) \vee \mathbf{t}^l \leq \mathbf{x}|\mathbf{X} \nleq \mathbf{b}_n\right) \to H(\mathbf{x}) = \frac{1}{-\log G(\mathbf{0})} \log \frac{G(\mathbf{x})}{G(\mathbf{x} \wedge \mathbf{0})}, \tag{3.6.24}$$

as $n \to \infty$, where $t_j^l \in [-\infty, \infty)$ denotes the lower endpoint of G_j. H is then the distribution function of the *multivariate generalized Pareto distribution.*

Based on (3.6.24), (3.6.17) and (3.6.19) we then obtain, when $\alpha_j > \gamma_j\mu_j$, $j = 1, \ldots, d$, that

$$H(\mathbf{x}) = \frac{l(\{1+\gamma_1\frac{x_1\wedge 0-\mu_1}{\alpha_1}\}^{-1/\gamma_1}, \ldots, \{1+\gamma_d\frac{x_d\wedge 0-\mu_d}{\alpha_d}\}^{-1/\gamma_d})}{l(\{1-\gamma_1\mu_1/\alpha_1\}^{-1/\gamma_1}, \ldots, \{1-\gamma_d\mu_d/\alpha_d\}^{-1/\gamma_d})}$$
$$- \frac{l(\{1+\gamma_1\frac{x_1-\mu_1}{\alpha_1}\}^{-1/\gamma_1}, \ldots, \{1+\gamma_d\frac{x_d-\mu_d}{\alpha_d}\}^{-1/\gamma_d})}{l(\{1-\gamma_1\mu_1/\alpha_1\}^{-1/\gamma_1}, \ldots, \{1-\gamma_d\mu_d/\alpha_d\}^{-1/\gamma_d})}.$$

Setting $\sigma_j = \alpha_j - \gamma_j\mu_j$ and $\zeta_j := (\frac{\sigma_j}{\alpha_j})^{-1/\gamma_j}$, $j = 1, \ldots, d$, we arrive at

$$H(\mathbf{x}) = \frac{l(\zeta_1\{1+\frac{\gamma_1(x_1\wedge 0)}{\sigma_1}\}^{-1/\gamma_1}, \ldots, \zeta_d\{1+\frac{\gamma_d(x_d\wedge 0)}{\sigma_d}\}^{-1/\gamma_d})}{l(\zeta_1, \ldots, \zeta_d)}$$
$$- \frac{l(\zeta_1\{1+\frac{\gamma_1x_1}{\sigma_1}\}^{-1/\gamma_1}, \ldots, \zeta_d\{1+\frac{\gamma_dx_d}{\sigma_d}\}^{-1/\gamma_d})}{l(\zeta_1, \ldots, \zeta_d)} \tag{3.6.25}$$

for $\mathbf{x} \in \mathbb{R}^d$ such that $\boldsymbol{\sigma} + \boldsymbol{\gamma}\mathbf{x} > \mathbf{0}$. Finally, when $\mathbf{x} \geq \mathbf{0}$ we obtain that

$$\bar{H}(\mathbf{x}) = \frac{l\left(\zeta_1\left\{1+\frac{\gamma_1 x_1}{\sigma_1}\right\}^{-1/\gamma_1}, \dots, \zeta_d\left\{1+\frac{\gamma_d x_d}{\sigma_d}\right\}^{-1/\gamma_d}\right)}{l(\zeta_1, \dots, \zeta_d)}$$

and $\zeta_j = \bar{H}_j(0)$ $(j = 1, \dots, d)$. Further note that with properties (L1) and (L2) for STDFs, we obtain in case $\mathbf{x} \geq \mathbf{0}$

$$\begin{aligned}\bar{H}_j(x_j) &= \frac{l(0, \dots, 0, \zeta_j\{1+\frac{\gamma_j x_j}{\sigma_j}\}^{-1/\gamma_j}, 0, \dots, 0)}{l(\zeta_1, \dots, \zeta_d)} \\ &= \zeta_j\left\{1+\frac{\gamma_j x_j}{\sigma_j}\right\}^{-1/\gamma_j} \frac{l(\mathbf{e}_j)}{l(\zeta_1, \dots, \zeta_d)} \\ &= \zeta_j\left\{1+\frac{\gamma_j x_j}{\sigma_j}\right\}^{-1/\gamma_j} / l(\zeta_1, \dots, \zeta_d),\end{aligned}$$

with $\zeta_j = \bar{H}_j(0)$. Imposing the constraint $l(\zeta_1, \dots, \zeta_d) = 1$ we then have

$$\bar{H}(\mathbf{x}) = l(\bar{H}_1(x_1), \dots, \bar{H}_d(x_d)). \tag{3.6.26}$$

For further properties concerning multivariate generalized Pareto distributions see Rootzén and Tajvidi [654] and Kiriliouk et al. [488].

4

Statistics for Claim Sizes

If there is any suspicion of heavy-tailed distributions, then it is advisable that the actuary should make a number of different data plots. Modelling of large claims is quite an uncertain undertaking, and hence the more graphs considered the better in order to make a balanced conclusion.

As a baseline distribution one might depart from the exponential distribution and inspect for HTE tails. If the right tail of the distribution is obviously heavier than any exponential distribution, then Weibull, log-normal or Pareto quantile plots offer potential improvements. Such a first step can be performed using different kinds of quantile plots (exponential, log-normal, Weibull or Pareto) and their derivative plots.

After this large claims modelling using extreme value methodology comes into play. Here the maximum likelihood methodology applied to the *peaks over threshold* (POT) approach plays the central role. We also emphasize methods based on quantile plotting in order to allow for graphical validation of the models and results. We first discuss the classical case of independence and identically distributed data, followed by regression settings, censored and multivariate data. In reinsurance, the development of large claims can take several years. When evaluating a portfolio, not all the claims are fully developed and the indexed payments at the last available development year are an underestimation of the real final indexed payment. When historical incurred information per claim is available, this should assist in the estimation of the tails of the payment distribution.

It remains desirable to construct a distribution with an appropriate tail fit but which at the same time has enough parameters to fit also in the medium range. An early reference here is Albrecht [32], who pointed out that claim size data are often well described by a Pareto distribution for large claims, while the log-normal distribution provides a good fit for medium-sized claims. For a general review on the construction of mixture models with tail components, see Scarott and MacDonald [667]. Here we discuss the method of splicing different distributions in more detail, and in particular we propose combining a mixed Erlang distribution with a tail fit.

All of this material will be illustrated using the data sets introduced in Chapter 1. While the automobile liability data and the Dutch fire insurance data will be used throughout, we end the chapter by analysing the Austrian storm risk, European flood risk data, the Groningen earthquake data, and the Danish fire insurance case in order to illustrate statistical methods for tail estimation.

Reinsurance: Actuarial and Statistical Aspects, First Edition.
Hansjörg Albrecher, Jan Beirlant and Jozef L. Teugels.

For a more general survey and statistical methods of extreme value theory see Embrechts et al. [329], Reiss et al. [645], Coles [221], and Beirlant et al. [100]. These references also contain more technical details that are omitted here.

4.1 Heavy or Light Tails: QQ- and Derivative Plots

As discussed in Section 3.4 the mean excess function offers a first tool to discriminate between HTE and LTE tails. In practice, based on a sample $X_1, X_2, \ldots, X_n$, the mean excess function $e(t) = \mathbb{E}(X - t|X > t)$ can be naively estimated when replacing the expectation by its empirical counterpart:

$$\hat{e}_n(x) = \frac{\sum_{i=1}^{n} X_i 1_{(t,\infty)}(X_i)}{\sum_{i=1}^{n} 1_{(t,\infty)}(X_i)} - t,$$

where for any set A, $1_A(X_i)$ equals 1 if $X_i \in A$, and 0 otherwise. The value t is often taken equal to one of the data points, say the $(k+1)$-largest observation $X_{n-k,n}$ for some $k = 1, 2, \ldots, n-1$. We then obtain

$$e_{k,n} = \hat{e}_n(X_{n-k,n}) = \frac{1}{k}\sum_{j=1}^{k} X_{n-j+1,n} - X_{n-k,n}. \tag{4.1.1}$$

The mean excess values $e_{k,n}$ can be plotted as a function of the threshold $x_{n-k,n}$ or as a function of the inverse rank k.

There is an interesting link between the values $e_{k,n}$ and exponential QQ-plots. For an exponential distribution the quantile values $y_j = Q\left(\frac{j}{n+1}\right)$ stand in linear relationship to the corresponding quantiles of the standard exponential distribution $x_j = Q(j/(n+1)) = -\log(1 - j/(n+1))$:

$$y_j = \frac{1}{\lambda} x_j, \; j = 1, \ldots, n.$$

Hence, when estimating $y_j = Q\left(\frac{j}{n+1}\right)$ by the empirical quantiles $X_{j,n}$, we have that the *exponential QQ-plot*, defined by

$$\left(-\log\left(1 - \frac{i}{n+1}\right), X_{i,n}\right), \quad i = 1, \ldots, n,$$

should exhibit a linear pattern which passes through the origin for the exponential model to be a plausible model. An estimator of the slope can then also be used as an estimator of $1/\lambda$.

Now $e_{k,n}$ can be viewed as an estimate of the slope $1/\lambda_k$ of the exponential QQ-plot to the right of an anchor point $\left(-\log \frac{k+1}{n+1}, X_{n-k,n}\right)$, and hence $(x_{n-k,n}, e_{k,n})$ or $(k, e_{k,n})$ for $k = 1, \ldots, n$, can be interpreted as a derivative plot of the exponential QQ-plot.

When fitting a regression line which passes through the anchor point using least squares regression minimizing

$$\sum_{j=1}^{k}\left(\log\frac{k+1}{j}\right)^{-1}\left(X_{n-j+1,n}-\left[X_{n-k,n}+\lambda_k^{-1}\left(\log\frac{n+1}{j}-\log\frac{n+1}{k+1}\right)\right]\right)^2$$

with respect to $1/\lambda_k$, one indeed obtains

$$1/\hat{\lambda}_k=\frac{e_{k,n}}{\frac{1}{k}\sum_{j=1}^{k}\log\frac{k+1}{j}},$$

so that $e_{k,n}\approx 1/\hat{\lambda}_k$ using the approximation $\frac{1}{k}\sum_{j=1}^{k}\log\frac{k+1}{j}\approx 1$, which is sharp even for small k.

Also, when the data come from a distribution with a tail heavier than exponential, the exponential QQ-plot will ultimately be convex and ultimately upcross the fitted regression line for every k, so that the slopes $e_{k,n}$ will increase always with increasing $X_{n-k,n}$ (or decreasing k), while for a tail lighter than exponential, the QQ-plot will ultimately be concave, ultimately appearing under the fitted regression line for every k, and the slopes will decrease with increasing $X_{n-k,n}$ (or decreasing k).

When modelling reinsurance claim data we expect convex exponential QQ-plots linked with increasing mean excess plots $(x_{n-k,n}, e_{k,n})$. A popular second step is to inspect log-normal or Pareto QQ-plots. Note that the mean excess plots of a Pareto-type distribution ultimately will be linear increasing with slope $1/(\alpha-1)$, as follows from (3.4.12). Again, log-normal, respectively Pareto, tail fits appear appropriate when the right upper end of the corresponding QQ-plot is linear from some point on. It is advisable to accompany the QQ-plot with the corresponding derivative plots.

- Since $\log X$ is exponentially distributed with $\lambda = \alpha$ when X is strict Pareto(α) distributed, the *Pareto QQ-plot* is defined as

  $$\left(-\log\left(1-\frac{i}{n+1}\right),\log X_{i,n}\right),\quad i=1,\dots,n,$$

  with derivative plot

  $$\left(\log x_{n-k,n}, H_{k,n}\right)\text{ or }\left(k, H_{k,n}\right)$$

  where

  $$H_{k,n}=\frac{1}{k}\sum_{j=1}^{k}\log X_{n-j+1,n}-\log X_{n-k,n}. \tag{4.1.2}$$

  $H_{k,n}$ is the estimator of $1/\alpha$ introduced by Hill [442]. Indeed, if the data come from a Pareto distribution, then the Pareto QQ-plot is linear and the derivative plot is horizontal at the level $1/\alpha$.

- The normal QQ-plot based on the logarithms of the data provides the *log-normal QQ-plot*

$$\left(\Phi^{-1}\left(\frac{i}{n+1}\right), \log X_{i,n}\right), \quad i = 1, \dots, n,$$

where Φ^{-1} denotes the standard normal quantile function. The derivative plot is then given by

$$\left(\log x_{n-k,n}, \frac{H_{k,n}}{N_{k,n}}\right) \text{ or } \left(k, \frac{H_{k,n}}{N_{k,n}}\right)$$

with

$$N_{k,n} = \frac{n+1}{k+1}\varphi\left(\Phi^{-1}\left(1-\frac{k+1}{n+1}\right)\right) - \Phi^{-1}\left(1-\frac{k+1}{n+1}\right),$$

since, with φ denoting the standard normal density,

$$\begin{aligned}
&\frac{1}{k}\sum_{j=1}^{k}\Phi^{-1}\left(1-\frac{j}{n+1}\right) - \Phi^{-1}\left(1-\frac{k+1}{n+1}\right) \\
&\approx \int_0^1 \Phi^{-1}\left(1-u\frac{k+1}{n+1}\right)du - \Phi^{-1}\left(1-\frac{k+1}{n+1}\right) \\
&= N_{k,n}.
\end{aligned}$$

- The quantile function of the Weibull distribution is given by

$$Q(p) = \left(-\frac{1}{\lambda}\log(1-p)\right)^{1/\tau}, \quad 0 < p < 1,$$

so that for this model $\log Q(p) = \frac{1}{\tau}\log[-\log(1-p)] + \frac{1}{\tau}\log\frac{1}{\lambda}$. Again taking $p = \frac{i}{n+1}$ $(i = 1, \dots, n)$ and estimating $Q(i/(n+1))$ by $X_{i,n}$ leads to the definition of the *Weibull QQ-plot*

$$\left(\log\left[-\log\left(1-\frac{i}{n+1}\right)\right], \log X_{i,n}\right), \quad i = 1, \dots, n.$$

The derivative plot is then given by

$$\left(\log x_{n-k,n}, \frac{H_{k,n}}{W_{k,n}}\right) \text{ or } \left(k, \frac{H_{k,n}}{W_{k,n}}\right)$$

with

$$W_{k,n} = \frac{1}{k}\sum_{j=1}^{k}\log\log\frac{n+1}{j} - \log\log\frac{n+1}{k+1}.$$

Insurance claim data often exhibit different statistical behavior over various subsets of the outcome set which can be observed in mean excess plots, starting with components in the center of the data followed by a Pareto tail. Sometimes such Pareto tails then turn out to be upper-truncated, as defined in Section 3.3.1.1.

Case studies. In Figures 4.1–4.3 the exponential, Pareto, log-normal, and Weibull QQ-plot together with the corresponding derivative plots are given for the Dutch fire insurance data, and the ultimate values for the car liability insurance from Companies A and B. In Figure 4.1 the regression lines based on the top 100 Dutch fire claim observations and passing through the corresponding anchor point at $k = 100$ are given. The corresponding slope estimate can be traced back in the derivative plot through the vertical coordinate of the anchor point in the QQ-plot, which then is the horizontal coordinate of the slope estimate in the derivative plot.

The Dutch fire insurance data show a heavy-tailed behavior since the exponential QQ-plot is convex, which is consistent with the mean excess plot being increasing over the whole data range. However, in total at least three components can be detected with different slopes in the QQ- and derivative plots, with Pareto behavior for $\log x \leq 16$, a decreasing Pareto derivative plot for $\log x \in (16, 18)$, and ultimately a heavy tail piece at $\log x > 18$ (approximately). Note the horizontal behavior of the $H_{k,n}$ plot for $\log x$ between 14 and 16, followed by constant Weibull derivatives for $\log x \in (16, 18)$. However, ultimately at the largest data points again Pareto behavior appears.

With the ultimate data values for Company A a three-component spliced distribution can be observed in Figure 4.2, starting with a component with decreasing derivative plots for $\log x \leq 12$, followed by a Pareto component when $\log x \in (12, 13)$ and a HTE tail piece for $x \in (13, 15)$. Finally, there is an ultimate section using the top eight data points which shows a strong downward trend in each derivative plot, which could indicate upper-truncation near some high value T. So here possible model candidates for tail fits are log-normal, Pareto or an upper-truncated tail.

Finally, the ultimate data from Company B also show three components, $\log x \leq 12$, $\log x \in (12, 13.5)$ and $\log x \in (13.5, 14.5)$, ending with a short Pareto piece appearing at the top 10 data points which follows from a linear increasing mean excess plot in that area.

These QQ- and derivative plots give first indications which then should be studied further using the extreme value and splicing methods developed next. □

4.2 Large Claims Modelling through Extreme Value Analysis

4.2.1 EVA for Pareto-type Tails

In order to model large claims, Pareto tail modelling is probably the most common approach. Here we use the subset of models with tails heavier than exponential for which the EVI γ is positive, as discussed in Section 3.3.2.2, which in fact equals the set of *Pareto-type models* that can be defined through tail functions $1 - F$, quantile functions Q, or tail quantile functions $U(x) = Q(1 - 1/x)$. Indeed

$$U(x) = x^{\gamma} \ell_U(x), \; x \uparrow \infty, \qquad (4.2.3)$$

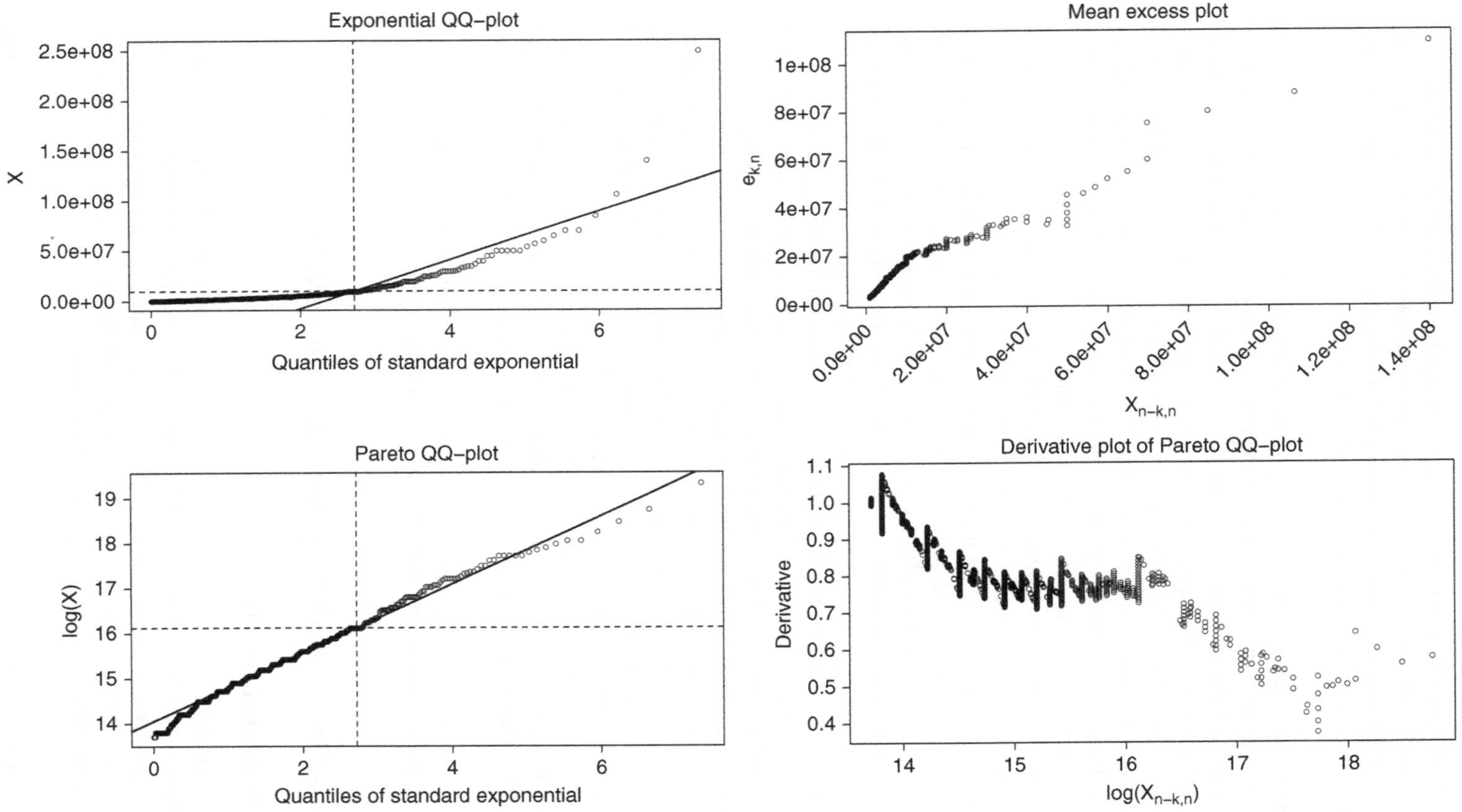

Figure 4.1 Dutch fire insurance data: exponential QQ-plot and mean excess plot $(x_{n-k,n}, e_{k,n})$ (top); Pareto QQ-plot and Hill plot $(\log x_{n-k,n}, H_{k,n})$ (second line); log-normal QQ-plot and derivative plot $(\log x_{n-k,n}, H_{k,n}/N_{k,n})$ (third line); Weibull QQ-plot and derivative plot $(\log x_{n-k,n}, H_{k,n}/W_{k,n})$ (bottom). For each QQ-plot the regression line through $X_{n-99,n}, \ldots, X_{n,n}$ is plotted.

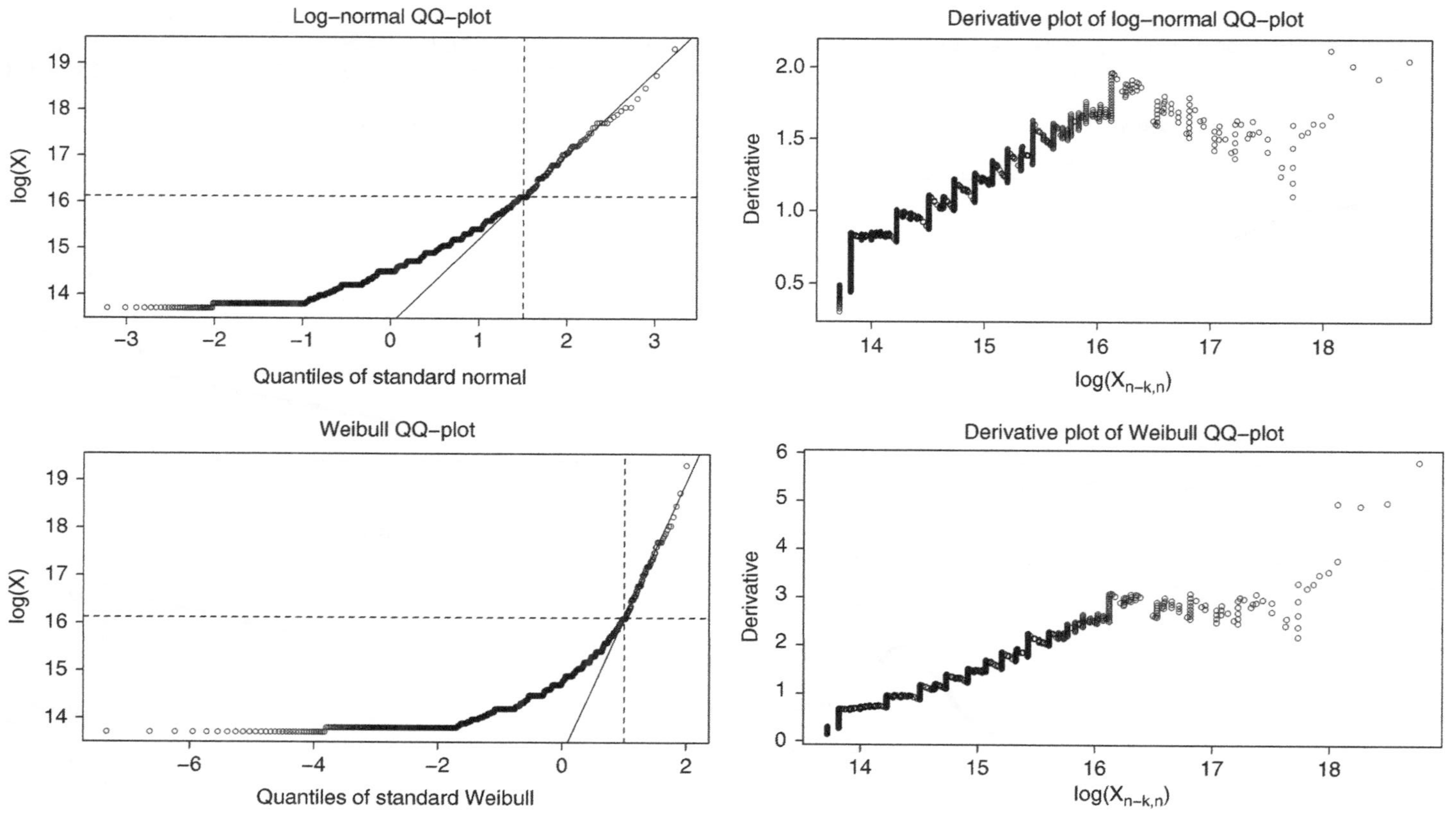

Figure 4.1 (Continued)

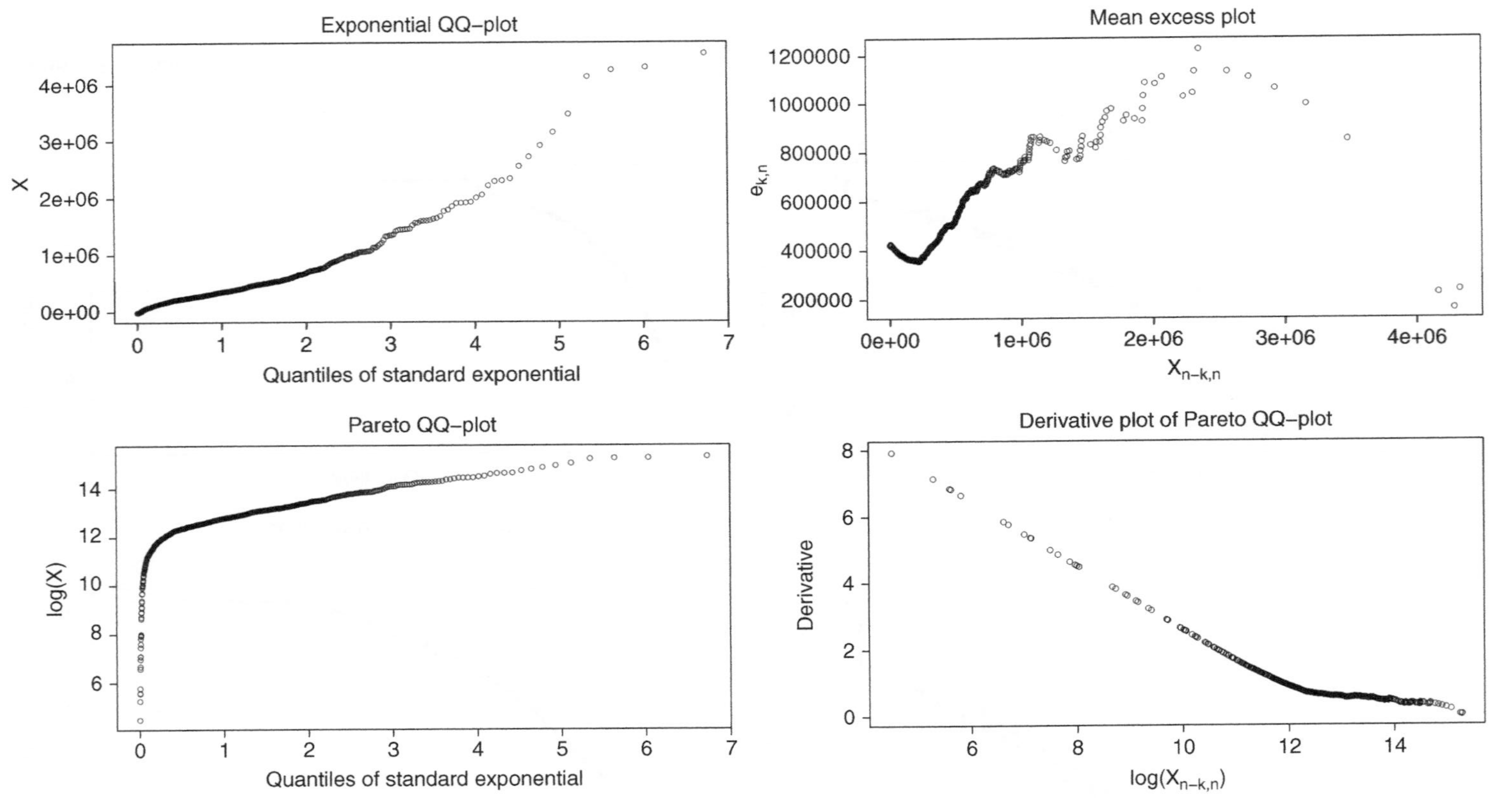

Figure 4.2 MTPL data for Company A, ultimate data values at evaluation: exponential QQ-plot and mean excess plot $(x_{n-k,n}, e_{k,n})$ (top); Pareto QQ-plot and Hill plot $(\log x_{n-k,n}, H_{k,n})$ (second line); log-normal QQ-plot and derivative plot $(\log x_{n-k,n}, H_{k,n}/N_{k,n})$ (third line); Weibull QQ-plot and derivative plot $(\log x_{n-k,n}, H_{k,n}/W_{k,n})$ (bottom).

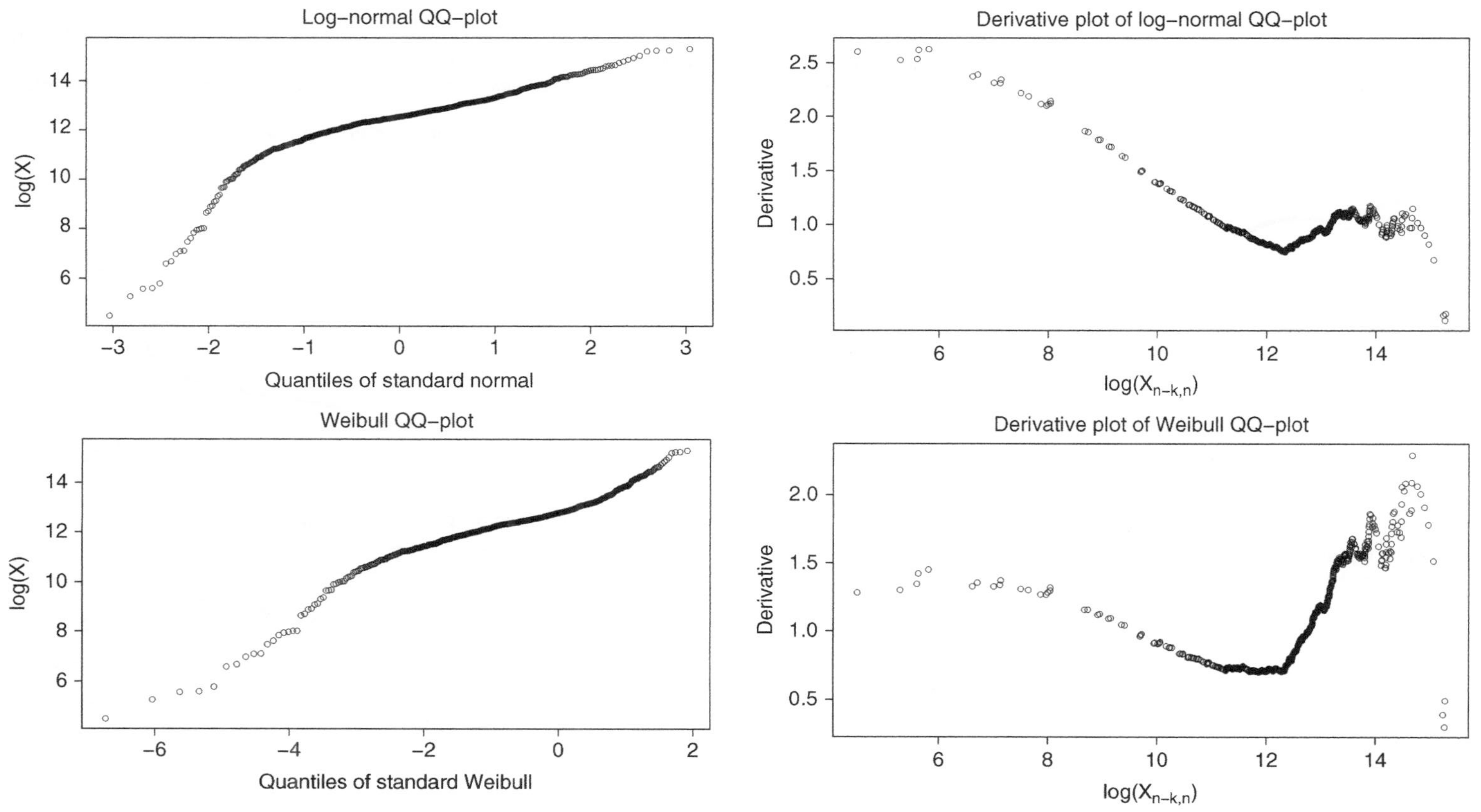

Figure 4.2 (Continued)

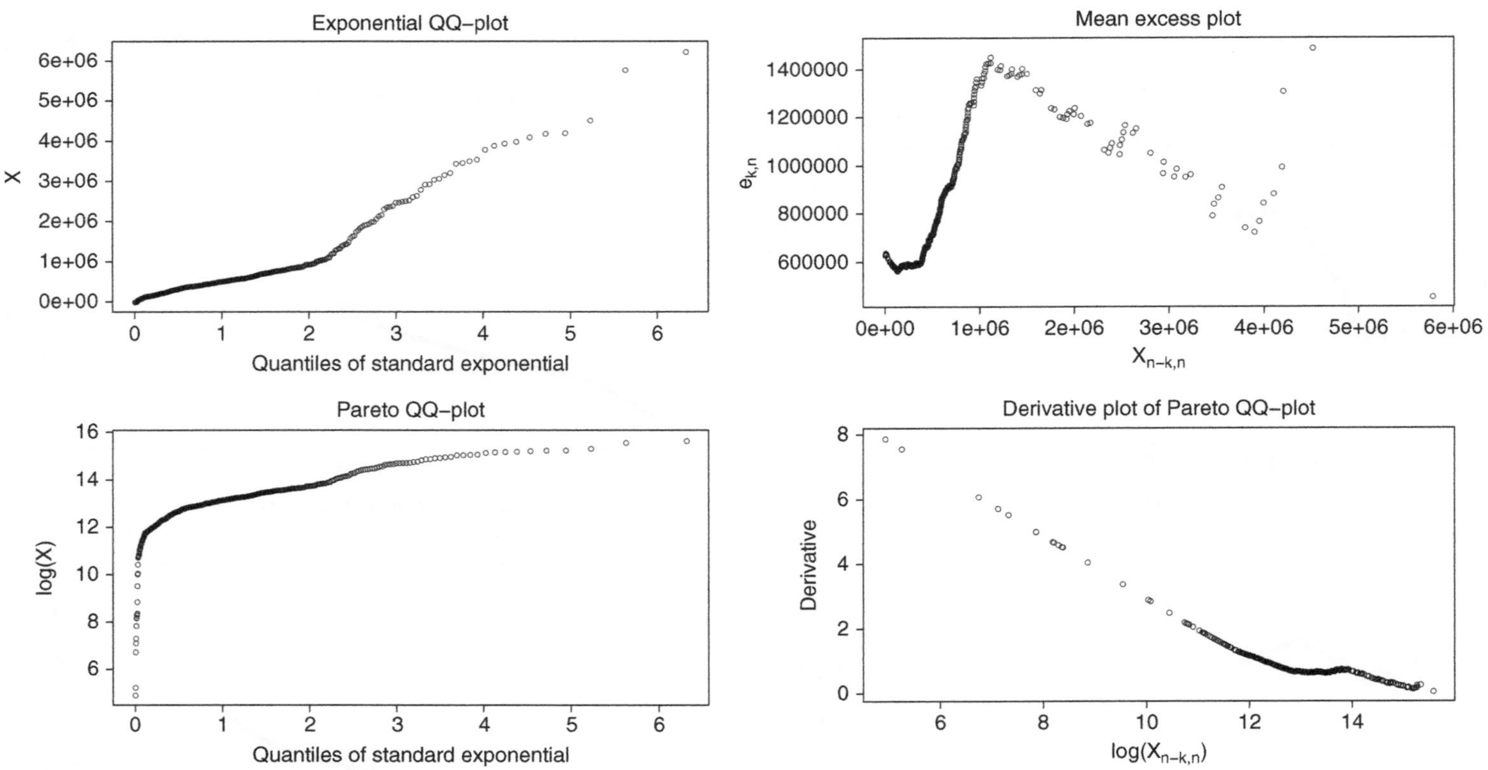

Figure 4.3 MTPL data for Company B, ultimate data values at evaluation: exponential QQ-plot and mean excess plot $(x_{n-k,n}, e_{k,n})$ (top); Pareto QQ-plot and Hill plot $(\log x_{n-k,n}, H_{k,n})$ (second line); log-normal QQ-plot and derivative plot $(\log x_{n-k,n}, H_{k,n}/N_{k,n})$ (third line); Weibull QQ-plot and derivative plot $(\log x_{n-k,n}, H_{k,n}/W_{k,n})$ (bottom).

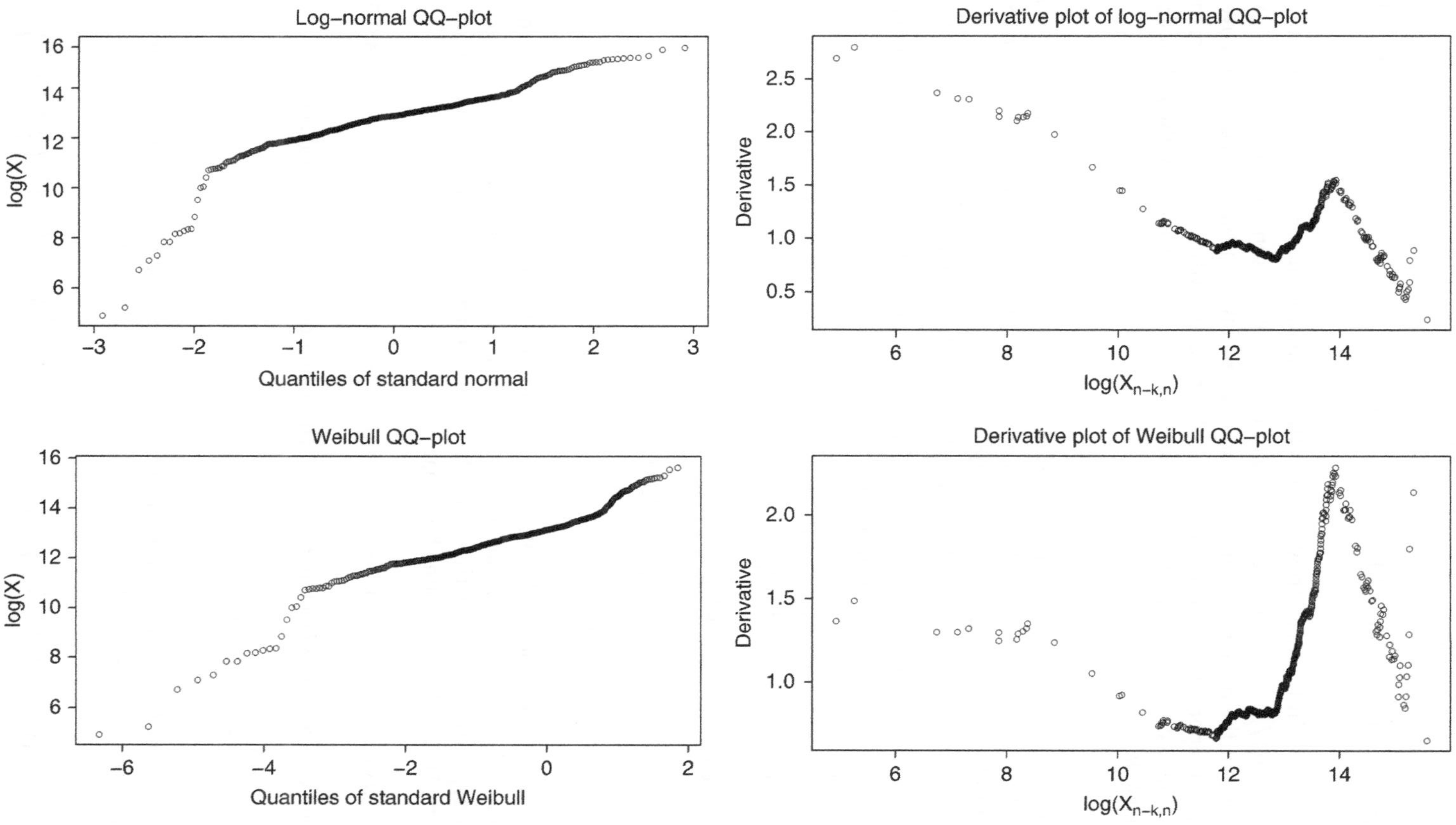

Figure 4.3 (Continued)

where $\gamma = 1/\alpha > 0$ and ℓ_U is a *slowly varying function*. Also (3.2.7) is equivalent to

$$\frac{\overline{F}(ut)}{\overline{F}(t)} = \mathbb{P}\left(\frac{X}{t} > u | X > t\right) \to u^{-1/\gamma}, \; t \uparrow \infty \tag{4.2.4}$$

for every $u > 0$. In this section we discuss the estimation of $\gamma = 1/\alpha$, large quantiles $Q(1-p) = U(1/p)$, and small tail probabilities $\overline{F}(x)$. We assume in this and the next subsection that the data are independent and identically distributed (i.i.d.). Moreover mathematical approximations of variances (AVar), bias (ABias), mean squared error (AMSE), and distributions of estimators using k largest observations of the n data will hold when $k, n \to \infty$ and $k/n \to 0$.

4.2.1.1 Estimating a Positive EVI

The most popular estimator for γ is given by the Hill estimator $H_{k,n}$ defined in (4.1.2), see [442]. In the preceeding section this estimator was retrieved through regression on the Pareto QQ-plot. Here, we also show how the maximum likelihood method based on the so-called POT approach leads to the same estimation method in the Pareto-type case.

- In Section 4.1 the Hill estimator was motivated as an estimator of the slope of a linear Pareto QQ-plot to the right of an anchor point $(\log((n+1)/(k+1)), \log X_{n-k,n})$. In fact, this interpretation can be carried over to the general case of Pareto-type distributions since then ultimately for $x \to \infty$ the Pareto QQ-plot is still linear with slope γ for a small enough k or, equivalently, for a large enough $X_{n-k,n}$. Indeed, under (4.2.3),

  $$\log U(x) \sim \gamma \log x + \log \ell_U(x), x \to \infty.$$

  It can now be shown that for every slowly varying function ℓ

  $$\log \ell(x) / \log x \to 0$$

  as $x \to \infty$. Hence, whereas Pareto QQ-plots are hardly ever completely linear, they are ultimately linear at some set of largest values. The speed at which the linearity sets in depends on the underlying slowly varying function. Like many publications, following Hall [419], we assume here that

  $$\ell_U(x) = C(1 + Dx^{-\beta}(1 + o(1))), \; x \to \infty, \tag{4.2.5}$$

  for some $C, \beta > 0$, and D a real constant. This can, however, be generalized to

  $$\frac{\ell_U(tx)}{\ell_U(x)} - 1 \sim b(x) h_{-\beta}(t),$$

  as $x \to \infty$ with b essentially a power function or, more correctly, a regularly varying function with index $-\beta$, and $h_{-\beta}(t) = \int_1^t u^{-\beta-1} du$. Under (4.2.5)

  $$\log U(x) - \log U\left(\frac{n+1}{k+1}\right) \sim \gamma \left(\log x - \log \frac{n+1}{k+1}\right) + D\left(x^{-\beta} - \left(\frac{n+1}{k+1}\right)^{-\beta}\right), \tag{4.2.6}$$

from which $H_{k,n}$ follows taking $x = (n+1)/j$ $(j = 1,\ldots,k)$, estimating $U((n+1)/j)$ by $X_{n-j+1,n}$ $(j = 1,\ldots,k+1)$, and taking the average of both sides of (4.2.6) over $j = 1,\ldots,k$ after deleting the last term on the right-hand side. Omitting this final term (or, equivalently, assuming a strict Pareto distribution with constant slowly varying function) causes a bias which will be more important with smaller β. Adverse situations for the Hill estimator are logarithmic slowly varying functions ℓ_U, as in case of the log-gamma distribution. Such cases exhibit $\beta = 0$.

- Alternatively, the Hill estimator is also a maximum likelihood estimator based on (4.2.4). Indeed, extreme value methodology proposes fitting the limiting Pareto distribution with distribution function $1 - x^{-1/\gamma}$ to the POT values $Y = X/t$ over a high threshold t conditionally on $X > t$. Note that the use of the mathematical limit in (4.2.4) to fit the exceedance data introduces an approximation error that leads to estimation bias. Let N_t denote the number of exceedances over t. Then the log-likelihood equals

$$\log L(\gamma | Y_1,\ldots,Y_{N_t}) = -N_t \log\gamma - \left(1 + \frac{1}{\gamma}\sum_{j=1}^{N_t} \log Y_j\right)$$

with

$$\frac{d\log L(\gamma)}{d\gamma} = -\frac{N_t}{\gamma} + \frac{1}{\gamma^2}\sum_{j=1}^{N_t} \log Y_j,$$

leading to the maximum likelihood estimator

$$\hat{\gamma} = \frac{1}{N_t}\sum_{j=1}^{N_t} \log Y_j.$$

Choosing an upper order statistic $X_{n-k,n}$ for the threshold t (so that $N_t = k$) we obtain $H_{k,n}$.

- From Section 4.1 it also follows that the Hill statistic can be interpreted as an estimator of the mean excess function of the log-transformed data, that is, $e_{\log X}(\log t) = \mathbb{E}(\log X - \log t | X > t)$, with the threshold value t substituted by $X_{n-k,n}$. As in (3.4.10) we here find

$$e_{\log X}(\log t) = \frac{\int_t^\infty \overline{F}(u) d\log u}{\overline{F}(t)}.$$

Estimating $F(u)$ using the *empirical distribution function*

$$\hat{F}_n(x) = \frac{1}{n}\sum_{i=1}^{n} 1_{\{X_i \le x\}} \qquad (4.2.7)$$

with value $1-\hat{F}_n(u) = j/n$ over the interval $[X_{n-j,n}, X_{n-j+1,n})$ we are led to the estimator

$$\frac{\int_{X_{n-k,n}}^{X_{n,n}} (1-\hat{F}_n(u))d\log u}{1-\hat{F}_n(X_{n-k,n})} = \frac{\sum_{j=1}^{k}(j/n)(\log X_{n-j+1,n} - \log X_{n-j,n})}{k/n}. \tag{4.2.8}$$

Using summation by parts one observes that this final expression equals the Hill estimator:

$$H_{k,n} = \frac{1}{k}\sum_{j=1}^{k} Z_j = \overline{Z}_k, \tag{4.2.9}$$

with

$$Z_j = j\left(\log X_{n-j+1,n} - \log X_{n-j,n}\right),\ j = 1,\dots,n$$

(with $\log X_{0,n} = 0$).

To deduce approximate expressions for the variance and bias of the Hill estimator it is helpful to consider the preceding interpretation in terms of the scaled log-spacings Z_j. Thanks to the Rényi representation $j(E_{n-j+1,n} - E_{n-j,n}) =_d E_j$ $(j = 1,\dots,n)$ concerning order statistics $E_{1,n} \le E_{2,n} \le \dots \le E_{n,n}$ from a random sample $E_1, E_2, \dots, E_n$ of n independent standard exponential random variables, we have *in case of a strict Pareto distribution* (i.e., with ℓ_U constant), that

$$Z_j =_d \gamma E_j, j = 1,\dots,n. \tag{4.2.10}$$

This representation is based on the memoryless property of the exponential distribution and the fact that $nE_{1,n}$ is standard exponentially distributed. From (4.2.9) and (4.2.10) we expect that, as $k \to \infty$,

$$\mathrm{AVar}(H_{k,n}) = \frac{\gamma^2}{k}(1+o(1)).$$

Concerning the bias due to the approximation error, we confine ourselves to the model (4.2.5). Then the theoretical analogue of the Hill estimator is given by

$$\begin{aligned} EH_{k,n} &:= \frac{1}{k}\sum_{j=1}^{k} \log U\left(\frac{n+1}{j}\right) - \log U\left(\frac{n+1}{k+1}\right) \\ &= \gamma\,\frac{1}{k}\sum_{j=1}^{k} \log\frac{n+1}{j} - \log\frac{n+1}{k+1} \\ &\quad + \frac{1}{k}\sum_{j=1}^{k} \log\left(1 + D\left(\frac{j}{n+1}\right)^{\beta}\right) - \log\left(1 + D\left(\frac{k+1}{n+1}\right)^{\beta}\right) \end{aligned}$$

$$
\begin{aligned}
&\approx \gamma \frac{1}{k}\sum_{j=1}^{k}\log\frac{k+1}{j} + D\left(\frac{k+1}{n+1}\right)^{\beta}\frac{1}{k}\sum_{j=1}^{k}\left(\left(\frac{j}{k+1}\right)^{\beta}-1\right)\\
&\approx \gamma - \frac{D\beta}{1+\beta}\left(\frac{k+1}{n+1}\right)^{\beta}\\
&= \gamma + \frac{b_{k,n}}{1+\beta},
\end{aligned}
$$

with $b_{k,n} := -D\beta\left(\frac{k+1}{n+1}\right)^{\beta}$. Hence, the approximate mean squared error is given by

$$
\mathrm{AMSE}(H_{k,n}) = \frac{\gamma^2}{k} + \left(\frac{D\beta}{1+\beta}\right)^2\left(\frac{k+1}{n+1}\right)^{2\beta},
$$

while in order to construct confidence bounds we have that, as $k, n \to \infty$ with $k/n \to 0$,

$$
\sqrt{k}\left(\frac{H_{k,n}}{\gamma} - 1\right) \approx_d \mathcal{N}(0,1). \tag{4.2.11}
$$

Case studies. In Figure 4.4 the Hill derivative values are plotted as a function of k for the Dutch fire insurance claims, and for the ultimate values from Companies A and B. Plotting $(k, H_{k,n})$ is the usual way to inspect the estimates of the extreme value index $\gamma > 0$. In the case of the Dutch fire insurance data at each k we also put the 95% confidence intervals for γ based on (4.2.11). From Figure 4.4 it follows that the ultimate tail behavior in each of these three cases is non-standard. It is clear that the interpretation of these plots – sometimes termed Hill horror plots – is difficult. This motivates the search for bias reduction techniques, combined with different plotting techniques, as discussed below. □

4.2.1.2 Estimating Large Quantiles and Small Tail Probabilities

One of the most important applications of EVA is the estimation of extreme quantiles $q_p = Q(1-p)$ with p small, also termed Value-at-Risk (VaR) in risk applications. Alternatively, the return period for a high claim amount x given by $r(x) = 1/\mathbb{P}(X > x)$ is another measure describing extreme risks.

The estimation of a high quantile under Pareto-type modelling can be performed by extrapolating along a fitted regression line on the Pareto QQ-plot through the point $(\log((n+1)/(k+1)), \log X_{n-k,n})$ with slope $H_{k,n}$. Following (4.2.6) with $x = 1/p$, estimating $U((n+1)/(k+1))$ by $X_{n-k,n}$ and γ by $H_{k,n}$, and omitting the second term on the right-hand side, that is, using

$$
\log Q(1-p) \approx \log X_{n-k,n} + H_{k,n}\left(\log\frac{1}{p} - \log\frac{n+1}{k+1}\right),
$$

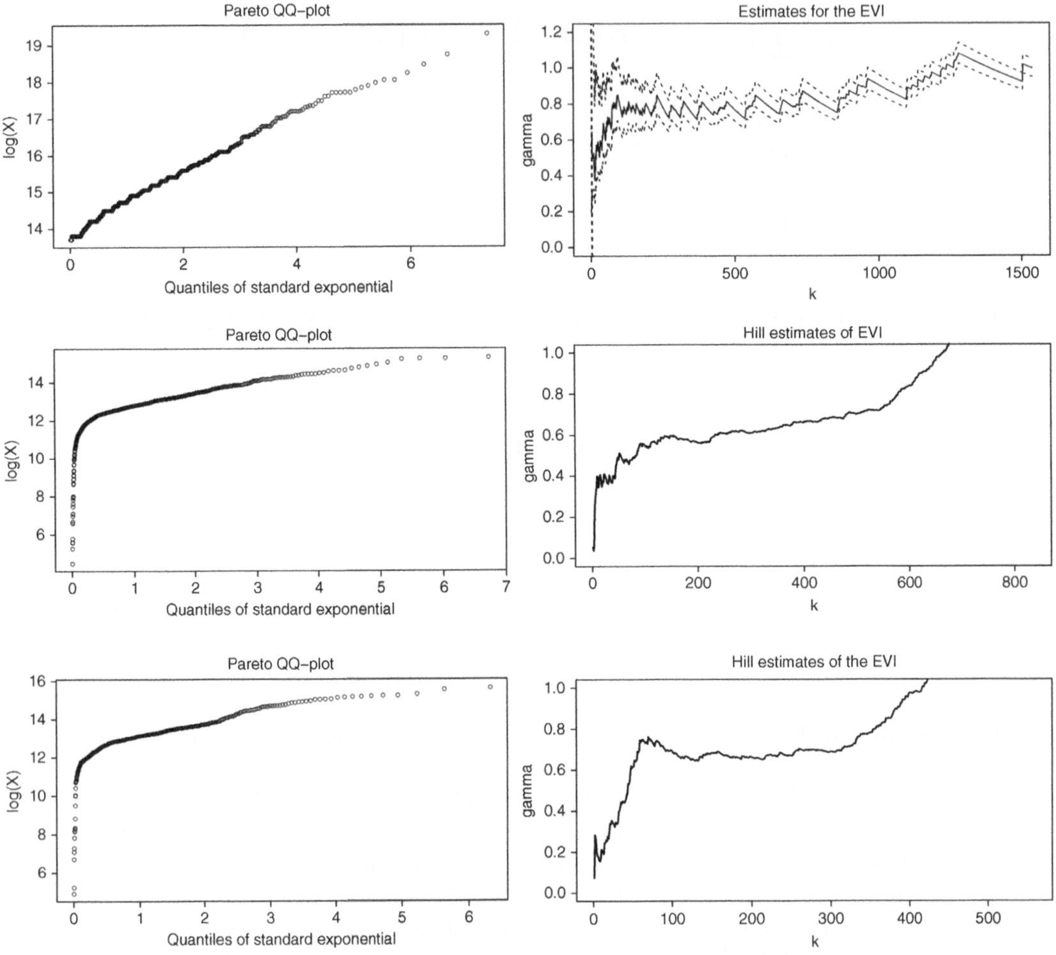

Figure 4.4 Pareto QQ-plot and Hill plot $(k, H_{k,n})$: Dutch fire insurance data with 95% confidence intervals for each k (top); MTPL data for Company A, ultimate values (middle); MTPL data for Company B, ultimate values (bottom).

we arrive at the estimator

$$\hat{q}^{+}_{k,p} = X_{n-k,n} \left(\frac{k+1}{(n+1)p} \right)^{H_{k,n}}, \tag{4.2.12}$$

which was first proposed by Weissman [777]. The estimator $\hat{q}^{+}_{k,p}$ can also be retrieved from (4.2.3), leading to the approximation $U(vx)/U(x) \approx v^{\gamma}$ for large values of x. Setting $vx = 1/p$, $x = (n+1)/(k+1)$ so that $v = (k+1)/((n+1)p)$, and estimating $U((n+1)/(k+1))$ by $X_{n-k,n}$ and γ by $H_{k,n}$, we obtain $\hat{q}^{+}_{k,p}$ again.

Estimation of return periods can be obtained using the inverse relationship on the Pareto QQ-plot:

$$\hat{r}^{+}_{k,x} = 1/\hat{p}^{+}_{k,x} = \frac{n+1}{k+1}\left(\frac{x}{X_{n-k,n}}\right)^{1/H_{k,n}}. \tag{4.2.13}$$

The expression for $\hat{p}^{+}_{k,x}$ can also be deduced from (4.2.4), leading to the approximation $\bar{F}(tu)/\bar{F}(t) \approx u^{-1/\gamma}$ for large values of t. Setting $tu = x$, $t = X_{n-k,n}$ so that $u = x/X_{n-k,n}$, and estimating $\bar{F}(t)$ by $(k+1)/(n+1)$ we obtain $\hat{p}^{+}_{k,x}$.

Approximate confidence bounds for such parameters have been derived based on asymptotic distributions of the estimators. In the case of the tail probability estimator we find with $p_x = \mathbb{P}(X > x)$

$$\sqrt{k}/\sqrt{1+\log^2(k/(np_x))}\left(\frac{\hat{p}^{+}_{k,x}}{p_x} - 1\right) \approx_d \mathcal{N}(0,1),$$

when $k, n \to \infty$, $k/n \to 0$, $np_x/k \to \tau \in [0,1)$ and $k^{-1/2}\log(np_x) \to 0$, while with $q_p = Q(1-p)$,

$$\sqrt{k}/\sqrt{1+\log^2(k/(np))}\left(\frac{\hat{q}^{+}_{k,p}}{q_p} - 1\right) \approx_d \mathcal{N}(0,1),$$

when $k, n \to \infty$, $k/n \to 0$, $np/k \to \tau \in [0,1)$ and $k^{-1/2}\log(np) \to 0$.

4.2.1.3 Bias Reduction

When constructing confidence intervals for risk measures such as p_x and q_p, again the approximation of the underlying conditional distribution by the simple Pareto distribution entails a bias for all the existing estimators, next to the bias induced by estimating γ. One approach to reduce the bias is to construct estimators based on regression models of the values Z_j. Indeed, under (4.2.5), using the approximation $j\log\frac{j+1}{j} \approx 1$ and with the mean value theorem on $x^{-\beta}$ at the points $j/(n+1)$ and $(j+1)/(n+1)$, the theoretical analogue of a Z_j random variable can be approximated by

$$j\left(\log U\left(\frac{n+1}{j}\right) - \log U\left(\frac{n+1}{j+1}\right)\right) \tag{4.2.14}$$

$$= \gamma\left(j\log\frac{j+1}{j}\right) + Dj\left[\left(\frac{n+1}{j}\right)^{-\beta} - \left(\frac{n+1}{j+1}\right)^{-\beta}\right]$$

$$\approx \gamma + b_{k,n}\left(\frac{j}{k+1}\right)^{\beta}. \tag{4.2.15}$$

An alternative representation, using $1 + u \approx e^u$ for small values of u, is then

$$j\left(\log U\left(\frac{n+1}{j}\right) - \log U\left(\frac{n+1}{j+1}\right)\right) \approx \gamma \exp\left(\gamma^{-1} b_{k,n}\left(\frac{j}{k+1}\right)^{\beta}\right).$$

The more accurate approximation

$$Z_j \approx \left(\gamma + b_{k,n}\left(\frac{j}{k+1}\right)^{\beta}\right) E_j,\ j = 1, \ldots, k,$$

where E_j denotes a sequence of independent standard exponentially distributed random variables, was derived in an asymptotic sense in Beirlant et al. [97]. For each k, model (4.2.15) can be considered as a non-linear regression model in which one can estimate the intercept γ, the slope $b_{k,n}$, and the power β with the covariates $j/(k+1)$. One can estimate these parameters jointly, or by using an external estimate for β, or using external estimation for β and $B = b_{k,n}\left(\frac{k+1}{n+1}\right)^{-\beta}$ on the regression model

$$Z_j = \gamma + B\left(\frac{k+1}{n+1}\right)^{\beta}\left(\frac{j}{k+1}\right)^{\beta},\quad j = 1, \ldots, k. \tag{4.2.16}$$

Gomes et al. found that external estimation for B and β should be based on k_1 extreme order statistics where $k = o(k_1)$ and $\hat{\beta} - \beta = o_p(1/\log n)$. Such an estimator for β was presented, for example, in Fraga Alves et al. [35]. Given an estimator $\hat{\beta}$ for β, an estimator for B was given in Gomes and Martins [399]:

$$\hat{B}_{k,n} = \left(\frac{k+1}{n+1}\right)^{\hat{\beta}} \frac{\left(\frac{1}{k}\sum_{j=1}^{k}(j/(k+1))^{\hat{\beta}}\right)\hat{C}_0 - \hat{C}_1}{\left(\frac{1}{k}\sum_{j=1}^{k}(j/(k+1))^{\hat{\beta}}\right)\hat{C}_1 - \hat{C}_2},$$

$$\hat{C}_m = \frac{1}{k}\sum_{j=1}^{k}(j/(k+1))^{m\hat{\beta}} Z_j,\quad m = 0, 1, 2.$$

When the three parameters are jointly estimated for each k, the asymptotic variance turns out to be $\gamma^2((1+\beta)/\beta)^4$, which is to be compared with the asymptotic variance γ^2 for the Hill estimator. Performing linear regression on $\left(\frac{j}{k+1}\right)^{\beta}$ importing an external estimator for β, the asymptotic variance drops down to $\gamma^2((1+\beta)/\beta)^2$. The original variance γ^2 is retained when using the external estimators for B and β in (4.2.16).

Bias reduction of the extreme quantile estimator $\hat{q}^{+}_{k,p}$ should not be based solely on replacing $H_{k,n}$ by a bias-reduced estimator for γ. Here we use the fact that $X_{n-k,n} =_d U(1/U_{k+1,n})$, where $U_{k+1,n}$ denotes the $(k+1)$th smallest order statistic from a uniform (0,1) sample of size n. Then we obtain from (4.2.3) and (4.2.5) with $x = 1/p$, approximating $U_{k+1,n}$ by its expected value $(k+1)/(n+1)$, and using $1+u \approx e^u$ for u small, that

$$\begin{aligned}
\frac{Q(1-p)}{X_{n-k,n}} &= \frac{p^{-\gamma}}{U_{k+1,n}^{-\gamma}} \frac{\ell_U(1/p)}{\ell_U(1/U_{k+1,n})} \\
&\approx \left(\frac{k+1}{(n+1)p}\right)^{\gamma} \frac{1+Dp^{\beta}}{1+D\left(\frac{k+1}{n+1}\right)^{\beta}} \\
&\approx \left(\frac{k+1}{(n+1)p}\right)^{\gamma} \exp\left(D(p^{\beta} - \left(\frac{k+1}{n+1}\right)^{\beta}\right) \\
&= \left(\frac{k+1}{(n+1)p}\right)^{\gamma} \exp\left(b_{k,n}\frac{1-((n+1)p/(k+1))^{\beta}}{\beta}\right),
\end{aligned}$$

so that a bias-reduced version of $\hat{q}^{+}_{k,p}$ is given by

$$\hat{q}^{BR}_{k,p} = X_{n-k,n}\left(\frac{k+1}{(n+1)p}\right)^{\hat{\gamma}} \exp\left(\hat{b}_{k,n}\frac{1-((n+1)p/(k+1))^{\hat{\beta}}}{\hat{\beta}}\right),$$

where $\hat{\gamma}$, $\hat{b}_{k,n}$, and $\hat{\beta}$ are bias-reduced estimators based on the regression model (4.2.15).

Bias-reduced estimators can also be obtained by improving on the approximation (4.2.4) of the POT distribution $\mathbb{P}(X/t > u|X > t)$ by the simple Pareto distribution, using an extension of the Pareto distribution as introduced in Beirlant et al. [102]. Indeed, when ℓ_U satisfies (4.2.5), then

$$\overline{F}(x) = C^{1/\gamma}x^{-1/\gamma}\left(1+\gamma^{-1}DC^{\beta/\gamma}x^{-\beta/\gamma}(1+o(1))\right). \tag{4.2.17}$$

The distribution of the POT's X/t $(X > t)$ can then be approximated using the expansion $(1+u)^b \approx 1+bu$ for u small:

$$\begin{aligned}
\frac{\overline{F}(tu)}{\overline{F}(t)} &\approx u^{-1/\gamma}\frac{1+\gamma^{-1}DC^{\beta/\gamma}t^{-\beta/\gamma}u^{-\beta/\gamma}}{1+\gamma^{-1}DC^{\beta/\gamma}t^{-\beta/\gamma}} \\
&\approx u^{-1/\gamma}\left(1+\gamma^{-1}DC^{\beta/\gamma}t^{-\beta/\gamma}(u^{-\beta/\gamma}-1)\right) \\
&\approx u^{-1/\gamma}\left(1+DC^{\beta/\gamma}t^{-\beta/\gamma}(1-u^{-\beta/\gamma})\right)^{-1/\gamma}.
\end{aligned}$$

This leads to the extended Pareto distribution (EPD) with distribution function

$$\overline{G}_{\gamma,\delta,\beta}(u) = \{u(1+\delta-\delta u^{\tau})\}^{-1/\gamma}, \quad u > 1, \tag{4.2.18}$$

$(\tau < 0,\ \delta > \max(-1, 1/\tau))$ with $\delta = \delta_t = DC^{\beta/\gamma}t^{-\beta/\gamma}$ and $\tau = -\beta/\gamma$. Note that for an EPD random variable Y with $\tau = -1$ and $\delta = \frac{\gamma}{\sigma} - 1$, it follows that $Y - 1$ is GPD distributed with parameters γ and σ.

Using the density of the EPD $g_{\gamma,\delta,\tau}(y) = \gamma^{-1}y^{-1/\gamma-1}\{1+\delta(1-y^{\tau})\}^{-1/\gamma-1}[1+\delta\{1-(1+\tau)y^{\tau}\}]$, maximum likelihood estimators are then derived through maximization of

$$\sum_{i=1}^{N_t} \log g_{\gamma,\delta,\tau}(Y_i),$$

with respect to γ, δ using an external estimator of τ through estimates of β and γ, where the values $Y_1, \ldots, Y_{N_t}$ denote the POT values over the threshold t.

Bias-reduced estimation of return periods is then obtained using $X_{n-k,n}$ again as a threshold t:

$$\hat{r}^{BR}_{k,x} = \frac{n+1}{k+1} / \overline{G}_{\hat{\gamma},\hat{\delta},\hat{\tau}}\left(\frac{x}{X_{n-k,n}}\right).$$

4.2.1.4 Estimating the Scale Parameter

Finally, note that the scale parameter C in (4.2.5), or $A = C^{1/\gamma}$ in (4.2.17), can be estimated with

$$\hat{A}_{k,n} = \frac{k+1}{n+1} X_{n-k,n}^{1/H_{k,n}}, \tag{4.2.19}$$

$$\hat{C}_{k,n} = X_{n-k,n}\left(\frac{k+1}{n+1}\right)^{H_{k,n}} \tag{4.2.20}$$

which follows, for instance, from (4.2.17) replacing x by $X_{n-k,n}$ and estimating $\overline{F}(x)$ by the empirical probability $(k+1)/(n+1)$.

The estimator $\hat{C}_{k,n}$ can also be retrieved using least squares regression on the k top points of the log–log plot

$$\left(\log \frac{n+1}{j}, \log X_{n-j+1,n}\right), \; j = 1, \ldots, k$$

minimizing

$$\sum_{j=1}^{k} (\log X_{n-j+1,n} - \gamma \log((n+1)/j) - \log C)^2. \tag{4.2.21}$$

Substituting $H_{k,n}$ for γ and taking the derivative with respect to $\log C$ indeed gives

$$\begin{aligned}
\log \hat{C} &= \frac{1}{k}\sum_{j=1}^{k} \log X_{n-j+1,n} - H_{k,n}\frac{1}{k}\sum_{j=1}^{k} \log \frac{n+1}{j} \\
&= H_{k,n}\left(1 - \frac{1}{k}\sum_{j=1}^{k} \log \frac{n+1}{j}\right) + \log X_{n-k,n} \\
&\approx -H_{k,n} \log \frac{n+1}{k+1} + \log X_{n-k,n} \\
&= \log \hat{C}_{k,n}.
\end{aligned}$$

In Beirlant et al. [104] it is shown that $\hat{A}_{k,n}$ is asymptotically normally distributed with asymptotic variance $(k\xi^2)^{-1}\log^2 U((n+1)/(k+1))$ and asymptotic bias $-b_{k,n}\{1/\beta + \log U((n+1)/(k+1))/(\gamma(1+\beta))\}$.

A bias-reduced estimator of the scale parameter A is then given by

$$\hat{A}^{BR}_{k,n} = \frac{k+1}{n+1} X^{1/\hat{\gamma}}_{n-k,n} \exp\left(\hat{b}_{k,n}/\hat{\beta}\right),$$

where $\hat{\gamma}^{BR}$ is a bias-reduced estimator of γ, and $\hat{b}_{k,n}$, $\hat{\beta}$ estimators of $b_{k,n}$ and β.

Case studies. The Hill and bias-reduced estimators of a positive EVI are plotted in Figure 4.5 as a function of k and $\log k$. Estimators for extreme quantiles and return periods are given in Figure 4.6, while the scale estimates can be found in Figure 4.7.

For the Dutch fire insurance data set a level $\gamma \approx 0.8$ is visible for $k > 150$ using the bias-reduced estimators, while for the smallest k, values between 0.4 and 0.5 appear when plotting the estimates as a function of $\log k$. These plots are to be compared with the plots on the second line in Figure 4.1, where the $H_{k,n}$ values are plotted against the data values.

Concerning the estimation of the quantile $Q(0.999)$ again two levels become apparent, namely around 1.5×10^8 at $k < 150$ and 3×10^8 at $k > 150$. Correspondingly, for the return period $\hat{r}^{+}_{200\text{ million}}$ two values $e^{7.5}$ when $k < 150$ and $e^{6.5}$ for $k > 150$ are detected. These components are found back again in the scale plots of Figure 4.7 with values around 9 and 11. These two values correspond to extrapolating on the Pareto QQ-plot using only the 150 largest values, compared to setting the anchor much deeper in the QQ-plot. Of course this last choice leads to a much more conservative tail extrapolation and eventually higher reinsurance premiums, as discussed below. Note that the scale parameter "compensates" for the lower EVI value for $k < 150$ with a larger value for the scale.

Concerning the ultimate values of Company A, note the three γ levels appearing from Figure 4.5 (middle): $\hat{\gamma} \approx 0.6$ for $k \geq 600$, $\hat{\gamma} \approx 0.3$ when $k \in (300; 600)$, ending with $\hat{\gamma} \approx 0.2$ when $k < 200$. In fact, two Pareto components are also visible in the mean excess plot in Figure 4.2 with two linear pieces with different slopes. Finally, the estimates drop down to 0 when $k \to 1$, which could be due to upper-truncation. For the estimates of $Q(0.999)$ notice an overall stable bias-reduced value at 5 million, with some slightly higher value at $k \in (300; 600)$. Note, however, that this value could be too large in view of the possible upper-truncation. Concerning the return period for values over 4 million, again we observe two levels: a return period close to e^6 for smaller values of k and a value somewhat larger than e^5 for $k \in (300; 600)$. We revisit the estimation of this tail using a truncated Pareto model below. Note that the three segments are also visible in the scale estimates in Figure 4.7 (middle).

For the ultimate values of Company B, a Pareto component with $\hat{\gamma} \approx 0.7$ is clearly visible for $k > 250$ from Figure 4.5 (bottom). After a systematic decrease for k down to 100, a level $\hat{\gamma} \approx 0.2$ is reached for $k \in (1; 50)$. This corresponds to the graphs from Figure 4.3, where a Pareto component is followed by a light tail component, ending with an ultimate Pareto section at the top data. The two Pareto levels are also visible at the estimators of the quantile $Q(0.999)$ with levels 6 and 30 million. This lowest level is of

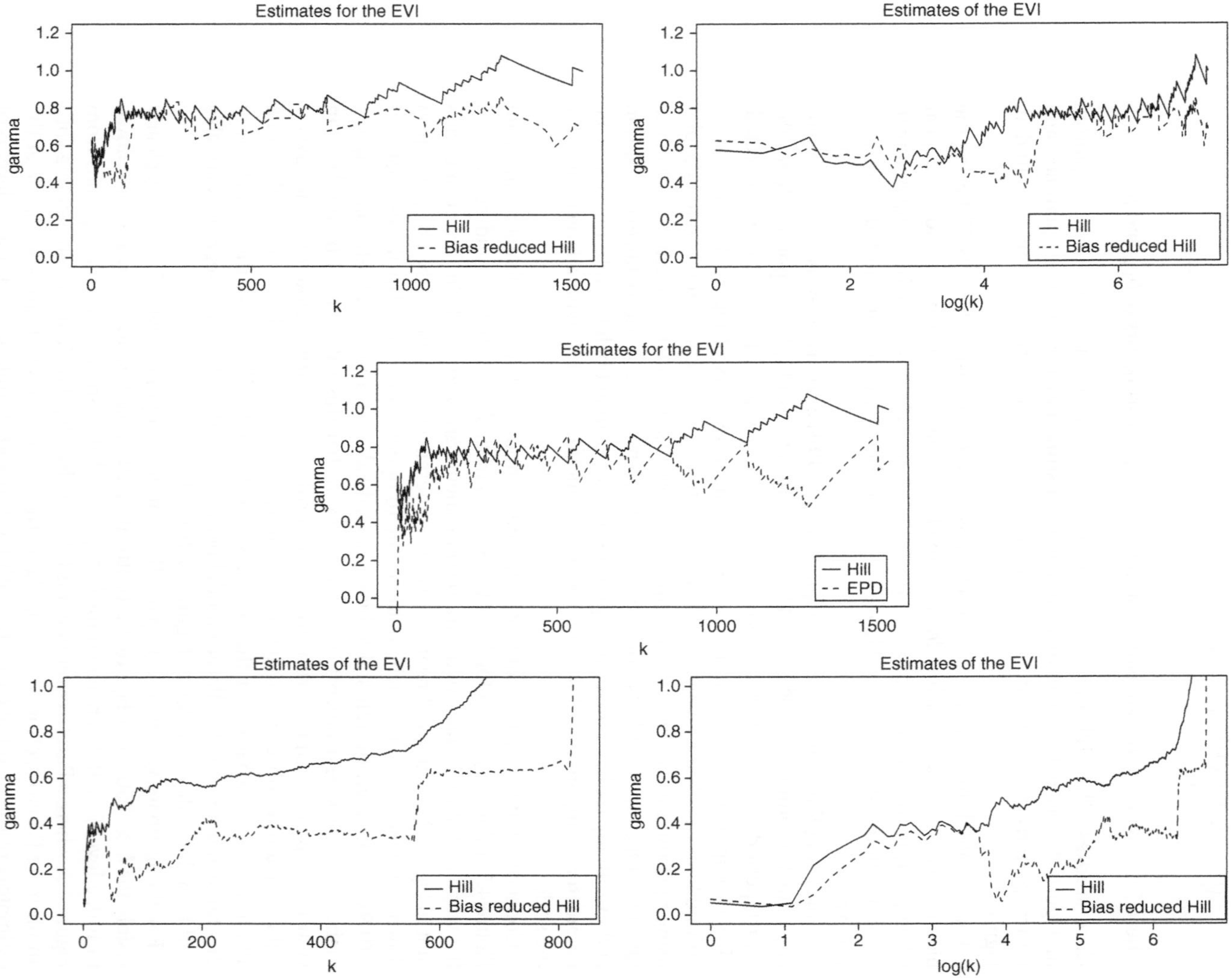

Figure 4.5 Hill estimates and bias-reduced versions, regression approach as a function of k (left) and $\log k$ (right), and EPD approach (middle) as a function of k: Dutch fire insurance data (top); MTPL data for Company A, ultimate values (middle); MTPL data for Company B, ultimate values (bottom).

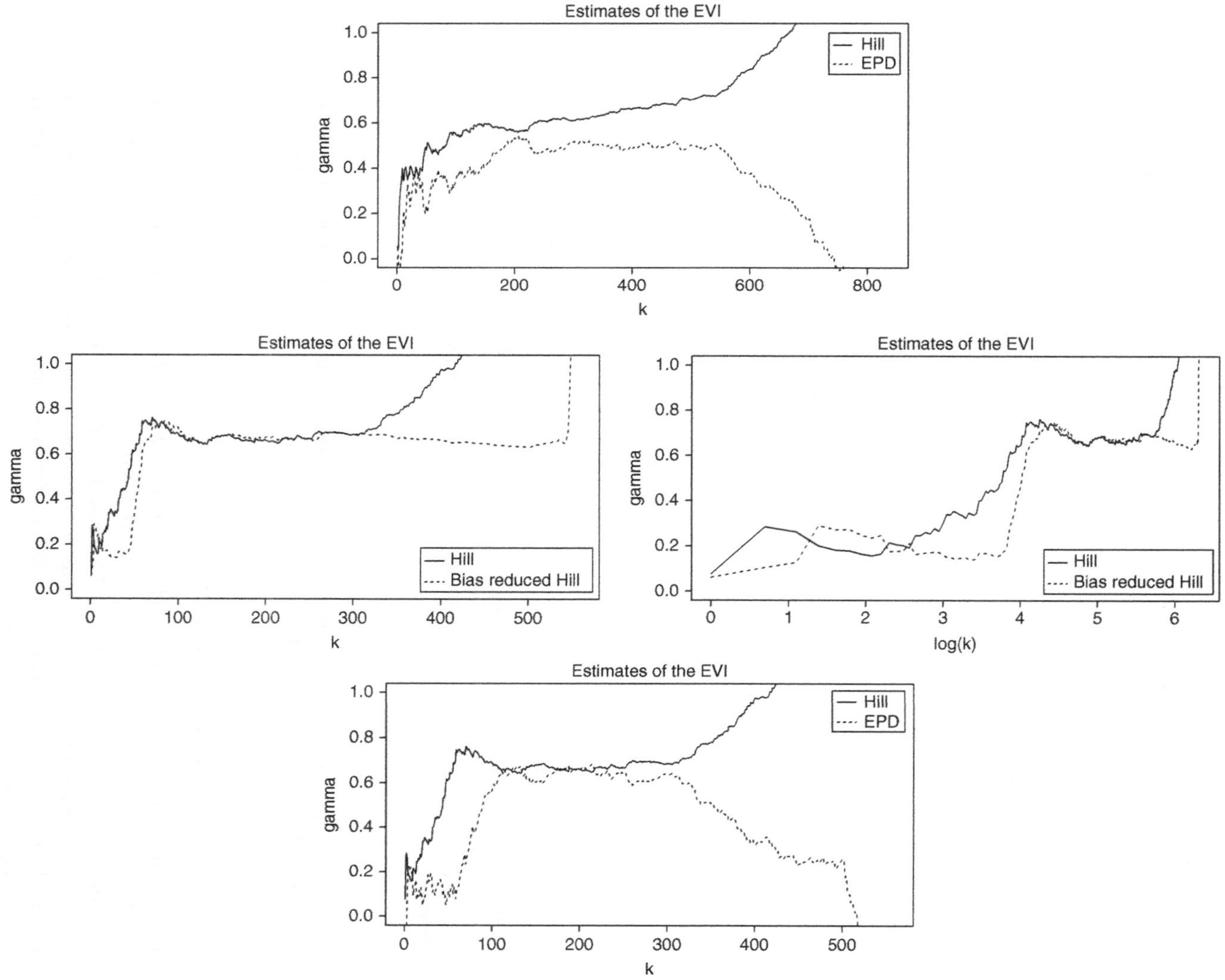

Figure 4.5 (Continued)

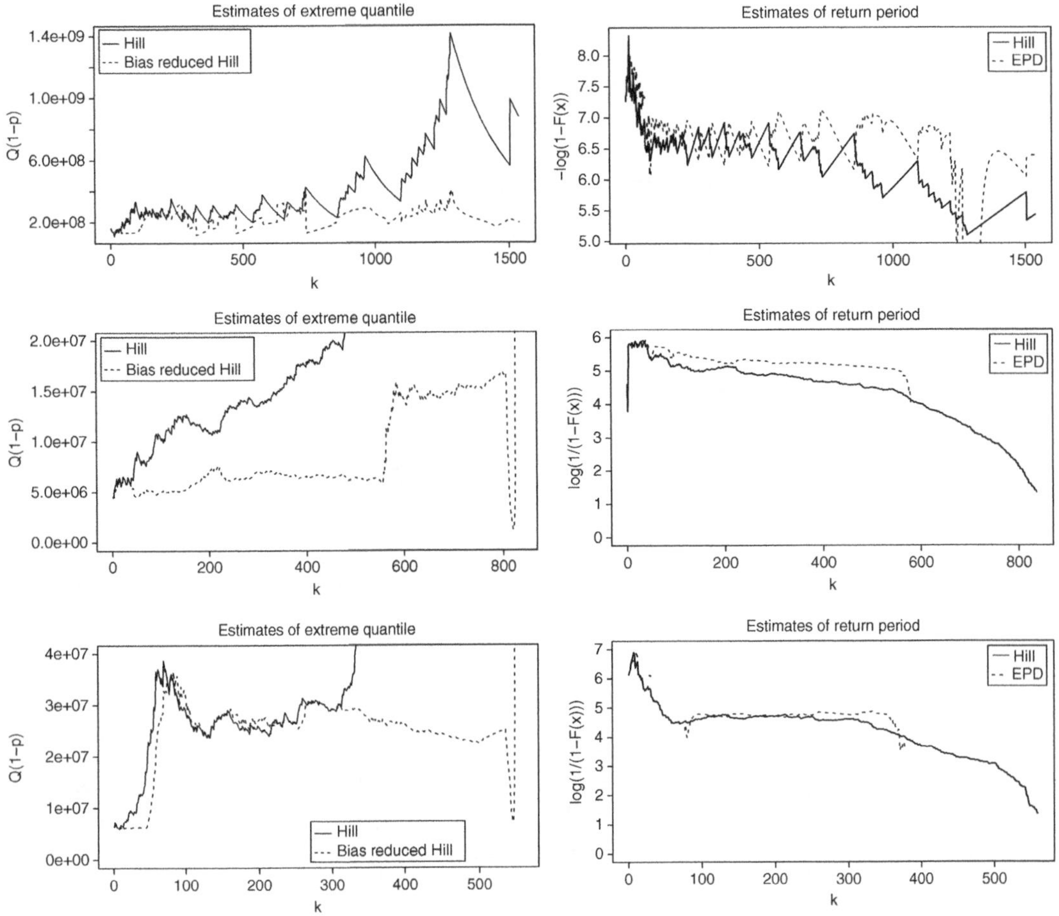

Figure 4.6 Quantile estimates $\hat{q}^{+}_{k,0.001}$ and $\hat{q}^{BR}_{k,0.001}$ (left) and log-return periods $\log \hat{r}^{+}_{k,x}$ and $\log \hat{r}^{BR}_{k,x}$ (right), as a function of k: Dutch fire insurance data ($x = 200$ million, top); MTPL data for Company A, ultimate values ($x = 4$ million, middle); MTPL data for Company B, ultimate values ($x = 6$ million, bottom).

course only based on a few top observations. Finally, the return period over 6 million is estimated at $e^{4.5}$ when $k \in (100, 400)$ and a value around e^{7} when using only a few exceedances. □

4.2.2 General Tail Modelling using EVA

In order to allow tail modelling with log-normal or Weibull tails, one has to incorporate the case where the EVI γ can be 0, next to positive values. Estimation of γ, extreme quantiles and return periods under the max-domains of attraction conditions C_{γ} in (3.2.5) or (3.2.6), with as few restrictions on the value of γ as possible, is the next step in tail modelling. Again we have two possible approaches: using quantile plotting or using a likelihood approach on POT values.

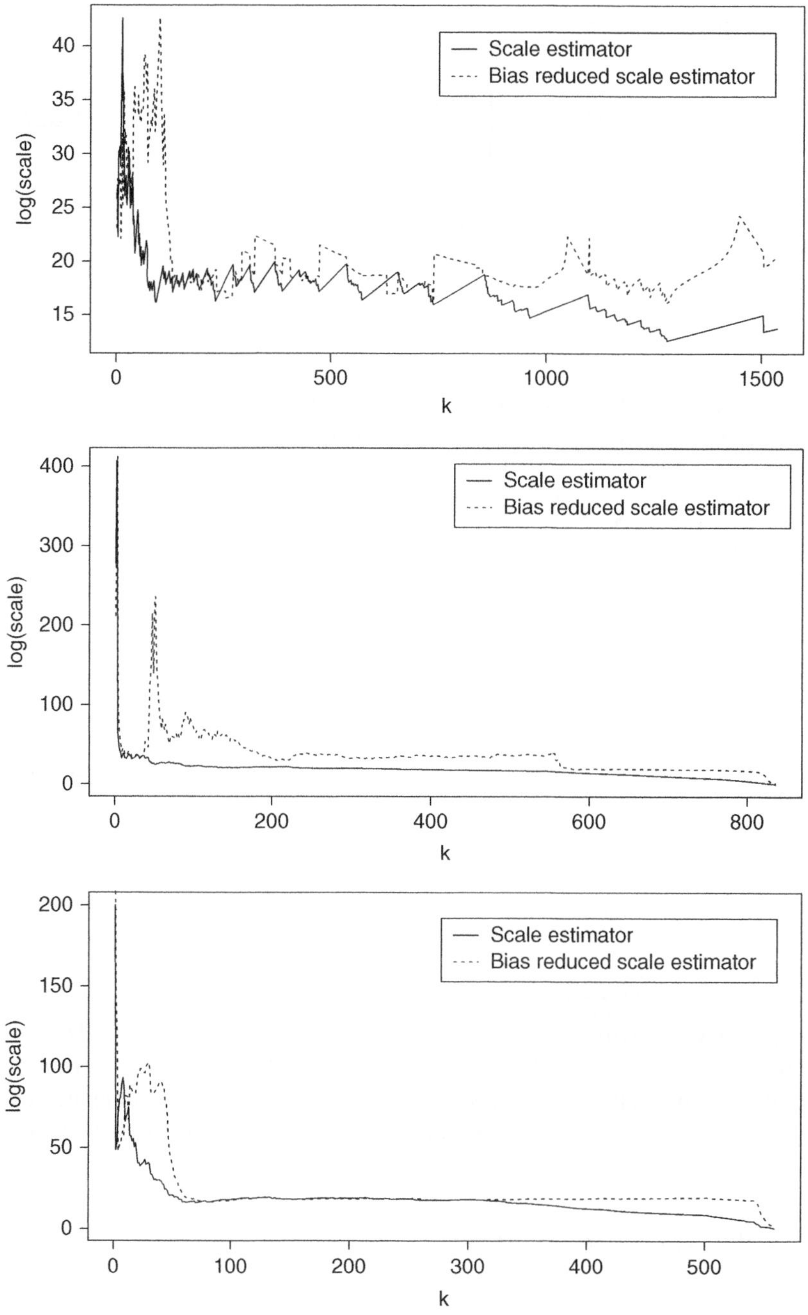

Figure 4.7 Scale estimates $\hat{A}_{k,n}$ and $\hat{A}^{BR}_{k,n}$ as a function of k: Dutch fire insurance data (top); MTPL data for Company A, ultimate values (middle); MTPL data for Company B, ultimate values (bottom).

- Here, several existing estimators start from the following condition, which follows from (3.2.5): for all $u \geq 1$ as $x \to \infty$

$$\frac{U(x)}{a(x)} \{\log U(xu) - \log U(x)\} \to \begin{cases} \log u, & \gamma \geq 0, \\ \frac{u^{\gamma}-1}{\gamma}, & \gamma < 0. \end{cases} \tag{4.2.22}$$

From this it follows with $\gamma^- = \min(\gamma, 0)$, that as $k, n \to \infty$, $k/n \to 0$

$$\begin{aligned} EH_{k,n} &= \frac{1}{k}\sum_{j=1}^{k} \log U\left(\frac{n+1}{j}\right) - \log\left(U\left(\frac{n+1}{k+1}\right)\right) \\ &= \frac{1}{k}\sum_{j=1}^{k} \log U\left(\frac{n+1}{k+1}\frac{k+1}{j}\right) - \log U\left(\frac{n+1}{k+1}\right) \\ &\approx \frac{a\left(\frac{n+1}{k+1}\right)}{U\left(\frac{n+1}{k+1}\right)} \frac{1}{k}\sum_{j=1}^{k} \left\{\left(\frac{j}{k+1}\right)^{-\gamma^-} - 1\right\} / \gamma^- \\ &\approx \frac{a\left(\frac{n+1}{k+1}\right)}{U\left(\frac{n+1}{k+1}\right)} (1-\gamma^-)^{-1}. \end{aligned} \tag{4.2.23}$$

Hence estimating $EH_{k,n}$ by $H_{k,n}$, and $U\left(\frac{n+1}{k+1}\right)$ by $X_{n-k,n}$, we find that for any estimator $\hat{\gamma}_{k,n}$ of γ

$$\hat{a}\left(\frac{n+1}{k+1}\right) = H_{k,n} X_{n-k,n}(1 - \hat{\gamma}^-_{k,n}), \tag{4.2.24}$$

leads to an estimator for $a\left(\frac{n+1}{k+1}\right)$.

Since a regularly varies with index $\gamma \in \mathbb{R}$, it also follows from (4.2.23) that $U((n+1)/(k+1))EH_{k,n} = ((n+1)/(k+1))^{\gamma}\ell((n+1)/(k+1))$ for some slowly varying function ℓ. Hence the approach using linear regression and extrapolation on linear tail patterns on a QQ-plot can be generalized to the case of a real-valued EVI using the *generalized QQ-plot*

$$\left(\log \frac{n+1}{k+1}, \log\left(X_{n-k,n} H_{k,n}\right)\right), \quad k = 1, \ldots, n-1,$$

which ultimately for smaller values of k will be linear with slope γ, whatever the sign or values of γ. Hence if a generalized QQ-plot is ultimately horizontal, then tail modelling using a distribution in the Gumbel domain of attraction is appropriate. An ultimately decreasing generalized QQ-plot indicates a negative EVI, which can occur, for instance, for truncated heavy-tailed distributions.

A *generalized Hill estimator* of γ estimating the slopes at the last k points on the generalized QQ-plot is then given by

$$\hat{\gamma}_{k,n}^{GH} = \frac{1}{k}\sum_{j=1}^{k} \log UH_{j,n} - \log UH_{k+1,n}$$

$$= H_{k+1,n} + \frac{1}{k}\sum_{j=1}^{k} \left(\log H_{j,n} - \log H_{k+1,n}\right),$$

where $UH_{j,n} = X_{n-j,n}H_{j,n}$.

Another generalization of the Hill estimator to real-valued EVI was given in Dekkers et al. [268], termed the *moment estimator*:

$$\hat{\gamma}_{k,n}^{M} = H_{k,n} + 1 - \frac{1}{2}\left(1 - \frac{H_{k,n}^2}{H_{k,n}^{(2)}}\right)^{-1},$$

where

$$H_{k,n}^{(2)} = \frac{1}{k}\sum_{j=1}^{k} \left(\log X_{n-j+1,n} - \log X_{n-k,n}\right)^2.$$

- Condition (3.2.6) for a distribution to belong to a domain of attraction of an extreme value distribution means that the generalized Pareto law is the limit distribution of the distribution of POT values $X - t$ given $X > t$ when $t \to x_+$: setting $h(t) = \sigma_t$

$$\mathbb{P}(X-t > u\sigma_t | X > t) \to_{t\to x_+} (1+\gamma u)^{-1/\gamma}, \text{ for all } u \text{ such that } 1+\gamma u > 0. \quad (4.2.25)$$

Hence, we are led to modelling the tail function $\mathbb{P}(Y > y)$ of POT values $Y = X-t$ with $X > t$ using the GPD with survival function $\left(1 + y\gamma/\sigma\right)^{-1/\gamma}$. Denoting the number of exceedances over t again by N_t, the log-likelihood is given by

$$\log L(\gamma, \sigma | Y_1, \ldots, Y_{N_t}) = -N_t \log \sigma - \left(\frac{1}{\gamma} + 1\right)\sum_{i=1}^{N_t} \log\left(1 + Y_i \frac{\gamma}{\sigma}\right).$$

Using a reparametrization $(\gamma, \sigma) \to (\gamma, \tau)$ with $\tau = \gamma/\sigma$, leads to the likelihood equations

$$\begin{cases} \frac{1}{\hat{\gamma}_t + 1} = \frac{1}{N_t}\sum_{i=1}^{N_t} \frac{1}{1+\hat{\tau}_t Y_i}, \\ \hat{\gamma}_t = \frac{1}{N_t}\sum_{i=1}^{N_t} \log\left(1 + \hat{\tau}_t Y_i\right). \end{cases}$$

Replacing t by an intermediate order statistic $X_{n-k,n}$ again gives

$$\begin{cases} \frac{1}{\hat{\gamma}^{ML}_{k,n}+1} = \frac{1}{k}\sum_{i=1}^{k} \frac{1}{1+\hat{\tau}^{ML}_{k,n}(X_{n-j+1,n}-X_{n-k,n})} \\ \hat{\gamma}^{ML}_{k,n} = \frac{1}{k}\sum_{i=1}^{k} \log\left(1+\hat{\tau}^{ML}_{k,n}(X_{n-j+1,n}-X_{n-k,n})\right). \end{cases}$$

In order to assess the goodness-of-fit of the GPD when modelling the POT values $Y = X - t$ for a given threshold t, one can use the transformation

$$R = \begin{cases} \frac{1}{\gamma}\log\left(1+\frac{\gamma}{\sigma}Y\right), & \text{if } \gamma \neq 0, \\ Y\frac{1}{\sigma}, & \text{if } \gamma = 0, \end{cases}$$

so that R is standard exponentially distributed if the POT values do follow a GPD, and the fit can be validated inspecting the overall linearity of the exponential QQ-plot

$$\left(-\log\left(1-\frac{i}{N_t+1}\right), \hat{R}_{i,N_t}\right), i = 1, \ldots, N_t$$

where $\hat{R}_{i,N_t}$ $(i = 1, \ldots, N_t)$ denote the ordered values of

$$\hat{R}_i = \begin{cases} \frac{1}{\hat{\gamma}}\log\left(1+\frac{\hat{\gamma}}{\hat{\sigma}}Y_i\right), & \text{if } \hat{\gamma} \neq 0, \\ Y_i/\hat{\sigma}, & \text{if } \hat{\gamma} = 0. \end{cases}$$

For all these estimators $\hat{\gamma}^{\bullet}_{k,n} - \gamma$ is asymptotically normal under some regularity conditions on the underlying distributions when $k, n \to \infty$ and $k/n \to 0$, with mean 0 if k is not too large (or, equivalently, if the threshold $t = x_{n-k,n}$ is not too small), and asymptotic variances (or covariance matrix for $(\hat{\gamma}^{ML}_{k,n}, \hat{\sigma}^{ML}_{k,n})$) given by

$$\mathrm{Var}(\hat{\gamma}^{GH}_{k,n}) \sim \begin{cases} \frac{1+\gamma^2}{k} & \text{if } \gamma \geq 0, \\ \frac{(1-\gamma)(1+\gamma+2\gamma^2)}{(1-2\gamma)k} & \text{if } \gamma < 0, \end{cases}$$

$$\mathrm{Var}(\hat{\gamma}^{M}_{k,n}) \sim \begin{cases} \frac{1+\gamma^2}{k} & \text{if } \gamma \geq 0, \\ \frac{(1-\gamma)^2(1-2\gamma)(1-\gamma+6\gamma^2)}{(1-3\gamma)(1-4\gamma)k} & \text{if } \gamma < 0, \end{cases}$$

$$\Sigma_{(\hat{\gamma}^{ML}_{k,n}, \hat{\sigma}^{ML}_{k,n})} \sim \frac{(1+\gamma)}{k}\begin{pmatrix} 1+\gamma & -\sigma \\ -\sigma & 2\sigma^2 \end{pmatrix}.$$

Estimators for small tail probabilities or return periods can easily be constructed from the POT approach. In fact, the approximation of $P(X - t > y|X > t)$ by $(1 + (\gamma/\sigma)y)^{-1/\gamma}$ for $y > 0$, setting $t + y = x$, leads to

$$\hat{p}^{ML}_{x,k} = \frac{k+1}{n+1}\left\{1+\frac{\hat{\gamma}^{ML}_{k,n}}{\hat{\sigma}^{ML}_{k,n}}(x - X_{n-k,n})\right\}^{-1/\hat{\gamma}^{ML}_{k,n}}.$$

Inversion leads to an extreme quantile estimator

$$\hat{q}^{ML}_{p,k} = X_{n-k,n} + \frac{\hat{\sigma}^{ML}_{k,n}}{\hat{\gamma}^{ML}_{k,n}} \left\{ \left(\frac{k+1}{(n+1)p} \right)^{\hat{\gamma}^{ML}_{k,n}} - 1 \right\}.$$

The estimators for high quantiles based on the approach used in the construction of the moment estimator $\hat{\gamma}^{M}_{k,n}$ are defined by

$$\hat{q}^{M}_{p,k} = X_{n-k,n} + \frac{\hat{a}\left(\frac{n+1}{k+1}\right)}{\hat{\gamma}^{M}_{k,n}} \left\{ \left(\frac{k+1}{(n+1)p} \right)^{\hat{\gamma}^{M}_{k,n}} - 1 \right\},$$

with $\hat{a}$ defined in (4.2.24). Note that $X_{n-k,n}H_{k,n}\left(1-\hat{\gamma}^{-}_{k,n}\right)$ can be seen as an alternative estimator for σ when comparing the expressions of $\hat{q}^{ML}_{p,k}$ and $\hat{q}^{M}_{p,k}$. This then in turn leads to a moment tail probability estimator

$$\hat{p}^{M}_{x,k} = \frac{k+1}{n+1} \left\{ 1 + \frac{\hat{\gamma}^{M}_{k,n}}{\hat{a}^{M}_{k,n}} (x - X_{n-k,n}) \right\}^{-1/\hat{\gamma}^{M}_{k,n}}.$$

The asymptotic distributions of these tail estimators have been derived in the literature. For instance, for $\hat{q}^{M}_{p,k}$ we have under some regularity conditions that with $a_n = (k+1)/(p(n+1))$ as $np_n \to c \geq 0$, $(\log a_n)/\sqrt{k} \to 0$ one has

- for $\gamma > 0$

$$\frac{\gamma\sqrt{k}}{a(n/k)a_n^{\gamma}\log a_n}\left(\hat{q}^{M}_{p,k} - Q(1-p)\right) \approx_d \mathcal{N}(0,(1+\gamma)^2),$$

- for $\gamma < 0$

$$\frac{\sqrt{k}}{a(n/k)}\left(\hat{q}^{M}_{p,k} - Q(1-p)\right) \approx_d \mathcal{N}\left(0, \frac{(1-\gamma)^2(1-3\gamma+4\gamma^2)}{\gamma^4(1-2\gamma)(1-3\gamma)(1-4\gamma)}\right).$$

For further details see de Haan and Ferreira [258].

Case studies. The estimators of the EVI, extreme quantiles and return periods, which are consistent under all max-domains of attraction, are given in Figures 4.8 and 4.9.

In the case of the Dutch fire example two linear increasing parts are clearly visible in the generalized QQ-plot with a smaller slope at the largest values. These correspond again to the two γ levels in the Hill derivative estimates, namely 0.8 for higher values of k and 0.4 for $k \downarrow 1$. The ML-GPD estimators are somewhat lower. The quantile estimates $\hat{q}_{0.001}$ and return period estimates $\hat{r}_{200 \text{ million}}$ confirm two levels, as in Figure 4.6, but the quantile levels are somewhat lower than under the Pareto analysis.

In the case of the ultimate values from Company A, we find an ultimately decreasing generalized QQ-plot and correspondingly negative values of the EVI estimators at the

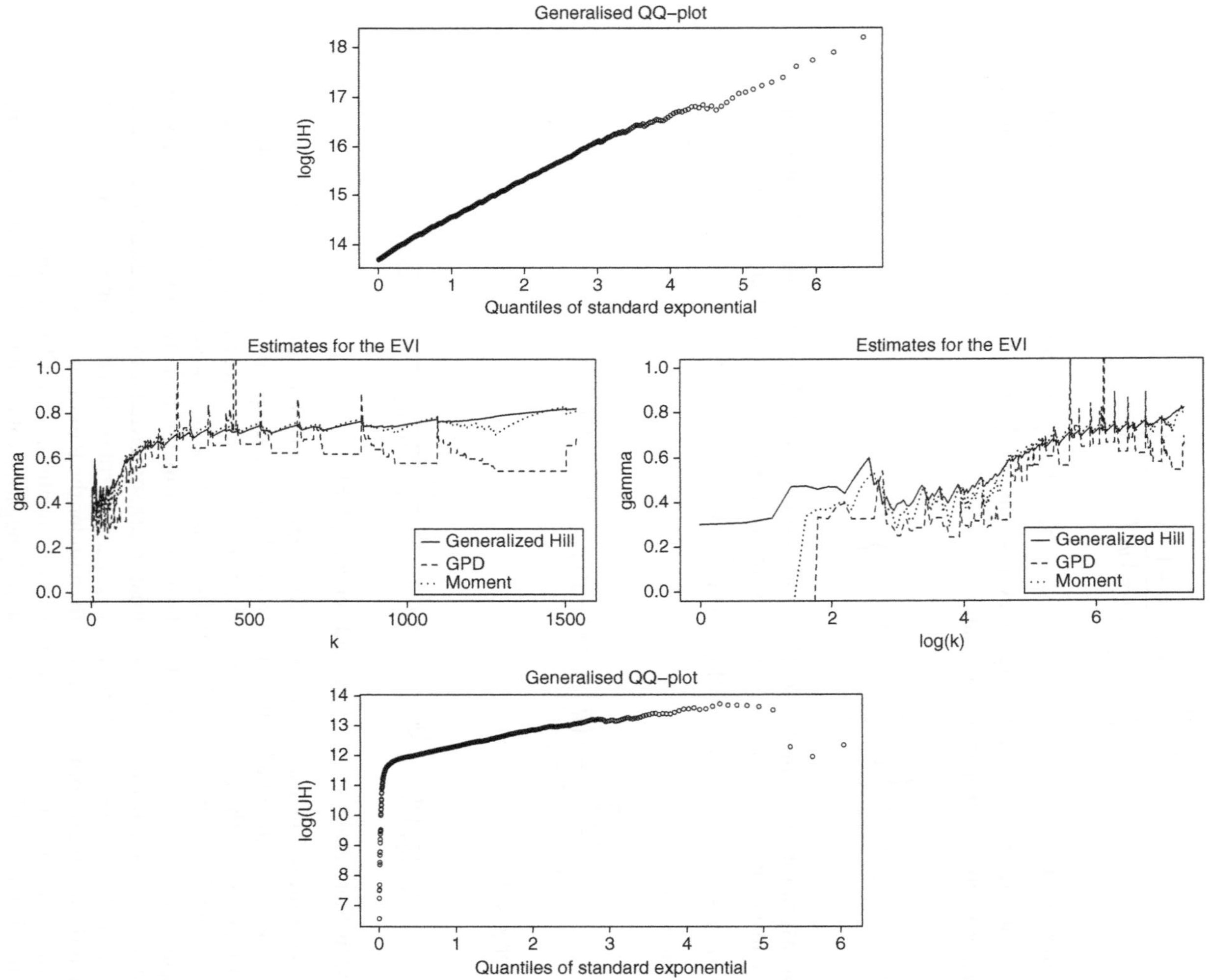

Figure 4.8 Generalized QQ-plot (middle) and estimators of EVI $\hat{\gamma}^{GH}_{k,n}$, $\hat{\gamma}^{M}_{k,n}$, $\hat{\gamma}^{ML}_{k,n}$ as a function of k (left) and $\log k$ (right): Dutch fire insurance data (first and second line); MTPL data for Company A, ultimate values (third and fourth line); MTPL data for Company B, ultimate values (bottom two lines).

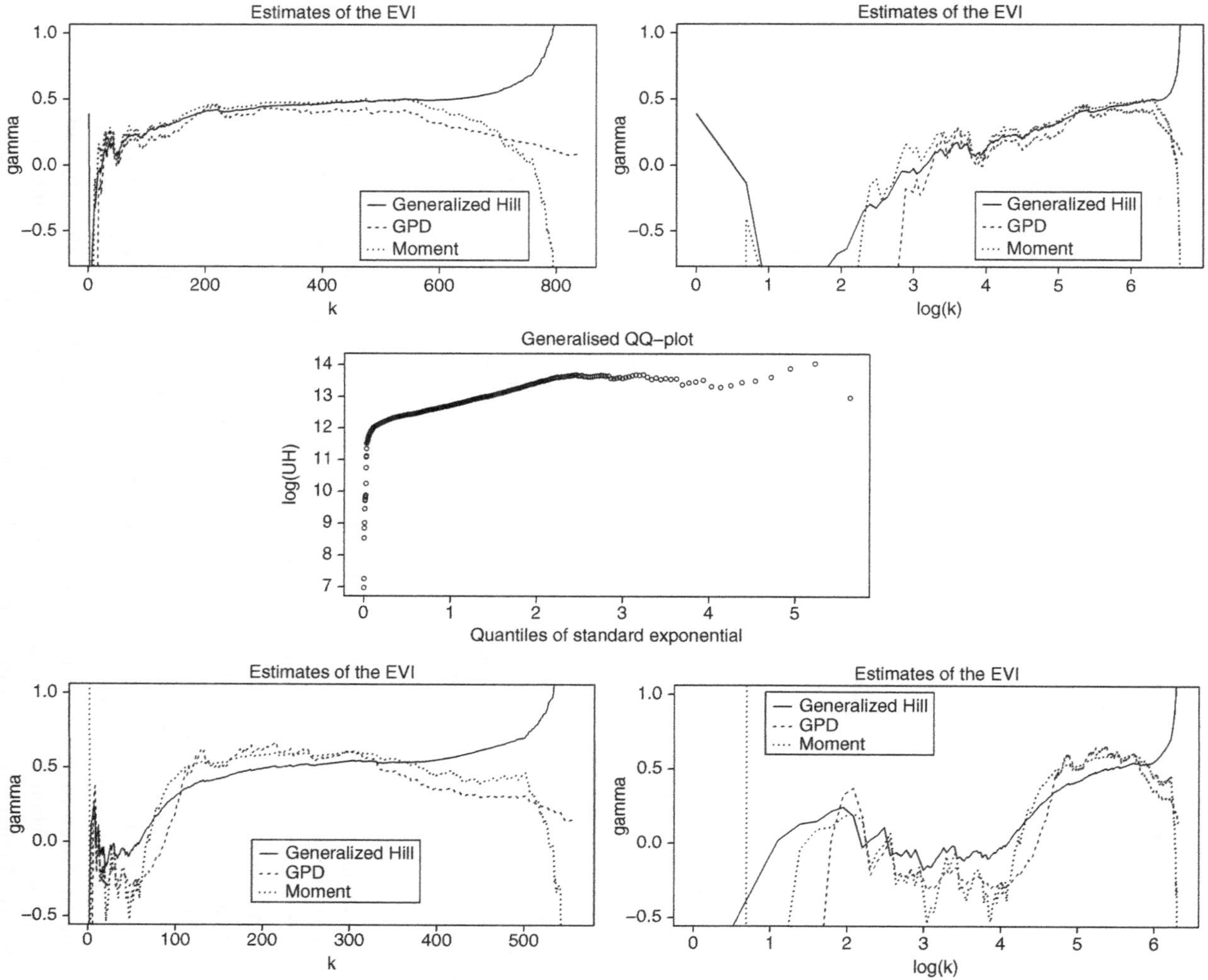

Figure 4.8 (Continued)

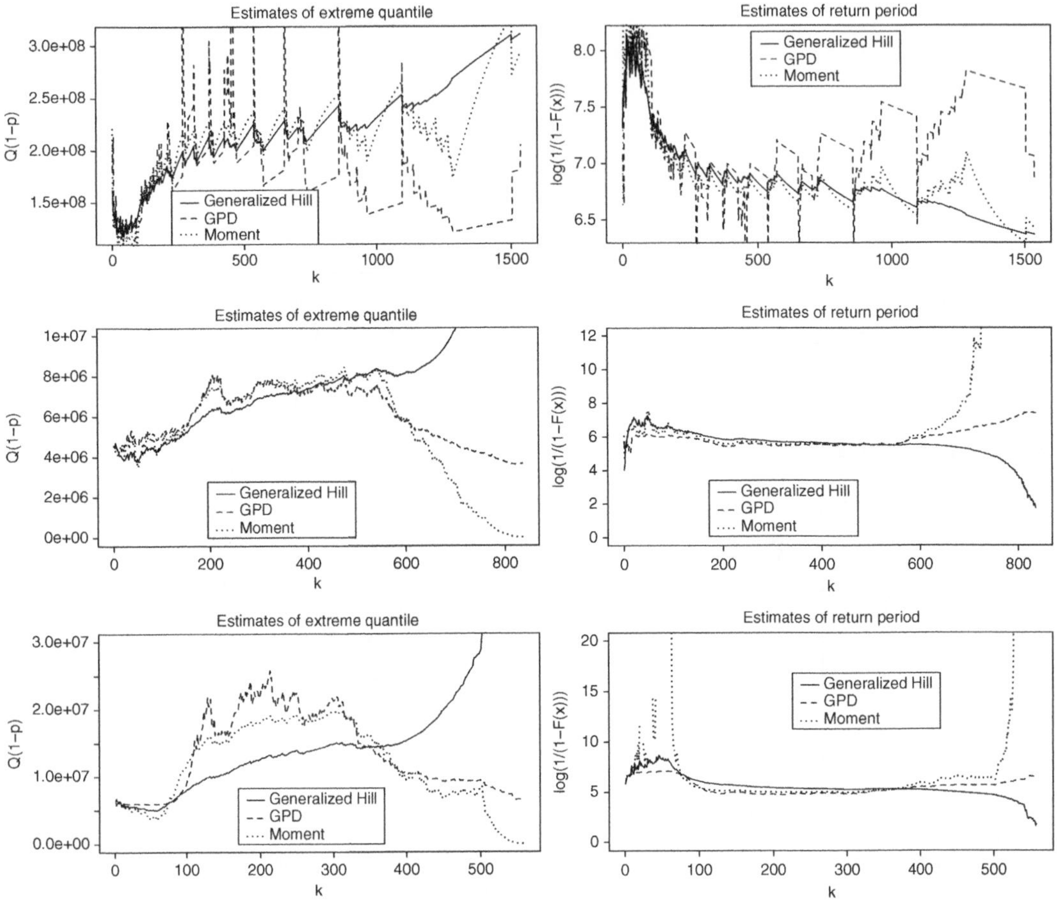

Figure 4.9 Large quantile estimators $\hat{q}^{GH}_{0.001,k}$, $\hat{q}^{M}_{0.001,k}$, $\hat{q}^{ML}_{0.001,k}$ (left) and log-return period estimators $-\log \hat{p}^{GH}_{x,k}$, $-\log \hat{p}^{M}_{x,k}$, $-\log \hat{p}^{ML}_{x,k}$ (right) as a function of k: Dutch fire insurance data ($x = 200$ million, top); MTPL data for Company A, ultimate values ($x = 4$ million, middle); MTPL data for Company B, ultimate values ($x = 6$ million, bottom).

smallest values of k. Concerning the quantile $Q(0.999)$ estimates, here no stable pictures appear for the estimates based on the generalized Hill estimator, with a decreasing plot as k decreases, ending at approximately 4 million Euros. The quantile level at 5 million appearing in the Pareto analysis is confirmed here for the smallest k values. The return period corresponding to 4 million is higher than e^6 in contrast to the Pareto analysis, which hints at a value just below e^6.

In the case of the ultimate values from Company B, the segment with $k \in (100, 250)$, which was already visible through a decreasing mean excess plot in Figure 4.3, corresponds with a slightly decreasing generalized QQ-plot and negative EVI estimates in that region. Given the ultimate Pareto tail at the top data in this case, the values of $\hat{\gamma}$ return positive when $k \downarrow 1$. In Figure 4.9 (bottom) the estimate of the quantile level $Q(0.999)$ of 5 million found from a Pareto analysis at the smallest k values is found back here. Note finally that the return periods for values over 6 million yield similar results as with Pareto tail modelling. □

4.2.3 EVA under Upper-truncation

Practical problems can arise when using the strict Pareto distribution and its generalization to the Pareto-type model because some probability mass can still be assigned to loss amounts that are unreasonably large or even impossible. With respect to tail fitting of an insurance claim data, upper-truncation is of interest and can be due to the existence of a maximum possible loss. Such truncation effects are sometimes visible in data, for instance when an overall linear Pareto QQ-plot shows non-linear deviations at only a few top data. Let W be an underlying non-truncated distribution with distribution function F_W, quantile function Q_W and tail quantile function U_W. Upper-truncation of the distribution of W at some value T was defined in the preceding chapter through the conditioning operation $W|W < T$. Let F_T and U_T be the distribution and tail quantile function of this truncated distribution. In practice one does not always know if the data $X_1, \dots, X_n$ come from a truncated or non-truncated distribution, and hence the behavior of estimators should be evaluated under both cases, and a statistical test for upper-truncation is useful. This section is taken from Beirlant et al. [95].

Upper-truncation of the distribution of W at some truncation point T yields

$$\overline{F}_T(x) = \frac{P(W > x) - P(W > T)}{1 - P(W > T)} = \frac{\overline{F}_W(x) - \overline{F}_W(T)}{F_W(T)}.$$

The corresponding quantile function Q_T is then given by

$$Q_T(1-p) = Q_W(1 - [\overline{F}_W(T) + p(1 - \overline{F}_W(T))]) = Q_W(\{1 - \overline{F}_W(T)\}(1-p))$$

while the tail function U_T satisfies

$$U_T(y) = U_W\left(1/\left\{\overline{F}_W(T) + \frac{F_W(T)}{y}\right\}\right) = U_W\left(\frac{1}{\overline{F}_W(T)}\left[1 + \frac{F_W(T)}{y\overline{F}_W(T)}\right]^{-1}\right), \tag{4.2.26}$$

or

$$U_T(y) = U_W\left(\frac{1}{\overline{F}_W(T)}\left[1 + \frac{1}{yD_T}\right]^{-1}\right), \tag{4.2.27}$$

with $D_T := \overline{F}_W(T)/F_W(T)$ the odds of the truncated probability mass under the untruncated distribution W. Note also that for a fixed T, upper-truncation models are known to exhibit an EVI $\gamma = -1$. This follows from verifying (3.2.5) for U_T as given in (4.2.27). For instance when $U_W(x) = x^\gamma$, we find $(\overline{F}_W(T))^\gamma U_T(x) = \left(1 + \frac{1}{xD_T}\right)^{-\gamma} = \left(1 - \frac{\gamma}{xD_T}(1 + o(1))\right)$ as $x \to \infty$. This final expression satisfies (3.2.5) with $\gamma = -1$.

4.2.3.1 EVA for Upper-truncated Pareto-type Distributions

We restrict attention to tail estimation for upper-truncated Pareto-type distributions:

$$1 - F_W(w) = w^{-1/\gamma}\ell_F(w), \ \gamma > 0,$$

where ℓ_F is a slowly varying function at infinity or, with W/t denoting the peaks over a threshold t when $W > t$,

$$\mathbb{P}(W/t > y|W > t) \to y^{-1/\gamma} \text{ as } t \to \infty, \text{ for every } y > 1.$$

Upper-truncation of a Pareto-type distribution at a high value T then necessarily requires $t < T \to \infty$ and

$$\begin{aligned}\mathbb{P}(X/t > y|X > t) &= \mathbb{P}(W/t > y|t < W < T)\\ &= \frac{(yt)^{-1/\gamma}\ell_F(yt) - T^{-1/\gamma}\ell_F(T)}{t^{-1/\gamma}\ell_F(t) - T^{-1/\gamma}\ell_F(T)}\\ &= \frac{y^{-1/\gamma}\frac{\ell_F(yt)}{\ell_F(t)} - \left(\frac{T}{t}\right)^{-1/\gamma}\frac{\ell_F(T)}{\ell_F(t)}}{1 - \left(\frac{T}{t}\right)^{-1/\gamma}\frac{\ell_F(T)}{\ell_F(t)}}.\end{aligned}$$

One can now consider two cases as $t, T \to \infty$:

- *Rough upper-truncation with the threshold t* when $T/t \to \beta > 1$ and

$$\mathbb{P}(X/t > y|X > t) \to \frac{y^{-1/\gamma} - \beta^{-1/\gamma}}{1 - \beta^{-1/\gamma}}, \quad 1 < y < \beta. \tag{4.2.28}$$

  This corresponds to situations where the deviation from the Pareto behavior due to upper-truncation at a high value will be visible in the data from the threshold t onwards, and an adaptation of the above Pareto tail extrapolation methods appears appropriate.
- *Light (or no) upper-truncation with the threshold t*: when $T/t \to \infty$

$$\mathbb{P}(X/t > y|X > t) \to y^{-1/\gamma}, \quad y > 1, \tag{4.2.29}$$

  and hardly any truncation is visible in the data from the threshold t onwards, and the Pareto-type model without truncation and the corresponding extreme value methods for Pareto-type tails appear appropriate when restricted to excesses over t.

Under rough upper-truncation we have

$$\mathbb{P}(X > x|X > t) \approx \frac{(x/t)^{-1/\gamma} - \beta^{-1/\gamma}}{1 - \beta^{-1/\gamma}}, \quad t < x < t\beta,$$

with density

$$f_X(x) = \frac{1}{\gamma}\left(\frac{x}{t}\right)^{-1-1/\gamma}\left(1 - \beta^{-1/\gamma}\right)^{-1}.$$

Estimating T by $X_{n,n}$ and taking $t = X_{n-k,n}$ so that $\widehat{\beta}^{-1} = R_{k,n} = X_{n-k,n}/X_{n,n}$, we obtain the following log-likelihood:

$$\log L_{k,n}(\gamma) = \log \Pi_{j=1}^{k} \frac{\left(X_{n-j+1,n}/X_{n-k,n}\right)^{-1-1/\gamma}}{\gamma\left[1-\left(X_{n,n}/X_{n-k,n}\right)^{-1/\gamma}\right]}$$
$$= -k\log\gamma - \left(1+\frac{1}{\gamma}\right)\sum_{j=1}^{k}\log\frac{X_{n-j+1,n}}{X_{n-k,n}}$$
$$- k\log\left(1-\left(\frac{X_{n,n}}{X_{n-k,n}}\right)^{-1/\gamma}\right).$$

Now

$$\frac{\partial \log L_{k,n}(\gamma)}{\partial\gamma} = -\frac{k}{\gamma} - \sum_{j=1}^{k}\log\frac{X_{n-j+1,n}}{X_{n-k,n}} - k\frac{\left(\frac{X_{n,n}}{X_{n-k,n}}\right)^{-1/\gamma}\log\frac{X_{n,n}}{X_{n-k,n}}}{1-\left(\frac{X_{n,n}}{X_{n-k,n}}\right)^{-1/\gamma}},$$

which leads to the defining equation for the likelihood estimator $\hat{\gamma}_{k,n}^{T}$:

$$H_{k,n} = \hat{\gamma}_{k,n}^{T} + \frac{R_{k,n}^{1/\hat{\gamma}_{k,n}^{T}}\log R_{k,n}}{1-R_{k,n}^{1/\hat{\gamma}_{k,n}^{T}}}.$$

This estimator was first proposed in Aban et al. [6]. Beirlant et al. [95] showed that with $\kappa = \beta^{1/\gamma} - 1$

$$\text{AVar}(\hat{\gamma}_{k,n}^{T}) = \begin{cases} \frac{\gamma^2}{k}\left(1-\frac{1+\kappa}{\kappa^2}\log^2(1+\kappa)\right)^{-1}, & \text{as } T/t \to \beta, \\ \frac{\gamma^2}{k}, & \text{as } T/t \to \infty. \end{cases}$$

From (4.2.27) it is clear that the estimation of D_T is an intermediate step in important estimation problems following the estimation of γ, namely of extreme quantiles and of the endpoint T. When U_W satisfies (4.2.3) it follows from (4.2.27) that as $t, T \to \infty$ and $T/t \to \beta$

$$Q_T(1-p) = U_T(1/p) = T\left(1+\frac{p}{D_T}\right)^{-\gamma}(1+o_{t,T}(1)), \tag{4.2.30}$$

so that

$$\left(\frac{Q_T\left(1-\frac{k+1}{n+1}\right)}{Q_T\left(1-\frac{1}{n+1}\right)}\right)^{1/\gamma} \approx \frac{1+\frac{1}{(n+1)D_T}}{1+\frac{k+1}{(n+1)D_T}} = \frac{1}{k+1}\left(\frac{1+(n+1)D_T}{1+\frac{(n+1)D_T}{k+1}}\right). \tag{4.2.31}$$

Motivated by (4.2.31) and estimating $Q_T(1-(k+1)/(n+1))/Q_T(1-1/(n+1))$ by $R_{k,n}$, one arrives at

$$\hat{D}_T := \hat{D}_{T,k,n} = \frac{k+1}{n+1}\frac{R_{k,n}^{1/\hat{\gamma}_{k,n}^T} - \frac{1}{k+1}}{1-R_{k,n}^{1/\hat{\gamma}_{k,n}^T}} \tag{4.2.32}$$

as an estimator for D_T. In practice one makes use of the admissible estimator

$$\hat{D}_T^{(0)} := \max\left\{\hat{D}_T, 0\right\}$$

to make it useful for truncated and non-truncated Pareto-type distributions.

For $D_T > 0$, in order to construct estimators of T and extreme quantiles $q_p = Q_T(1-p)$, as in (4.2.31) we find that

$$\left(\frac{Q_T(1-p)}{Q_T\left(1-\frac{k+1}{n+1}\right)}\right)^{1/\gamma} = \frac{1+\frac{k+1}{(n+1)D_T}}{1+\frac{p}{D_T}} = \frac{D_T+\frac{k+1}{n+1}}{D_T+p}. \tag{4.2.33}$$

Then taking logarithms on both sides of (4.2.33) and estimating $Q_T(1-(k+1)/(n+1))$ by $X_{n-k,n}$ we find an estimator $\hat{q}_p^T := \hat{q}_{k,p}^T$ of q_p:

$$\log \hat{q}_{k,p}^T = \log X_{n-k,n} + \hat{\gamma}_{k,n}^T \log\left(\frac{\hat{D}_T+\frac{k+1}{n+1}}{\hat{D}_T+p}\right), \tag{4.2.34}$$

which equals the Weissman estimator $\hat{q}_{k,p}^W$ when $\hat{D}_T = 0$. An estimator $\hat{T}_{k,n}$ of T follows from letting $p \to 0$ in the above expressions for $\hat{q}_{p,k,n}^T$:

$$\log \hat{T}_{k,n} = \max\left\{\log X_{n-k,n} + \hat{\gamma}_{k,n}^T \log\left(1+\frac{k+1}{(n+1)\hat{D}_T}\right), \log X_{n,n}\right\}. \tag{4.2.35}$$

Here we take the maximum of $\log X_{n,n}$ and the value following from (4.2.34) with $p \to 0$ in order for this endpoint estimator to be admissible. It has been shown that $\hat{q}_{k,p}^T$ is superefficient under rough upper-truncation, which means that the asymptotic variance is $o(1/k)$ and the asymptotic bias is also smaller than, for instance, that of the moment quantile estimator $\hat{q}_{p,k}^M$.

However, $\hat{q}_{k,p}^T$ is not a consistent estimator for q_p under light upper-truncation and when $np_n \to 0$. In that case one should use

$$\log \hat{q}_{p,k,n}^{\infty} = \log X_{n-k,n} + \hat{\gamma}_{k,n}^T \log\left(\frac{k+1}{p(n+1)}\right). \tag{4.2.36}$$

The estimation of tail probabilities $p_x = P(X > x)$ can be based directly on (4.2.28) using $R_{k,n}$ as an estimator for $1/\beta$:

$$\hat{p}^T_{x,k} = \frac{k+1}{n+1} \frac{\left(\frac{x}{X_{n-k,n}}\right)^{-1/\hat{\gamma}^T_{k,n}} - R_{k,n}^{1/\hat{\gamma}^T_{k,n}}}{1 - R_{k,n}^{1/\hat{\gamma}^T_{k,n}}}. \tag{4.2.37}$$

Of course, in order to decide between (4.2.34) and (4.2.36) one should use a statistical test for deciding between rough and light upper-truncation.

4.2.3.2 Testing for Upper-truncated Pareto-type Tails

Aban et al. [6] proposed a test for $H_0^{(1)} : T = \infty$ versus $H_1^{(1)} : T < \infty$ under the strict Pareto model, rejecting H_0 at level $q \in (0,1)$ when

$$X_{n,n} < \left(\frac{nA}{-\log q}\right)^{\gamma} \tag{4.2.38}$$

for some $1 < k < n$ with A the scale parameter in the Pareto model. In (4.2.38), γ is estimated by $H_{k,n}$, the maximum likelihood estimator under H_0, while A is estimated using $\hat{A}_{k,n}$ from (4.2.19). Note that the rejection rule (4.2.38) can be rewritten as

$$T_{A,k,n} := kR_{k,n}^{1/H_{k,n}} > \log \frac{1}{q}, \tag{4.2.39}$$

and the P-value is given by $\exp\left(-kR_{k,n}^{1/H_{k,n}}\right)$.

Considering the testing problem

$$H_0^{(2)} : X \text{ satisfies (4.2.29) } \textit{versus } H_1^{(2)} : X \text{ satisfies (4.2.28)}$$

under the upper-truncated Pareto-type model, Beirlant et al. [95] propose to reject $H_0^{(2)}$ when an appropriate estimator of $(n+1)D_T/(k+1)$ is significantly different from 0. Here we construct such an estimator generalizing $R_{k,n}^{1/\gamma}$ with an average of ratios $(X_{n-k,n}/X_{n-j+1,n})^{1/\gamma}$, $j = 1, \dots, k$, which then possesses an asymptotic normal distribution under the null hypothesis. Observe that with (4.2.30) under $H_0^{(2)}$ as $k \to \infty$

$$\begin{aligned} \overline{E}_{k,n} = \frac{1}{k}\sum_{j=1}^{k} \left(\frac{Q_T\left(1-\frac{k+1}{n+1}\right)}{Q_T\left(1-\frac{j}{n+1}\right)}\right)^{1/\gamma} &\approx \frac{1}{k}\sum_{j=1}^{k} \frac{1+\frac{j}{k+1}\frac{k+1}{(n+1)D_T}}{1+\frac{k+1}{(n+1)D_T}} \\ &\approx \frac{1+\frac{1}{2}\frac{k}{nD_T}}{1+\frac{k}{nD_T}}. \end{aligned}$$

Estimating $\overline{E}_{k,n}$ by

$$\overline{R}_{k,n}(\gamma) = \frac{1}{k}\sum_{j=1}^{k}\left(\frac{X_{n-k,n}}{X_{n-j+1,n}}\right)^{1/\gamma}$$

leads now to

$$L_{k,n}(\hat{\gamma}) := \frac{\overline{R}_{k,n}(\hat{\gamma}) - \frac{1}{2}}{1 - \overline{R}_{k,n}(\hat{\gamma})} \tag{4.2.40}$$

as an estimator of $(n+1)D_T/(k+1)$, with $\hat{\gamma}$ an appropriate estimator of γ. Under $H_0^{(2)}$, the Hill estimator $H_{k,n}$ is an appropriate estimator of γ. Moreover, it can be shown that under some regularity assumptions on the underlying Pareto-type distribution, we have under $H_0^{(2)}$ for $k, n \to \infty$ and $k/n \to 0$, that $\sqrt{k}L_{k,n}(H_{k,n})$ is asymptotically normal with mean 0 and variance 1/12. It is then also shown under rough upper-truncation as $k, n, T \to \infty$, $k/n \to 0$ that $L_{k,n}(H_{k,n})$ tends to a negative constant so that an asymptotic test based on $L_{k,n}(H_{k,n})$ rejects $H_0^{(2)}$ on level q when

$$T_{B,k,n} := \sqrt{12\,k}L_{k,n}(H_{k,n}) < -z_q, \tag{4.2.41}$$

with $P(\mathcal{N}(0,1) > z_q) = q$. The P-value is then given by $\Phi(\sqrt{12\,k}L_{k,n}(H_{k,n}))$.

Case study. Given the fact that the mean excess function of the ultimate values from Company A are ultimately decreasing at the largest values, an upper-truncated Pareto model is a possible tail model. This is also clear from the Hill plot in Figure 4.10, which systematically decreases to 0 as $k \downarrow 1$, and from the plots of the generalized Hill, moment and POT estimators of γ, which decrease to -1 as $k \downarrow 1$. The P-values of the T_B test for upper-truncation are lower than 0.05 in areas around $k = 100$ and $k = 250$. This means that, at the corresponding thresholds $t = Q_T(1 - k/n)$, the upper-truncated Pareto model in (4.2.28) yields a more appropriate model than the strict Pareto model to fit the distribution of the exceedances X/t. This is illustrated on the Pareto QQ-plot in Figure 4.10 overlaying this upper-truncated Pareto fit over the top 250 points. While the strict Pareto fit corresponds to a regression line on these points, the concave curve provided by modelling a truncation effect appears to provide a better fit.

The estimates $\hat{\gamma}_k^T$, $-\log \hat{p}^T_{k,4\text{million}}$ and even more $\hat{q}^T_{k,0.001}$ are quite stable as a function of k leading to the approximate values, respectively $\hat{\gamma}^T \approx 0.65$, $-\log \hat{p}^T_{4\text{million}}$ just above 6, and $\hat{q}^+_{k,0.001}$ just below 5 million. This is a bit lower than the earlier estimates of the 0.999 quantile, and leads to a return period corresponding with 4 million, which is close to the values returned by the GPD-ML method. Note that the endpoint T is estimated here at around 5 million at $k = 100$ and $k = 250$. □

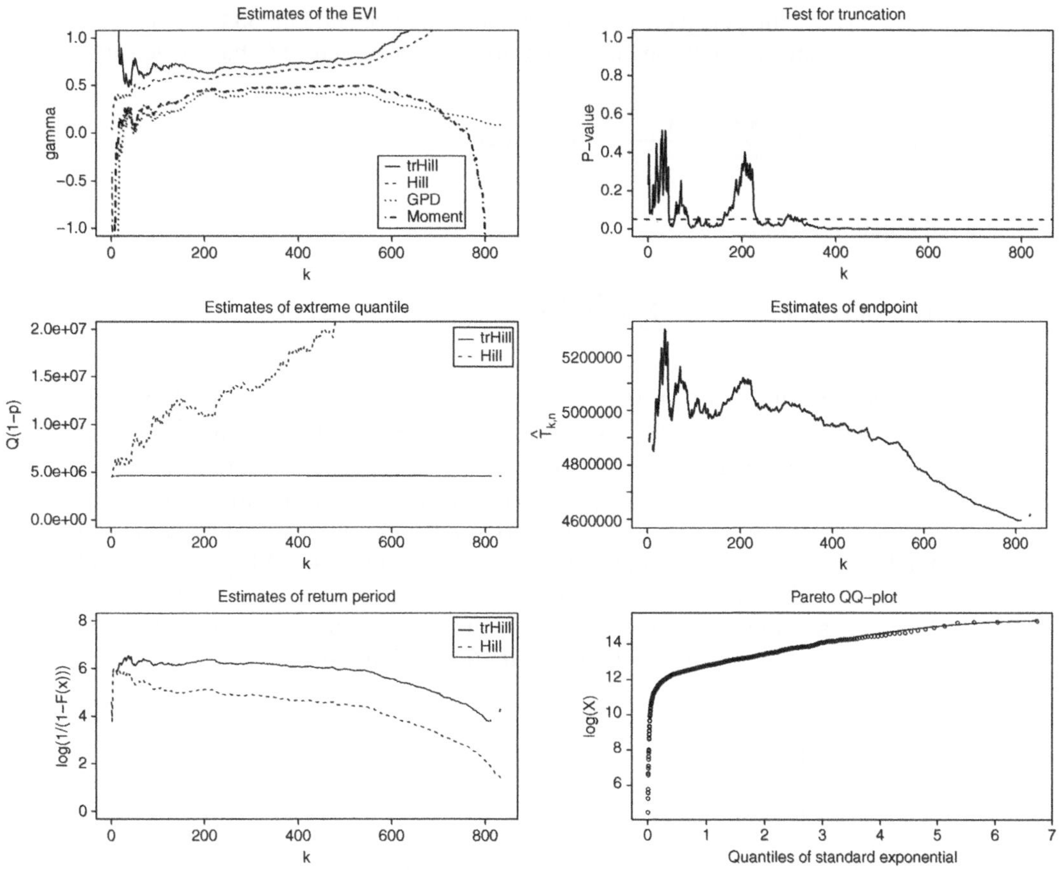

Figure 4.10 MTPL data for Company A, ultimate estimates: $\hat{\gamma}_k^T$ and $H_{k,n}$ as a function of k (top left), P-value of $T_{B,k}$ as a function of k (top right), $\hat{q}^T_{k,0.001}$ and $\hat{q}^+_{k,0.001}$ as a function of k (middle left), estimates of endpoint $\hat{T}_k$ (middle right), $-\log \hat{p}^T_{4\,\text{million},k}$ as a function of k (bottom left), truncated Pareto fit to Pareto QQ-plot based on top $k = 250$ observations (bottom right).

4.3 Global Fits: Splicing, Upper-truncation and Interval Censoring

Given an appropriate tail fit, the ultimate goal consists of fitting a distribution with a global satisfactory fit. Rather than trying to splice specific parametric models such as log-normal or Weibull models for the modal part of the distribution, one can rely on fitting a mixed Erlang (ME) distribution, as discussed in Verbelen et al. [758]. We also consider this set-up in the presence of truncation and censoring.

4.3.1 Tail-mixed Erlang Splicing

The Erlang distribution has a gamma density

$$f_E(x; r, \lambda) = \frac{\lambda^r}{(r-1)!} x^{r-1} e^{-\lambda x}, \; x > 0,$$

where r is a positive integer shape parameter. Following Lee and Lin [534] we consider mixtures of M Erlang distributions with common scale parameter $1/\lambda$ having density

$$f_{\mathrm{ME}}(x;\mathbf{r},\alpha,\lambda)=\sum_{j=1}^{M}\alpha_j\frac{\lambda^{r_j}}{(r_j-1)!}x^{r_j-1}e^{-\lambda x},\quad x>0,$$

and tail function

$$1-F_{\mathrm{ME}}(x;\mathbf{r},\alpha,\lambda)=e^{-\lambda x}\sum_{j=1}^{M}\alpha_j\sum_{n=0}^{r_j-1}\frac{(\lambda x)^n}{n!},$$

where the positive integers $\mathbf{r}=(r_1,\dots,r_M)$ with $r_1<r_2<\dots<r_M$ are the shape parameters of the Erlang distributions, and $\alpha=(\alpha_1,\dots,\alpha_M)$ with $\alpha_j>0$ and $\sum_{j=1}^{M}\alpha_j=1$ are the weights in the mixture. Tijms [745] showed that the class of mixtures of Erlang distributions with a common scale $1/\lambda$ is dense in the space of positive continuous distributions on $\mathbb{R}^+$. Moreover this class is also closed under mixtures, convolution and compounding. Hence aggregate risks calculations are simple, and XL premiums and risk measures based on quantiles can also be evaluated in a rather straightforward way.

For instance, a composite ME generalized Pareto distribution can be built using (3.5.14), that is, a *two-component spliced distribution* with density

$$f_{\mathrm{ME,GP}}(x)=\begin{cases}\pi\frac{f_{\mathrm{ME}}(x;\mathbf{r},\alpha,\lambda)}{F_{\mathrm{ME}}(t;\mathbf{r},\alpha,\lambda)}, & 0<x\le t,\\ (1-\pi)\frac{1}{\sigma}\left(1+\frac{\gamma}{\sigma}(x-t)\right)^{-1-1/\gamma}, & x>t.\end{cases}$$

If a continuity requirement at t were imposed, this would lead to

$$\pi=\frac{F_{\mathrm{ME}}(t;\mathbf{r},\alpha,\lambda)}{F_{\mathrm{ME}}(t;\mathbf{r},\alpha,\lambda)+\sigma f_{\mathrm{ME}}(t;\mathbf{r},\alpha,\lambda)}.$$

The survival function of this spliced distribution is given by

$$1-F_{\mathrm{ME,GP}}(x)=\begin{cases}1-\pi\frac{F_{\mathrm{ME}}(x;\mathbf{r},\alpha,\lambda)}{F_{\mathrm{ME}}(t;\mathbf{r},\alpha,\lambda)}, & 0<x\le t,\\ (1-\pi)\left(1+\frac{\gamma}{\sigma}(x-t)\right)^{-1/\gamma}, & x>t.\end{cases}$$

Alternatively, one can take $\hat{\pi}=1-k_*/n$, where k_* is an appropriate number of top order statistics corresponding to an extreme value threshold $t=x_{n-k_*,n}$.

Fitting ME distributions through direct likelihood maximization is difficult. A first algorithm was proposed by Tijms [745], but it turns out to be slow and can lead to overfitting. Lee and Lin [534] use the *expectation-maximization* (EM) algorithm proposed by Dempster et al. [271] to fit the ME distribution. Model selection criteria, such as the Akaike information criterion (AIC) and Bayesian information criterion (BIC) information criteria, are then used to avoid overfitting. Verbelen et al. [758] extend this approach to censored and/or truncated data. The need for the EM algorithm follows from the data incompleteness due to mixing and censoring.

The EM algorithm is used to compute the maximum likelihood estimator (MLE) for incomplete data where direct maximization is impossible. It consists of two steps that are put in an iteration until convergence:

- **E-step**: Compute the conditional expectation of the log-likelihood given the observed data and previous parameter estimates.
- **M-step**: Determine a subsequent set of parameter estimates in the parameter range through maximization of the conditional expectation computed in the E-step.

Rather than proposing a data-driven estimator of the splicing point t, we use an expert opinion on the splicing point t based on EVA as outlined above. Then, π can be estimated by the fraction of the data not larger than t. Similarly, T is deduced from the EVA. The extreme value index γ is estimated in the algorithm, starting from the value obtained from the EVA at the threshold t. The final estimates for γ always turned out to be close to the EVA estimates. Next, the ME parameters (α, λ) are estimated using the EM algorithm as developed in Verbelen et al. [758]. The number of ME components M is estimated using a backward stepwise search, starting from a certain upper value, whereby the smallest shape is deleted if this decreases an information criterion such as AIC or BIC. Moreover, for each value of M, the shapes $\mathbf{r}$ are adjusted based on maximizing the likelihood starting from $\mathbf{r} = (s, 2s, \dots, ..., M \times s)$, where s is a chosen spread factor.

Of course tail splicing of an ME can also be performed using a simple Pareto fit, or an EPD fit, whether or not adapted for truncation. For instance, splicing an ME with an upper-truncated Pareto approximation leads to

$$f_{\mathrm{ME,TPa}}(x) = \begin{cases} \pi \frac{f_{\mathrm{ME}}(x;\mathbf{r},\alpha,\lambda)}{F_{\mathrm{ME}}(t;\mathbf{r},\alpha,\lambda)}, & 0 < x \le t, \\ (1-\pi)\frac{\frac{1}{t\gamma}\left(\frac{x}{t}\right)^{-1-1/\gamma}}{1-\left(\frac{T}{t}\right)^{-1/\gamma}}, & t < x < T. \end{cases} \tag{4.3.42}$$

Case study: MTPL data for Company A. When applying the splicing technique to the ultimate data from Company A, we noted in the discussion of Figure 4.5 (middle) that also in this case two Pareto tail pieces appear. When trying to splice one tail piece with an ME, the algorithm here also leads to a three-component ME fit coupled with a Pareto fit, with parameters $t^l = 0$, $\pi = 0.966$, $k = 20$, and $\gamma = 0.327$, while $M = 3$, $\alpha = (0.201, 0.714, 0.085)$, $\mathbf{r} = (1, 3, 10)$, and $\lambda^{-1} = 103\,917$, see Figure 4.11. However here the tail fit is satisfactory. Of course, splicing with two Pareto components and an ME is another option. □

4.3.2 Tail-mixed Erlang Splicing under Censoring and Upper-truncation

In reinsurance data left truncation appears at some point, denoted here by t^l, which can be a deductible or a percentage of the retention u from an XL contract. Claims leading to a cumulative payment below t^l at a given stage during development are then left truncated. Such a claim constitutes an IBNR claim. As discussed above, an upper-truncation mechanism at some point T can appear.

We denote the ME density and distribution function by f_{ME} and F_{ME}, and similarly f_{EV} and F_{EV} for the EVA distribution. We then define, omitting the model parameters from the notation for the moment,

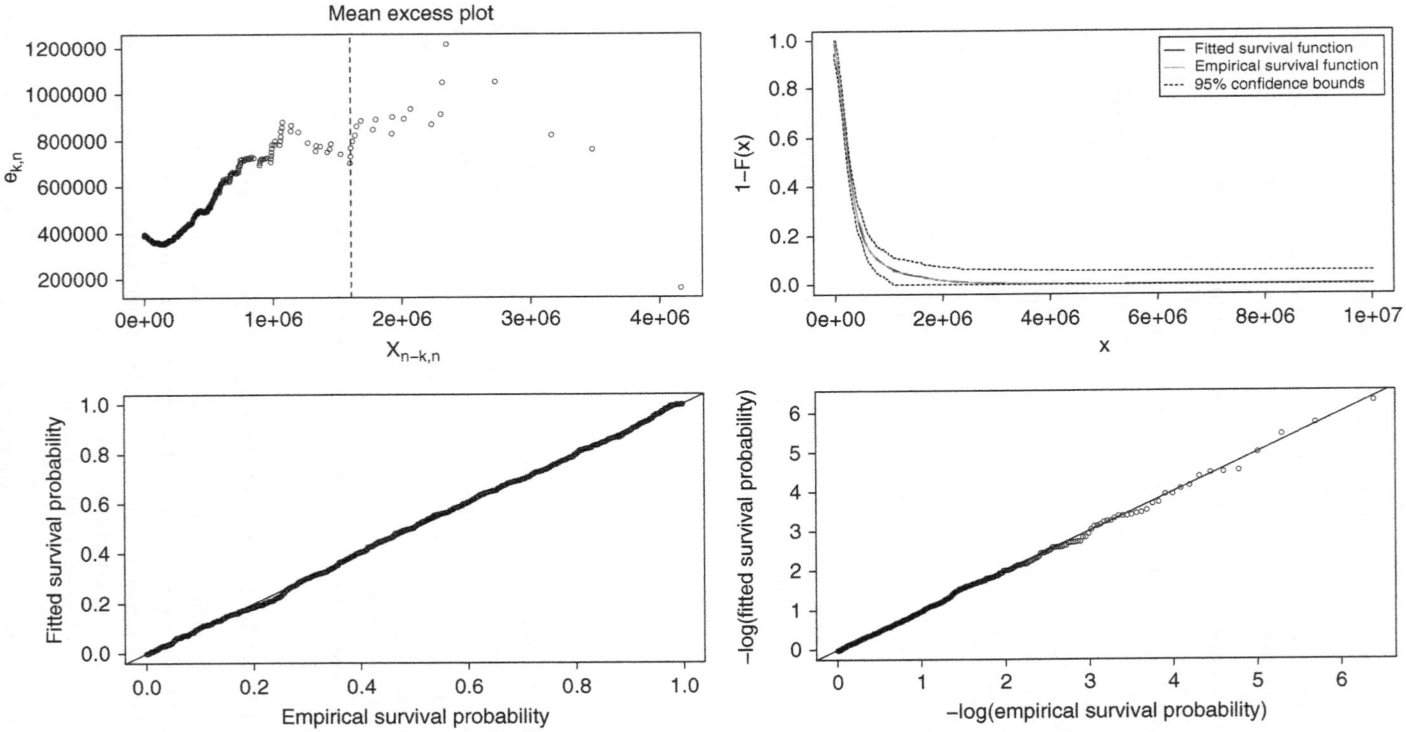

Figure 4.11 MTPL data for Company A ultimates: fit of spliced model with a mixed Erlang and a Pareto component with thresholds t indicated on mean excess plot (top left); empirical and model survival function (top right); PP plot of empirical survival function against splicing model RTF (bottom left); idem with $-\log$ transformation (bottom right).

$$f_1(x) = \begin{cases} \frac{f_{ME}(x)}{F_{ME}(t)-F_{ME}(t^l)} & \text{if } t^l \leq x \leq t, \\ 0 & \text{otherwise,} \end{cases}$$

$$f_2(x) = \begin{cases} \frac{f_{EV}(x)}{F_{EV}(T)-F_{EV}(t)} & \text{if } t \leq x \leq T, \\ 0 & \text{otherwise,} \end{cases}$$

with $0 \leq t^l < t < T$ where T can be equal to ∞. The densities f_1 and f_2 are then valid densities on the intervals $[t^l, t]$ and $[t, T]$, respectively. For the first density, this means that it is lower truncated at t^l and upper truncated at t, and the second density is lower truncated at t and upper truncated at T. The corresponding distribution functions are

$$F_1(x) = \begin{cases} 0 & \text{if } x \leq t^l, \\ \frac{F_{ME}(x)-F_{ME}(t^l)}{F_{ME}(t)-F_{ME}(t^l)} & \text{if } t^l < x < t, \\ 1 & \text{if } x \geq t, \end{cases}$$

$$F_2(x) = \begin{cases} 0 & \text{if } x \leq t, \\ \frac{F_{EV}(x)-F_{EV}(t)}{F_{EV}(T)-F_{EV}(t)} & \text{if } t < x < T, \\ 1 & \text{if } x \geq T. \end{cases}$$

We consider the splicing density and distribution function

$$f(x) = \begin{cases} 0 & \text{if } x \leq t^l, \\ \pi f_1(x) & \text{if } t^l < x \leq t, \\ (1-\pi)f_2(x) & \text{if } t < x < T, \\ 0 & \text{if } x \geq T, \end{cases}$$

and

$$F(x) = \begin{cases} 0 & \text{if } x \leq t^l, \\ \pi F_1(x) & \text{if } t^l < x \leq t, \\ \pi + (1-\pi)F_2(x) & \text{if } t < x < T, \\ 1 & \text{if } x \geq T. \end{cases} \tag{4.3.43}$$

Next to truncation, censoring mechanisms occur in reinsurance[1]:

- *right censoring* occurs for instance when a claim has not been settled at the evaluation date (RBNS claims). See Chapter 1 for the case of motor liability data. The final claim amount x_i will be larger than the lower censoring value l_i
- *left censoring* occurs when only an upper bound u_i to the claim x_i is given

1 Note that the censoring definitions used here are different from others in the actuarial literature (e.g., see Klugman et al. [491]).

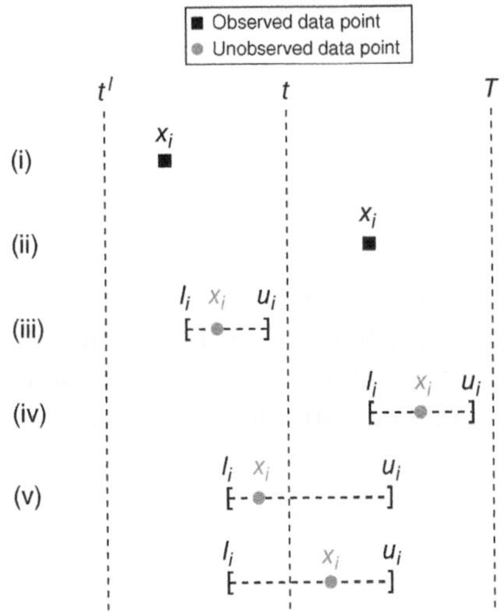

Figure 4.12 The different classes of censored observations.

- *interval censoring* means that the final claim value x_i is only known to be inside an interval $[l_i, u_i] \subset [t^l, T]$.

In the splicing context with an EVA component from a threshold t on, we have the following five classes of observations:

(i) Uncensored observations with $t^l \leq l_i = u_i = x_i \leq t < T$.
(ii) Uncensored observations with $t^l < t < l_i = u_i = x_i \leq T$.
(iii) Interval censored observations with $t^l \leq l_i < u_i \leq t < T$.
(iv) Interval censored observations with $t^l < t \leq l_i < u_i \leq T$.
(v) Interval censored observations with $t^l \leq l_i < t < u_i \leq T$.

These classes are shown in Figure 4.12. In the conditioning argument in the E-step of the algorithm, the fifth case is split into $x_i \leq t$ and $x_i > t$, as indicated in Figure 4.12.

For the Erlang mixture, the number M and the integer shapes $\mathbf{r}$ are fixed when estimating $\boldsymbol{\Theta}_1 = (\alpha, \lambda)$. Also, $\boldsymbol{\Theta}_2$ denotes the extreme value parameter γ (together with σ when using the GPD tail fit). The idea behind the EM algorithm in this context is to consider the censored sample $\mathcal{X}$ in contrast to the complete data $\mathcal{Y}$ which is not fully observed. Given a complete version of the data, we can construct a complete likelihood function as

$$\mathcal{L}_{\text{complete}}(\boldsymbol{\Theta}; \mathcal{Y}) = \prod_{i=1}^{n} \left(\pi f_1(X_i; t^l, t, \boldsymbol{\Theta}_1)\right)^{1_{\{X_i \leq t\}}} \times \prod_{i=1}^{n} \left((1-\pi) f_2(X_i; t, T, \boldsymbol{\Theta}_2)\right)^{1_{\{X_i > t\}}},$$

where $1_{\{X_i \leq t\}}$ is the indicator function for the event $\{X_i \leq t\}$. The corresponding complete data log-likelihood function is

$$\ell_{\text{complete}}(\boldsymbol{\Theta}; \mathcal{Y}) = \sum_{i=1}^{n} 1_{\{X_i \leq t\}} \left(\log \pi + \log f_1(X_i; t^l, t, \boldsymbol{\Theta}_1)\right) + \sum_{i=1}^{n} 1_{\{X_i > t\}} \Big(\log(1-\pi) + \log f_2(X_i; t, T, \boldsymbol{\Theta}_2)\Big).$$

As we do not fully observe the complete version $\mathcal{Y}$ of the data sample, it is not possible to optimize the complete data log-likelihood directly. The intuitive idea for obtaining parameter estimates in the case of incomplete data is to compute the expectation of the complete data log-likelihood and then use this expected log-likelihood function to estimate the parameters. However, taking the expectation of the complete data log-likelihood requires the knowledge of the parameter vector, and so the algorithm has to run iteratively. Starting from an initial guess for the parameter vector, the EM algorithm iterates between two steps. In the hth iteration of the E-step the expected value of the complete data log-likelihood is computed with respect to the unknown data $\mathcal{Y}$ given the observed data $\mathcal{X}$ and using the current estimate of the parameter vector $\Theta^{(h-1)}$ as true values,

$$\mathbb{E}\left(\ell_{\text{complete}}(\boldsymbol{\Theta}; \mathcal{Y}) | \mathcal{X}; \boldsymbol{\Theta}^{(h-1)}\right).$$

In the M-step, one maximizes the expected value of the complete data log-likelihood obtained in the E-step with respect to the parameter vector:

$$\boldsymbol{\Theta}^{(h)} = \arg\max_{\boldsymbol{\Theta}} \mathbb{E}\left(\ell_{\text{complete}}(\boldsymbol{\Theta}; \mathcal{Y}) | \mathcal{X}; \boldsymbol{\Theta}^{(h-1)}\right).$$

Both steps are iterated until convergence.

In the E-step we distinguish the five cases of data points again to determine the contribution of a data point to this expectation:

(i) $\log \pi + \mathbb{E}\left(\log f_1(X_i; t^l, t, \boldsymbol{\Theta}_1) | \, t^l \leq l_i = u_i \leq t < T; \boldsymbol{\Theta}_1^{(h-1)}\right)$

(ii) $\log(1-\pi) + \mathbb{E}\left(\log f_2(X_i; t, T, \boldsymbol{\Theta}_2) | t^l < t < l_i = u_i \leq T; \boldsymbol{\Theta}_2^{(h-1)}\right)$

(iii) $\log \pi + \mathbb{E}\left(\log f_1(X_i; t^l, t, \boldsymbol{\Theta}_1) | t^l \leq l_i < u_i \leq t < T; \boldsymbol{\Theta}_1^{(h-1)}\right)$

(iv) $\log(1-\pi) + \mathbb{E}\left(\log f_2(X_i; t, T, \boldsymbol{\Theta}_2) | t^l < t \leq l_i < u_i \leq T; \boldsymbol{\Theta}_2^{(h-1)}\right)$

(v) $\mathbb{E}\Big(\left[\log \pi + \log f_1(X_i; t^l, t, \boldsymbol{\Theta}_1)\right] 1_{\{X_i \leq t\}} + \left[\log(1-\pi) + \log f_2(X_i; t, T, \boldsymbol{\Theta}_2)\right] 1_{\{X_i > t\}} | t^l \leq l_i < t < u_i \leq T; \boldsymbol{\Theta}^{(h-1)}\Big)$

Note that the event $\{t^l \leq l_i = u_i \leq t < T\}$ indicates that we know t^l, $l_i = u_i$, t and T, and that the ordering $t^l \leq l_i = u_i \leq t < T$ holds. Similar reasonings hold for the other conditional arguments in the expectations. Then, using the law of total probability, the final case can be rewritten as

$$\mathbb{E}\left(\log \pi + \log f_1(X_i; t^l, t, \mathbf{\Theta}_1) | t^l \leq l_i < X_i < u_i \leq T; \mathbf{\Theta}_1^{(h-1)}\right)$$
$$\times P\left(X_i \leq t | t^l \leq l_i < t < u_i \leq T; \mathbf{\Theta}^{(h-1)}\right)$$
$$+ \mathbb{E}\left(\log(1-\pi) + \log f_2(X_i; t, T, \mathbf{\Theta}_2) | t^l \leq l_i \leq t < X_i < u_i \leq T; \mathbf{\Theta}_2^{(h-1)}\right)$$
$$\times P\left(X_i > t | t^l \leq l_i < t < u_i \leq T; \mathbf{\Theta}^{(h-1)}\right),$$

where $\{t^l \leq l_i < X_i \leq t < u_i \leq T\}$ denotes that t^l, l_i, t, u_i and T are known, that the ordering $t^l \leq l_i < t < u_i \leq T$ holds, and that $\{X_i \leq t\}$. Using (4.3.43) we find that the probability in the first term is then given by

$$\mathbb{P}\left(X_i \leq t | t^l \leq l_i < t < u_i \leq T; \mathbf{\Theta}^{(h-1)}\right)$$
$$= \frac{F\left(t; t^l, t, T, \mathbf{\Theta}^{(h-1)}\right) - F\left(l_i; t^l, t, T, \mathbf{\Theta}^{(h-1)}\right)}{F\left(u_i; t^l, t, T, \mathbf{\Theta}^{(h-1)}\right) - F\left(l_i; t^l, t, T, \mathbf{\Theta}^{(h-1)}\right)}$$
$$= \frac{\pi^{(h-1)} - \pi^{(h-1)} F_1\left(l_i; t^l, t, \mathbf{\Theta}_1^{(h-1)}\right)}{\pi^{(h-1)} + (1-\pi^{(h-1)}) F_2\left(u_i; t, T, \mathbf{\Theta}_2^{(h-1)}\right) - \pi^{(h-1)} F_1\left(l_i; t^l, t, \mathbf{\Theta}_1^{(h-1)}\right)},$$

and similarly for the second term. The M-step with maximization with respect to π, $\mathbf{\Theta}_1$ and $\mathbf{\Theta}_2$, and the choice of the initial values, is discussed in detail in Reynkens et al. [647].

EVA is not available in the literature for interval censored data. The role of the empirical survival and quantile functions in the construction of a tail analysis for complete data (i.e., setting $X_{n-j+1,n}$ as an estimator of $Q(1 - j/(n+1))$, $j = 1, \dots, n$) is taken over by the Turnbull [747] estimator $1 - \hat{F}_n^{TB}$ as an estimator of $1 - F$ and the corresponding quantile function $\hat{Q}_n^{TB}$. The Turnbull estimator is an extension to interval censoring of the Kaplan–Meier estimator or product-limit estimator [482], that is, when $u_i = \infty$.

- The *Kaplan–Meier estimator* $1 - \hat{F}_n^{KM}$ of $1 - F$ is defined as follows: letting $0 = \tau_0 < \tau_1 < \tau_2 < \dots < \tau_N$ (with $N < n$) denote the observed possible censored data, N_j the number of observations $X_i \geq \tau_j$, and d_j the number of values l_i equal to τ_j, then

$$1 - \hat{F}_n^{KM}(t) = \Pi_{\tau_j < t}\left[1 - \frac{d_j}{N_j}\right].$$

This expression is motivated from the fact that

$$1 - F(\tau_i) = P(X > \tau_i | X > \tau_{i-1}) P(X > \tau_{i-1} | X > \tau_{i-2})$$
$$\dots P(X > \tau_2 | X > \tau_1) P(X > \tau_1 | X > \tau_0) P(X > \tau_0).$$

- Turnbull's algorithm is then constructed as follows:
Let $0 = \tau_0 < \tau_1 < \ldots < \tau_m$ denote here the grid of all points l_i, u_i, $i = 1, 2, \ldots, n$. Define δ_{ij} as the indicator whether the observation in the interval $(l_i, u_i]$ could be equal to τ_j, $j = 1, \ldots, m$. δ_{ij} equals 1 if $(\tau_{j-1}, \tau_j] \subset (l_i, u_i]$ and 0 otherwise. Initial values are assigned to $1 - F(\tau_j)$ by distributing the mass $1/n$ for the ith individual equally to each possible $\tau_j \in (l_i, u_i]$. The algorithm is given as:
  1. *Compute the probability p_j that an observation equals τ_j by $p_j = F(\tau_j) - F(\tau_{j-1})$, $j = 1, \ldots, m$.*
  2. *Estimate the number of observations at τ_j by*

$$d_j = \sum_{i=1}^{n} \frac{\delta_{ij} p_j}{\sum_k^m \delta_{ik} p_k}.$$

  3. *Compute the estimated number of data with $l_i \geq \tau_j$ by $N_j = \sum_{k=j}^m d_k$.*
  4. *Update the product-limit estimator using the values of d_j and N_j found in the two preceding steps. Stop the iterative process if the new and old estimate of $1 - F$ for all τ_j do not differ too much.*

In case of interval censored data we can then estimate the mean excess function e (see (3.4.10)) substituting $1 - F$ by the Turnbull estimator $1 - \hat{F}_n^{TB}$:

$$e_n^{TB}(x) := \frac{\int_x^\infty (1 - \hat{F}_n^{TB}(u))du}{1 - \hat{F}_n^{TB}(x)}. \tag{4.3.44}$$

As discussed in Section 4.2.1, the mean excess function based on the log-data leads to an estimator of a positive extreme value index γ. As in (4.2.8), using the Turnbull estimator rather than the classical empirical distribution we obtain an estimator of $\gamma > 0$ in the case of incomplete data:

$$H_n^{TB}(x) := \frac{\int_x^\infty (1 - \hat{F}_n^{TB}(u))d\log u}{1 - \hat{F}_n^{TB}(x)}. \tag{4.3.45}$$

We then compute these statistics at the positions $x = \hat{Q}_n^{TB}(1 - (k+1)/(n+1))$, $k = 1, \ldots, n-1$. Such plots will assist in choosing an appropriate threshold t and estimates of the extreme value index γ to validate the tail component in the splicing.

Case study: Dutch fire insurance data. In this case no censoring is present, while the EVA did not indicate any upper-truncation effect. However, there is a left truncation point $t^l = 900\,000$. Fitting (4.3.42) with $T = \infty$, $\pi = 0.925$ and $t = 9\,075\,878$ on the basis of the mean excess plot, setting $\gamma = 0.784$ to be compared with the Hill estimator at the threshold t, combined with an ME component with $M = 2$, $\boldsymbol{\alpha} = (0.901, 0.099)$, $\mathbf{r} = (1, 5)$, $\lambda^{-1} = 1\,038\,901$ leads to the fit presented in Figure 4.13.

Note, however, that the tail fit following from $\left(-\log\left(1 - \frac{j}{n+1}\right), -\log(1 - F_{\mathrm{ME},Pa}(X_{j,n}))\right)$ $(j = 1, \ldots, n)$ (Figure 4.13, bottom right) is unsatisfactory. This is expected from the EVA

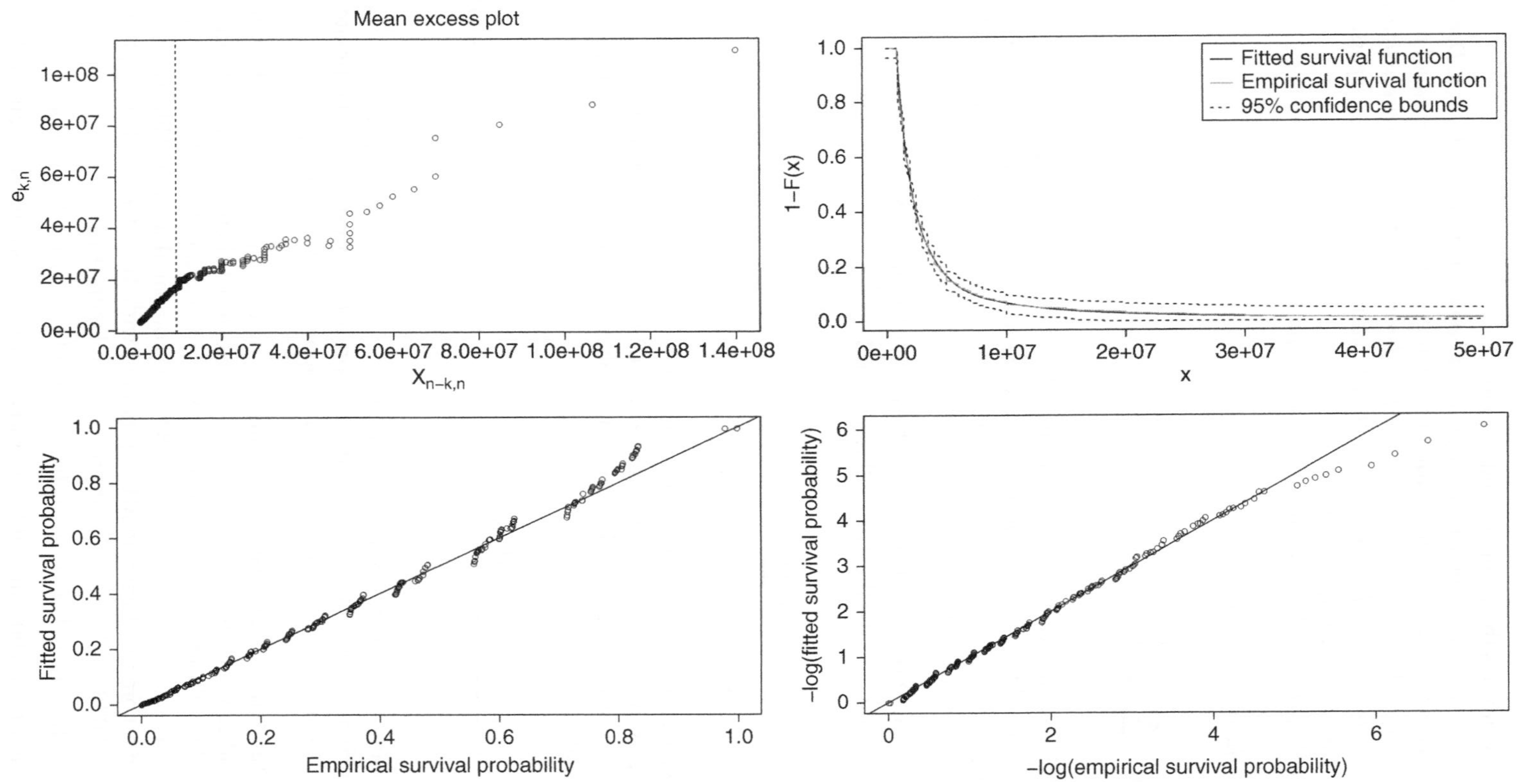

Figure 4.13 Dutch fire claim data: fit of spliced model mixed Erlang and Pareto with threshold t indicated on the mean excess plot (top left); empirical and model survival function (top right); PP plot of empirical survival function against splicing model RTF (bottom left); idem with $-\log$ transformation (bottom right).

discussed above, where we found two Pareto tail pieces with a lower EVI value following from the bias-reduced estimators $\hat{\gamma}^+$ at the top 1% of the data. This is also visible with the extreme quantile estimates of $Q(0.999)$ in Figure 4.6 (top). From this the following splicing model was fitted:

$$f_{\mathrm{ME}_3,Pa_1,Pa_2}(x) = \begin{cases} \pi_1 \frac{f_{\mathrm{ME}}(x;\mathbf{r},\alpha,\lambda)}{F_{ME}(t_1;\mathbf{r},\alpha,\lambda)-F_{ME}(t^l;\mathbf{r},\alpha,\lambda)} \\ \quad = \pi_1 \frac{e^{-\lambda x}\sum_{j=1}^{M}\alpha_j \frac{\lambda^{r_j}}{(r_j-1)!}x^{r_j-1}}{1-e^{-\lambda t_1}\sum_{j=1}^{M}\alpha_j \sum_{n=0}^{r_j-1}\frac{(\lambda t_1)^n}{n!}}, & t^l < x \leq t_1, \\ \pi_2 \frac{1}{t_1\gamma_1}\left(\frac{x}{t_1}\right)^{-1-1/\gamma_1} \Big/ \left[1-\left(\frac{t_2}{t_1}\right)^{-1/\gamma_1}\right], & t_1 < x \leq t_2, \\ (1-\pi_1-\pi_2)\frac{1}{t_2\gamma_2}\left(\frac{x}{t_2}\right)^{-1-1/\gamma_2}, & t_2 < x. \end{cases}$$

with $t^l = 900\,000$, $\pi_1 = 0.925$, $\pi_2 = 0.065$, $t_1 = 9\,075\,878$, $t_2 = 45\,000\,000$, $\gamma_1 = 0.947$ and $\gamma_2 = 0.427$, while $M = 2$, $\boldsymbol{\alpha} = (0.901, 0.099)$, and $\mathbf{r} = (1, 5)$, $\lambda^{-1} = 1\,038\,901$ (see Figure 4.14). □

Case study: MTPL data for Company A. The interval censoring approach is considered here with the indexed payments in 2010 as a lower bound and upper bounds for the non-closed claims which are derived from the indexed incurred values. Concerning the upper bounds two methods are applied here.

- First, in Figure 4.15 (top) we plot the percentage of incurred values which correctly act as upper bounds for the final payments of the closed claims as a function of the development year. From this we observe that from the sixth year of development the incurred values start to be reliable upper bounds with 90% confidence. We then restrict attention to the claims with at least 5 years of development, that is, with accident year before 2006. This restricted data set contains 596 claims of which 45% are censored.

  First we inspect the tails within the interval censoring approach on the basis of e_n^{TB} and H_n^{TB} from (4.3.45) with x taken in $\hat{Q}_n^{TB}(1 - (k + 1)/(n + 1))$. We conclude that this mean excess plot adapted for interval censoring based on (4.3.44) has a shape comparable to the mean excess plot based on the ultimate values, but with a different horizontal scale and with a Hill-type estimate $\hat{\gamma} \approx 0.45$ (see (4.3.45)) that is situated between the two levels found in Figure 4.5. We coupled a ME with a Pareto (ME-Pa). The parameters are:

$$\text{ME-Pa}: \pi = 0.873,\ t = 500\,000,\ \boldsymbol{\alpha} = (0.171, 0.829),\ \mathbf{r} = (1,4),\ 1/\lambda = 55\,227,$$
$$\gamma = 0.44,\ T = \infty.$$

  See Figure 4.16.
- Another approach follows from Figure 4.15 (bottom) where, for every development year d, we present the boxplots based on all claims, closed or non-closed in 2010, of the ratios $R_{i,d}$ of the final cumulative payment Z_i in 2010 over the incurred value

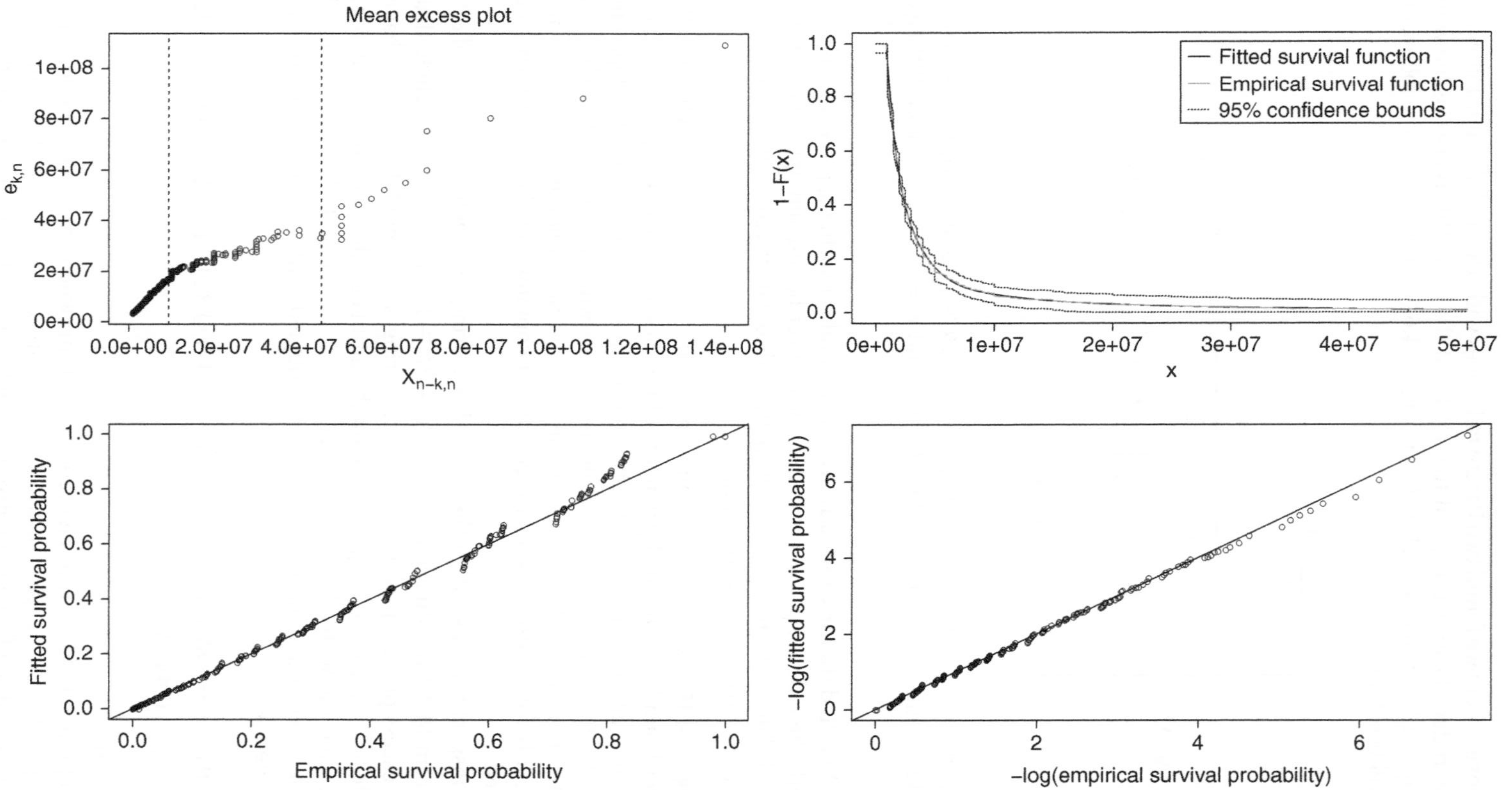

Figure 4.14 Dutch fire claim data: fit of spliced model with a mixed Erlang and two Pareto components with thresholds t_1 and t_2 as indicated on the mean excess plot (top left); empirical and model survival function (top right); PP plot of empirical survival function against splicing model RTF (bottom left); idem with $-\log$ transformation (bottom right).

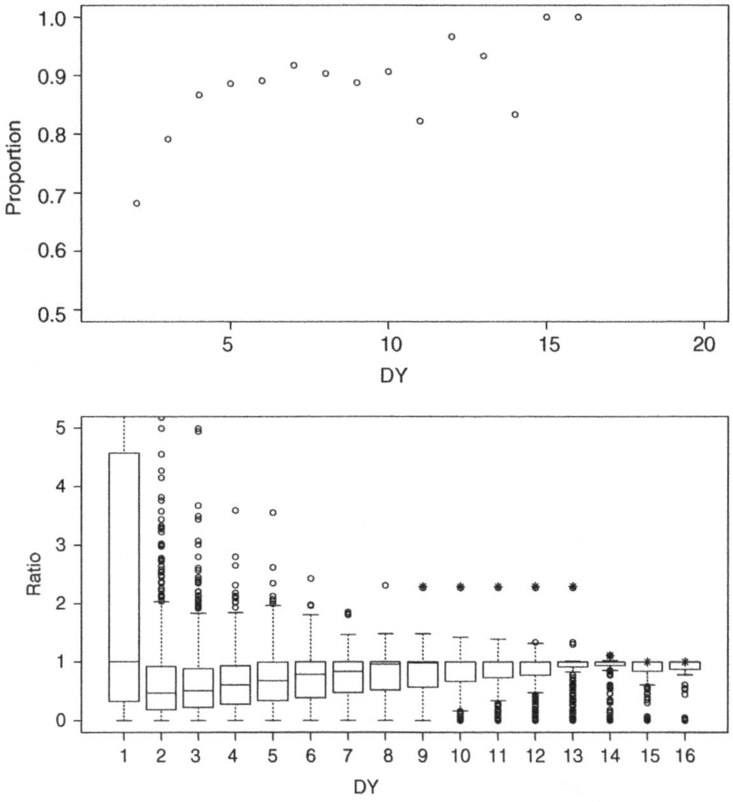

Figure 4.15 MTPL data for Company A: percentage of closed claims with incurred value being a correct upper bound for final payment as a function of the number of development years (DY) (top); boxplots of $R_{i,d}$ for every development year d and factor f_d used in the interval censoring approach (bottom).

$I_{i,d}$ for the given development year for claim i: $R_{i,d} = Z_i/I_{i,d}$. When a claim is closed before a particular development year d, the ratio for that claim in year d equals 1. This plot yields relevant information on the possibility of using the incurred values as an upper bound: if a ratio $R_{i,d}$ is larger than 1, the incurred value is smaller than the final available cumulative payment. The ratios $R_{i,d}$ are also right censored in case the cumulative payment is censored. Estimating the right endpoints of the distributions of the $R_{i,d}$ values per d using the methods developed in Einmahl et al. [315] then leads to factors f_d so that $\tilde{I}_{i,d} = f_d I_{i,d}$ provide more reliable upper bounds for the real final cumulative payments. We then still deleted the claims from 2010 since the upper bounds for these losses are still not reliable. In Figure 4.15 (bottom) we also plot the factors f_d.

We then inspect the tails again within the interval censoring approach on the basis of e^{TB} and H^{TB} from (4.3.45) with $\tilde{I}_{i,d}$ serving as an upper bound for the final cumulative payment of claim i. The corresponding tail fit and splicing results, given in Figure 4.17, compare well with the results in Figure 4.16. However, the confidence

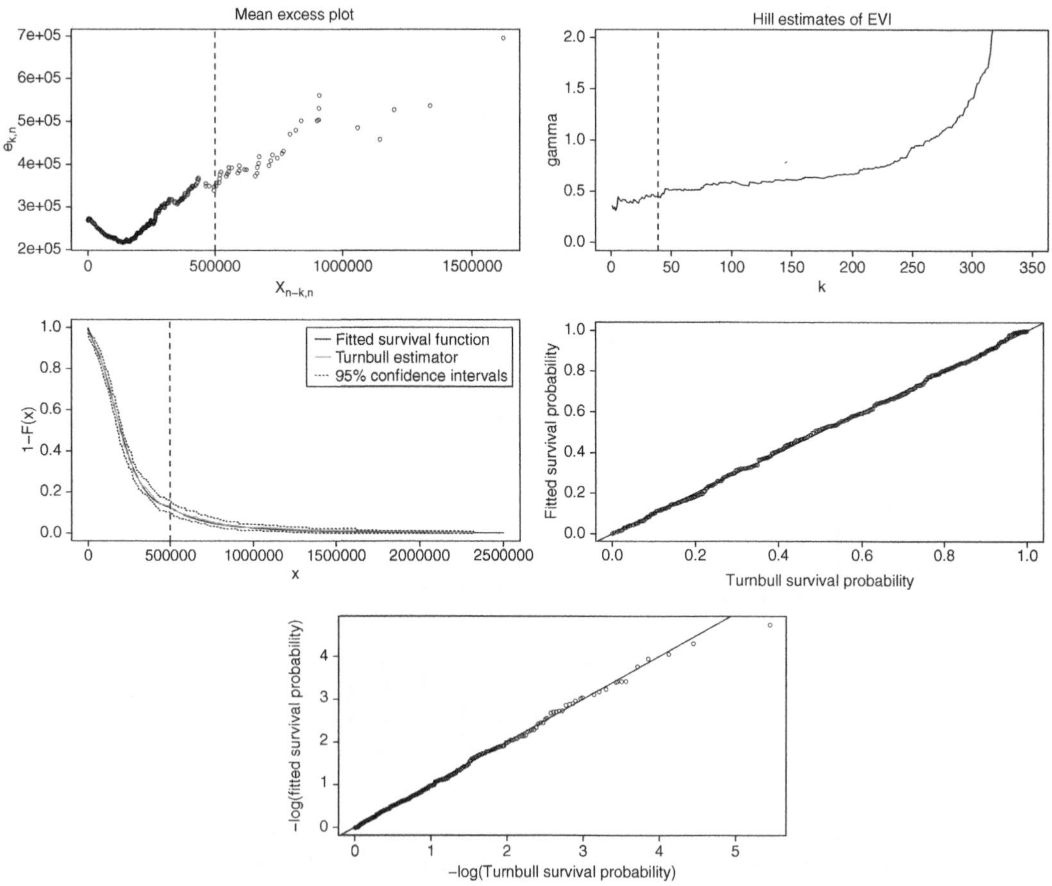

Figure 4.16 MTPL data for Company A: fit of spliced mixed Erlang and Pareto models with interval censoring based on upper bounds $I_{i,d}$, $i = 1, \ldots, 596$, $d = 5, \ldots, 16$, for non-closed claims: mean excess plot based on (4.3.44) (top left); Hill plot based on (4.3.45) (top right); empirical and model survival function (middle left); PP plot of empirical survival function against splicing model RTF (middle right); idem with $-\log$ transformation (bottom).

intervals based on the Turnbull estimator are wider when using the larger upper bounds $\tilde{I}_{i,d}$ for larger claim sizes (see Figure 4.17, bottom right).

The parameters of the splicing model here are:

$$\text{ME-Pa} : \pi = 0.777,\ t = 500\,000,\ \boldsymbol{\alpha} = (0.155, 0.845),\ \mathbf{r} = (1,4),\ 1/\lambda = 63\,410,$$
$$\gamma = 0.506,\ T = \infty.$$

Case study: MTPL data for Company B. Again, the interval censoring approach is considered here with the indexed payments in 2010 as a lower bound and the indexed incurred values as upper bound of the intervals for claims under development in 2010. Here we restrict attention to the claims with at least 5 years of development and use

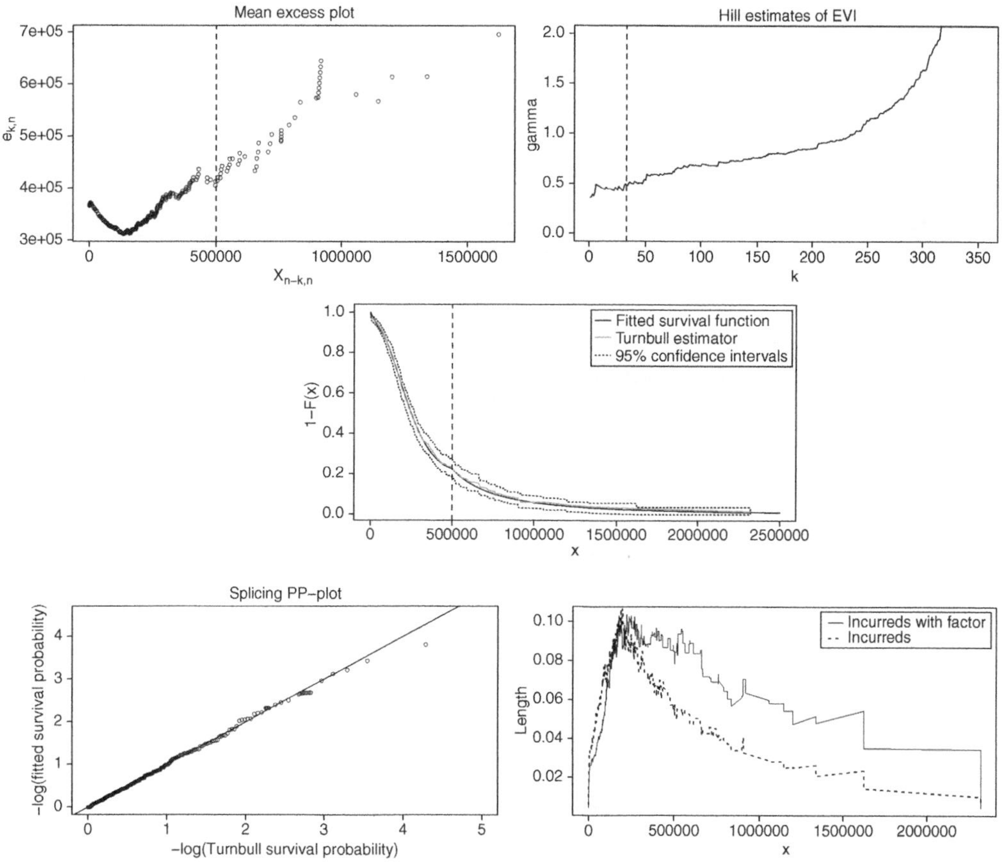

Figure 4.17 MTPL data for Company A (1995–2009): fit of spliced mixed Erlang and Pareto models with interval censoring based on upper bounds $\tilde{I}_{i,d}$, $i = 1, \dots, 849$, $d = 1, \dots, 15$ for non-closed claims: mean excess plot based on (4.3.44) (top left); Hill plot based on (4.3.45) (top right); empirical and model survival function (middle); PP plot of $-\log$ survival function, empirical against splicing mode (bottom left); size of confidence intervals using interval censoring with upper bounds $I_{i,d}$ and $\tilde{I}_{i,d}$ (bottom right).

the incurred values $I_{i,d}$ as upper bounds. This restricted data set contains 428 claims, of which 48% are censored.

On the basis of e^{TB} and H^{TB} from (4.3.44) and (4.3.45) with x in $\hat{Q}_n^{TB}(1 - (k+1)/(n+1))$, we conclude that the mean excess plot adapted for interval censoring has a shape comparable to the mean excess plot based on the ultimate values, but with a different horizontal scale, and with a stable plot of the Hill type estimates for γ at 0.5. This value can be seen as a compromise between the two levels found in Figure 4.5. We then splice an ME with a Pareto (ME-Pa) tail. The parameters are:

$$\text{ME-Pa}: \pi = 0.704,\ t = 360\,000,\ \boldsymbol{\alpha} = (0.098, 0.902),\ \mathbf{r} = (1, 5),\ 1/\lambda = 47\,796,$$
$$\gamma = 0.501,\ T = \infty.$$

See Figure 4.18. □

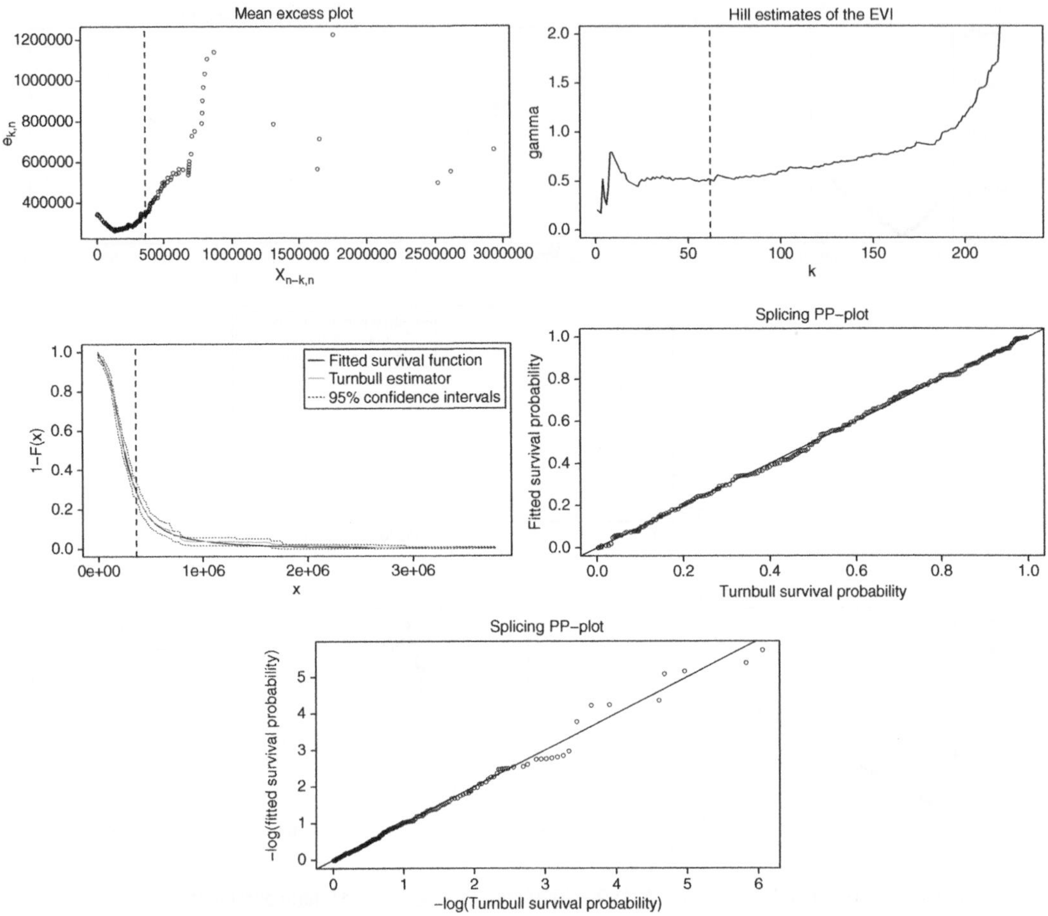

Figure 4.18 MTPL data for Company B: mean excess plot based on (4.3.44) (top left) and Hill plot based on (4.3.45)(top right) for interval censored data based on accidents from 1990 to 2005; fit of spliced model mixed Erlang and Pareto models: empirical and model survival function (middle left); PP plot of empirical survival function against splicing model RTF (middle right) left); idem with $-\log$ transformation (bottom).

If the upper bounds are put to ∞, that is, if one uses the right censoring framework, then, under the random right censoring assumption of independence between the real cumulative payment at closure of the claim and the censoring variable C which is observed in case the claim is right censored, estimators of $\gamma > 0$ have been proposed in Beirlant et al. [96, 101], Einmahl et al. [315], and Worms and Worms [796]. Using the likelihood approach, Beirlant et al. [101] proposed the estimator

$$H_{k,n}^{(c1)} = \frac{\frac{1}{k}\sum_{j=1}^{k} \log Z_{n-j+1,n} - \log Z_{n-k,n}}{\hat{p}_k}, \tag{4.3.46}$$

with $Z_i = \min(X_i, C_i)$ $(i = 1, \ldots, n)$ and $\hat{p}_k$ the proportion of non-censored data in the top k Z-data. Einmahl et al. [315] derived asymptotic results, while Beirlant et al. [96] proposed a bias-reduced version. Worms and Worms [796] derived a tail index estimator which is derived through the estimation of the mean excess function of the log-data, comparable with the estimator derived in (4.3.45):

$$H_{k,n}^{(c2)} = \frac{\sum_{j=1}^{k} \left(1 - \hat{F}_n^{KM}(Z_{n-j,n})\right)\left(\log Z_{n-j+1,n} - \log Z_{n-j,n}\right)}{1 - \hat{F}_n^{KM}(Z_{n-k,n})}, \tag{4.3.47}$$

where the Kaplan–Meier estimator can be written as

$$1 - \hat{F}_n^{KM}(x) = \Pi_{Z_{i,n} \le x} \left(1 - \frac{1}{n-i+1}\right)^{\Delta_{i,n}} = \Pi_{Z_{i,n} \le x} \left(1 - \frac{\Delta_{i,n}}{n-i+1}\right),$$

with $\Delta_{i,n}$ equal to 1 if the ith smallest observation $Z_{i,n}$ is non-censored, and 0 otherwise.

Case study: MTPL data for Company A. In the MTPL application the validity of the random right censoring assumption is questionable since censoring is informative here: censoring is more likely to occur with larger claim sizes, which are also related to larger development times. The estimates of γ using this approach also happen to be closer to 1, which indicates over-estimation. In case of Company A, the results for the estimators $H_{k,n}^{(c2)}$ and H_n^{TB} are compared in Figure 4.19. □

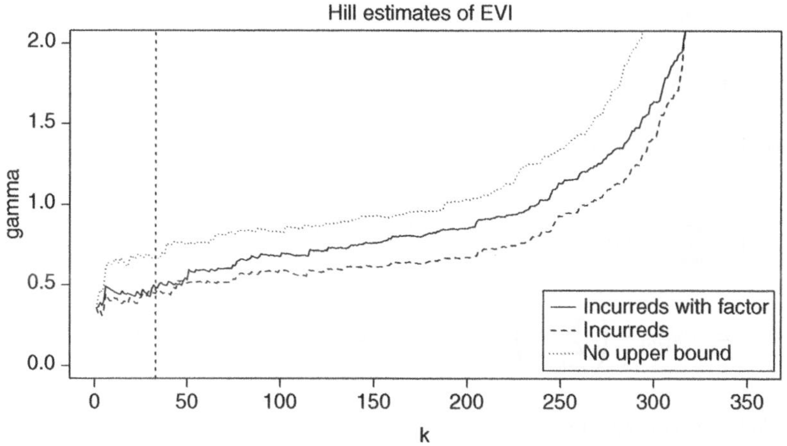

Figure 4.19 Hill plots adapted for interval censoring H_n^{TB} with upper bounds $I_{i,d}$ and $\tilde{I}_{i,d}$, and Hill estimates based on random right censoring $H_{k,n}^{(c2)}$ without upper bounds. The vertical line indicates the splicing threshold t used above.

4.4 Incorporating Covariate Information

In certain instances, the assumption of i.i.d. random variables, underlying the extreme value methods discussed above, may be violated. When analysing claim data from different companies, the tail fits may differ. Also, loss distributions may change over calendar years or along the number of development years. Sometimes considering covariates may remedy the situation. Let the covariate information, whether using continuous or indicator variables, be contained in a covariate vector $\mathbf{x} = (x_1, \dots, x_p)$. The extension of the POT approach based on (4.2.25) has been popular in literature, starting with the seminal paper by Davison and Smith [253]. However, there are also some methods available that focus on response random variables that exhibit Pareto-type tails. Here we denote the response variables Z_i (rather than X_i as in the preceding sections) with the corresponding exceedances or POTs $Y = Z/t$ or $Y = Z - t$ when $Z > t$.

4.4.1 Pareto-type Modelling

When modelling time dependence or incorporating any other covariate information in an independent data setting with Pareto-type distributed responses Z_i, the exceedances are defined through $Y_i = Z_i/t$ for some appropriate threshold t. Note that in many circumstances the threshold should then also be modelled along $x = x_i$, $i = 1, \dots, n$. As before, we assume that as $z \to \infty$

$$1 - F_{n,i}(z) = A_i z^{-1/\gamma_i}(1 + o_i(1)), \tag{4.4.48}$$

where $A_i, \gamma_i > 0$. Regression can be modelled through the scale parameter A and/or the extreme value index γ.

Changes in γ can be modelled in a parametric way using likelihood techniques. Suppose, for instance, that regression modelling of $\gamma > 0$ using an exponential link function appears appropriate in a given case study:

$$\gamma_i = \exp\left(\beta_0 + \beta_1 x_i\right),\ i = 1, \dots, n.$$

The log-likelihood function is then given by

$$\begin{aligned}\log L(\beta_0, \beta_1) &= \sum_{i=1}^{N_t} \log\left(\exp[-\beta_0 - \beta_1 x_i]\, Y_i^{-1-\exp(-\beta_0-\beta_1 x_i)}\right)\\ &= -N_t\beta_0 - \left(\sum_{i=1}^{N_t} x_i\right)\beta_1 - \sum_{i=1}^{N_t}\left(1 + \exp(-\beta_0 - \beta_1 x_i)\right)\log Y_i,\end{aligned}$$

leading to the likelihood equations

$$\begin{cases} \exp\beta_0 = \frac{1}{N_t}\sum_{i=1}^{N_t}(\log Y_i)e^{-\beta_1 x_i} \\ \frac{1}{N_t}\sum_{i=1}^{N_t} x_i(\log Y_i)e^{-\beta_1 x_i} = \left(\frac{1}{N_t}\sum_{i=1}^{N_t} x_i\right)\left(\frac{1}{N_t}\sum_{i=1}^{N_t}(\log Y_i)e^{-\beta_1 x_i}\right). \end{cases}$$

Beirlant and Goegebeur [99] propose to inspect the goodness-of-fit of such a regression model under constant scale parameter A on the basis of a Pareto QQ-plot using

$$\hat{R}_i := Y_i^{1/\hat{\gamma}_i}, \; i = 1, \ldots, N_t,$$

which are indeed approximately Pareto distributed with tail index 1, when the regression model is appropriate.

The case where γ does not depend on i, while A does depend on i, was formalized in Einmahl et al. [316] assuming that there exists a tail function $1 - F$ and a continuous, positive function A defined on $[0, 1]$ such that

$$\lim_{z\to\infty} \frac{1 - F_{n,i}(z)}{1 - F(z)} = A\left(\frac{i}{n}\right), \tag{4.4.49}$$

uniformly for all $n \in \mathbb{N}$ and all $i = 1, \ldots, n$ with $\int_0^1 A(s)ds = 1$. A is then called the *skedasis function*, which characterizes the trend in the extremes through the changes in the scale parameter A. Under (4.4.49), Einmahl et al. [316] showed that the Hill estimator $H_{k,n}$ is still a consistent estimator for γ. Assuming equidistant covariates $x_i = i/n, i = 1, \ldots, n$, as in (4.2.19),

$$\hat{A}_{1-F} = \frac{k+1}{n+1} Z_{n-k,n}^{1/H_{k,n}},$$

$$\hat{A}_{1-F_i} = \frac{1}{n+1}\left[\sum_{j=1}^{n} 1_{\{Z_j > Z_{n-k,n}\}} K_h\left(x_i - \frac{j}{n}\right)\right] Z_{n-k,n}^{1/H_{k,n}},$$

where $\sum_{j=1}^{n} 1_{\{Z_j > Z_{n-k,n}\}} K_h\left(x_i - \frac{j}{n}\right)$ denotes the number of Z values larger than the threshold $Z_{n-k,n}$ with covariate value x in a neighbourhood of x_i. The contribution of the observations to $\hat{A}_{1-F_i}$ is governed by a symmetric density kernel function K on $[-1, 1]$ and $K_h(x) = K(x/h)/h$, so that K gives more weight to the observations with covariates closer to x_i. We hence obtain

$$\hat{A}(s) = \frac{\hat{A}_{1-F_i}}{\hat{A}_{1-F}} = \frac{1}{k+1} \sum_{i=1}^{n} 1_{\{Z_i > Z_{n-k,n}\}} K_h\left(s - \frac{i}{n}\right).$$

Finally, estimators of small tail probabilities and large quantiles follow directly from (4.4.49), (4.2.12) and (4.2.13):

$$\widehat{\bar{F}}_i(z) = \hat{A}(x_i) \frac{k+1}{n+1} \left(\frac{z}{Z_{n-k,n}}\right)^{-1/H_{k,n}},$$

$$\hat{Q}_i(1-p) = Z_{n-k,n} \left(\frac{(k+1)\hat{A}(x_i)}{(n+1)p}\right)^{H_{k,n}},$$

4.4.2 Generalized Pareto Modelling

Let $Y_1, \ldots, Y_n$ be independent GPD random variables and let $\mathbf{x_i}$ denote the covariate information vector, that is,

$$\mathbb{P}(Y_i \leq y) = 1 - \left(1 + \frac{\gamma(\mathbf{x}_i)}{\sigma(\mathbf{x}_i)}(y - \mu(\mathbf{x}_i))\right)^{-1/\gamma(\mathbf{x}_i)}, \quad i = 1, \ldots, n,$$

where $\gamma(\mathbf{x}), \sigma(\mathbf{x}), \mu(\mathbf{x})$ denote admissible functions of $\mathbf{x}$, whether of *parametric* nature using three vectors of regression coefficients β_j ($j = 1, 2, 3$) of length p with $\gamma(\mathbf{x}) = \gamma_{\beta_1}(\mathbf{x})$, $\sigma(\mathbf{x}) = \sigma_{\beta_2}(\mathbf{x})$ and $\mu(\mathbf{x}) = \mu_{\beta_3}(\mathbf{x})$, or of *non-parametric* nature. Again this model is used as an approximation of the conditional distribution of excesses $Y(\mathbf{x}) = Z - \mu(\mathbf{x})$ over a high threshold $\mu(\mathbf{x})$ given that there is an exceedance. The choice of an appropriate threshold $\mu(\mathbf{x})$ is of course even more difficult than in the non-regression setting since the threshold can depend on the covariates in order to take the relative extremity of the observations into account.

- When *parametric functions* $\gamma(\mathbf{x}) = \gamma_{\beta_1}(\mathbf{x})$, $\sigma(\mathbf{x}) = \sigma_{\beta_2}(\mathbf{x})$ and $\mu(\mathbf{x}) = \mu_{\beta_3}(\mathbf{x})$ have been chosen, the estimators of β_j ($j = 1, 2, 3$) can be obtained by maximizing the log-likelihood function

$$\log L\left(\beta_1, \beta_2, \beta_3\right) = \sum_{i=1}^{N_\mu} \Bigg\{ -\log \sigma_{\beta_2}(\mathbf{x}) - \left(1 + \frac{1}{\gamma_{\beta_1}(\mathbf{x})}\right) \log\left(1 + \frac{\gamma_{\beta_1}(\mathbf{x})}{\sigma_{\beta_2}(\mathbf{x})} Y_i(\beta_3)\right)\Bigg\},$$

  where N_μ denotes the number of excesses over the threshold function $\mu(\mathbf{x})$.
- Alternatively, *non-parametric regression* techniques are available to estimate the parameter functions $\gamma(\mathbf{x}), \sigma(\mathbf{x})$. Consider independent random variables $Z_1, \ldots, Z_n$ and associated covariate information $\mathbf{x}_1, \ldots, \mathbf{x}_n$. Suppose we focus on estimating the tail of the distribution of Z at $\mathbf{x}^*$. Fix a high local threshold $\mu(\mathbf{x}^*)$ and compute the exceedances $Y_i = Z_j - \mu(\mathbf{x}^*)$, provided $Z_j > \mu(\mathbf{x}^*)$, $i = 1, \ldots, N_{\mu_{\mathbf{x}^*}}$. Here j is the index of the ith exceedance in the original sample, and $N_{\mu_{\mathbf{x}^*}}$ denotes the number of exceedances over the threshold $\mu_{\mathbf{x}^*}$. Then re-index the covariates in an appropriate way such that $\mathbf{x}_i$ denotes the covariate associated with exceedance Y_i.

  Using local polynomial maximum likelihood estimation, one approximates $\gamma(\mathbf{x})$ and $\sigma(\mathbf{x})$ by polynomials, centered at $\mathbf{x}^*$. Let h denote a bandwidth parameter and consider a univariate covariate x. Assuming γ, respectively σ, being m_1, respectively m_2, times differentiable one has for $|x_i - x^*| \leq h$,

$$\gamma(x_i) = \sum_{j=0}^{m_1} \beta_{1j}(x_i - x^*)^j + o(h^{m_1}),$$

$$\sigma(x_i) = \sum_{j=0}^{m_2} \beta_{2j}(x_i - x^*)^j + o(h^{m_2}),$$

where

$$\beta_{1j} = \frac{1}{j!}\frac{\partial^j \gamma(x)}{\partial x^j}\bigg|_{x=x^*} \quad \text{and } \beta_{2j} = \frac{1}{j!}\frac{\partial^j \sigma(x)}{\partial x^j}\bigg|_{x=x^*}.$$

The coefficients of these approximations can be estimated by local maximum likelihood fits of the GPD, with the contribution of each observation to the log-likelihood being governed by a kernel function K. The local polynomial maximum likelihood estimator $(\boldsymbol{\beta}_1, \boldsymbol{\beta}_2) = (\hat{\beta}_{10}, \ldots, \hat{\beta}_{1m_1}, \hat{\beta}_{20}, \ldots, \hat{\beta}_{2m_2})$ is then the maximizer of the kernel weighted log-likelihood function

$$\log L(\boldsymbol{\beta}_1, \boldsymbol{\beta}_2) = \sum_{i=1}^{N_{\mu(x^*)}} \log g\left(Y_i, \sum_{j=0}^{m_1} \beta_{1j}(x_i - x^*)^j, \sum_{j=0}^{m_2} \beta_{2j}(x_i - x^*)^j\right) K_h(x_i - x^*)$$

with respect to $\hat{\beta}_{10}, \ldots, \hat{\beta}_{1m_1}, \hat{\beta}_{20}, \ldots, \hat{\beta}_{2m_2}$, where $g(y; \mu, \sigma) = (1/\sigma)(1 + (\gamma/\sigma)y)^{-1-1/\gamma}$ is the density of the generalized Pareto distribution.

A more recent approach is using penalized log-likelihood optimization based on spline functions. Let the covariates $\mathbf{x}$ be one-dimensional within an interval $[a, b]$. The goal is to fit reasonably smooth functions h_γ and h_σ with $\gamma(x) = h_\gamma(x)$ and $\sigma(x) = h_\sigma(x)$ to the observations (Y_i, x_i), $i = 1, \ldots, N_\mu$. The penalized log-likelihood is then given by

$$\ell^p(h_\gamma, h_\sigma; Y_i, x_i) = \sum_{i=1}^{N_\mu} \left\{ \log \frac{1}{\sigma(x)} - \left(1 + \frac{1}{\gamma(x)}\right) \log\left(1 + \frac{\gamma(x)}{\sigma(x)} Y_i\right) \right\} - \lambda_\gamma \int_a^b (h''_\gamma(t))^2 dt - \lambda_\sigma \int_a^b (h''_\sigma(t))^2 dt.$$

The introduction of the penalty terms is a standard technique to avoid over-fitting when one is interested in fitting smooth functions (see Hastie and Tibshirani [428] or Green and Silverman [408]). Next • stands for γ or σ. Intuitively the penalty functions $\int_a^b (h''_\bullet(t))^2 dt$ measure the roughness of twice-differentiable curves and the smoothing parameters $\lambda_\bullet$ are chosen to regulate the smoothness of the estimates $\hat{h}_\bullet$: larger values of these parameters lead to smoother fitted curves.

Let $a = s_0 < s_1 < \ldots < s_m < s_{m+1} = b$ denote the ordered and distinct values among $\{x_1, \ldots, x_{N_\mu}\}$. A function h defined on $[a, b]$ is a cubic spline with the above knots if the following conditions are satisfied:

- on each interval $[s_i, s_{i+1}]$, h is a cubic polynomial
- at each knot s_i, h and its first and second derivatives are continuous.

A cubic spline is a natural cubic spline if in addition to the two latter conditions it satisfies the natural boundary condition that the second and third derivatives of h at a and b are zero. It follows from Green and Silverman [408] that for a natural cubic spline h with knots $s_1, \ldots, s_m$ one has

$$\int_a^b (h''_{\bullet}(t))^2 dt = \mathbf{h}_{\bullet}^t K \mathbf{h}_{\bullet},$$

where $\mathbf{h}_{\bullet} = (h_{\bullet}(s_1), \dots, h_{\bullet}(s_m))$, and K is a symmetric $m \times m$ matrix of rank $m-2$ only depending on the knots $s_1, \dots, s_m$. Hence

$$\ell^p(\mathbf{h}_\gamma, \mathbf{h}_\sigma; Y_i, x_i) = \sum_{i=1}^{N_\mu} \left\{ \log \frac{1}{\sigma(x)} - \left(1 + \frac{1}{\gamma(x)}\right) \log \left(1 + \frac{\gamma(x)}{\sigma(x)} Y_i(x_i)\right) \right\} - \lambda_\gamma \mathbf{h}_\gamma^t K \mathbf{h}_\gamma - \lambda_\sigma \mathbf{h}_\sigma^t K \mathbf{h}_\sigma.$$

In order to assess the validity of a chosen regression model one can generalize the exponential QQ-plot of generalized residuals defined before in the non-regression case:

$$\left(-\log\left(1 - \frac{i}{N_\mu + 1}\right), \hat{R}_{i,N_\mu}\right),$$

with

$$\hat{R}_i = \begin{cases} \frac{1}{\hat{\gamma}(\mathbf{x}_i)} \log\left(1 + \frac{\hat{\gamma}(\mathbf{x}_i)}{\hat{\sigma}(\mathbf{x}_i)} Y_i(\mathbf{x}_i)\right), & \hat{\gamma}(\mathbf{x}_i) \neq 0, \\ Y_i(\mathbf{x}_i) \frac{1}{\hat{\sigma}(\mathbf{x}_i)}, & \hat{\gamma}(\mathbf{x}_i) = 0, i = 1, \dots, N_\mu. \end{cases}$$

Finally, given regression estimators for $(\gamma(x), \sigma(x))$ using an appropriate threshold function $\mu(x)$, extreme quantile estimators are given by

$$\hat{Q}_x(1-p) = \begin{cases} \mu(x) + \frac{\hat{\sigma}(x)}{\hat{\gamma}(x)} \left(\left[\frac{p}{\hat{\bar{F}}_{Z|x}(\mu(x))} \right]^{-\hat{\gamma}(x)} - 1 \right), & \hat{\gamma}(x) \neq 0, \\ \mu(x) - \hat{\sigma}(x) \log\left(\frac{p}{\hat{\bar{F}}_{Z|x}(\mu(x))} \right), & \hat{\gamma}(x) = 0, \end{cases}$$

where $\hat{\bar{F}}_{Z|x}$ can, for instance, be taken to be equal to the Nadaraya–Watson estimator

$$\hat{\bar{F}}_{Z|x}(u) = \frac{\sum_{i=1}^n 1_{\{Z_i > u\}} K_h(x_i - x)}{\sum_{i=1}^n K_h(x_i - x)}.$$

For more details and other non-parametric methods, refer to Davison and Ramesh [252], Hall and Tajvidi [420], Chavez-Demoulin and Davison [200], Daouia et al. [249] [248], Gardes and Girard [368, 369], Gardes and Stupfler [370], Goegebeur, Guillou and Osmann [393], and Stupfler [713], as well as Chavez-Demoulin et al. [201] for other non-parametric extreme value regression methods and applications.

Case study: Austrian storm claim data. We consider here the modelling of the normalized historical losses of residential buildings from Section 1.3.3 in Vienna and the Upper Austria provinces as a function of the building value weighted wind index W.

We model the conditional extreme value index $\gamma = \beta_1$ constant in W, while $\log \sigma_W$ is considered to be linear in W:

$$\log \sigma_W = \beta_{2,1} + \beta_{2,2} W.$$

Finally, we take here $\mu_W = 0$, that is, we take all the data, since the data can already be considered as exceedances. Hence the model is

$$\text{normalized loss} \sim GPD\left(\beta_1, e^{\beta_{2,1}+\beta_{2,2}W}, 0\right).$$

The results from a maximum likelihood analysis are

- Upper Austria: $\beta_1 = 0.445,\ \beta_{2,1} = -7.2,\ \beta_{2,2} = 0.046$;
- Vienna: $\beta_1 = 0.337,\ \beta_{2,1} = -10.8,\ \beta_{2,2} = 0.065$

(cf. Figure 4.20). We also plot the estimates of the quantile $Q(0.97|W)$ using parametric and non-parametric fits, jointly with the residual QQ-plots. From the residual QQ-plot for the Vienna province we deduce that the storm with $w = 59$ is an outlier. Deleting that storm from the data set leads to $\beta_1 = -0.163,\ \beta_{2,1} = -10.5$ and $\beta_{2,2} = 0.064$. Hence this particular storm has a high influence on the analysis. □

4.4.3 Regression Extremes with Censored Data

In Section 4.3 we discussed the problem when estimating the distribution of the final payments based on censored data using the Kaplan–Meier estimator of the distribution of the payment data. Here we propose to consider regression modelling of the final payments given the development time at the closure of a claim. Note, however, that both the final payments and development periods are right censored, both variables being censored (or not censored) at the same time. We again use the notation Z_i $(i = 1, \dots, n)$ for the observed cumulative payment at the end of the study from that section, and similarly $nDY_{e,i}$ for the observed number of development years at the end of 2010. Again $\Delta_{i,n}$ denotes the indicator of non-censoring corresponding to the ith smallest observed value payment $Z_{i,n}$. Akritas and Van Keilegom [11] proposed the following non-parametric estimator of the conditional distribution of X given a specific value of nDY assuming that X and the censoring variable C (see Section 4.3) are conditionally independent given nDY:

$$1 - \hat{F}_{X|nDY}(x|d) = \prod_{Z_i \le x} \left(1 - \frac{W_i(d;h)}{\sum_{j=1}^{n} W_j(d;h) 1_{\{Z_j \ge Z_i\}}}\right)^{\Delta_i}$$

with weights

$$W_i(d;h) = \begin{cases} K\left(\frac{d - nDY_{e,i}}{h}\right) / \sum_{\Delta_j=1} K\left(\frac{d-nDY_{e,j}}{h}\right) & \text{if } \Delta_i = 1, \\ 0 & \text{if } \Delta_i = 0. \end{cases}$$

Denoting the weight W corresponding to the ith smallest Z value $Z_{i,n}$ with $W_{i,n}$ we then arrive at the following Hill-type estimator of the conditional extreme value given $nDY = d$, generalizing the unconditional Worms and Worms estimator $H_{k,n}^{(c2)}$ defined in (4.3.47):

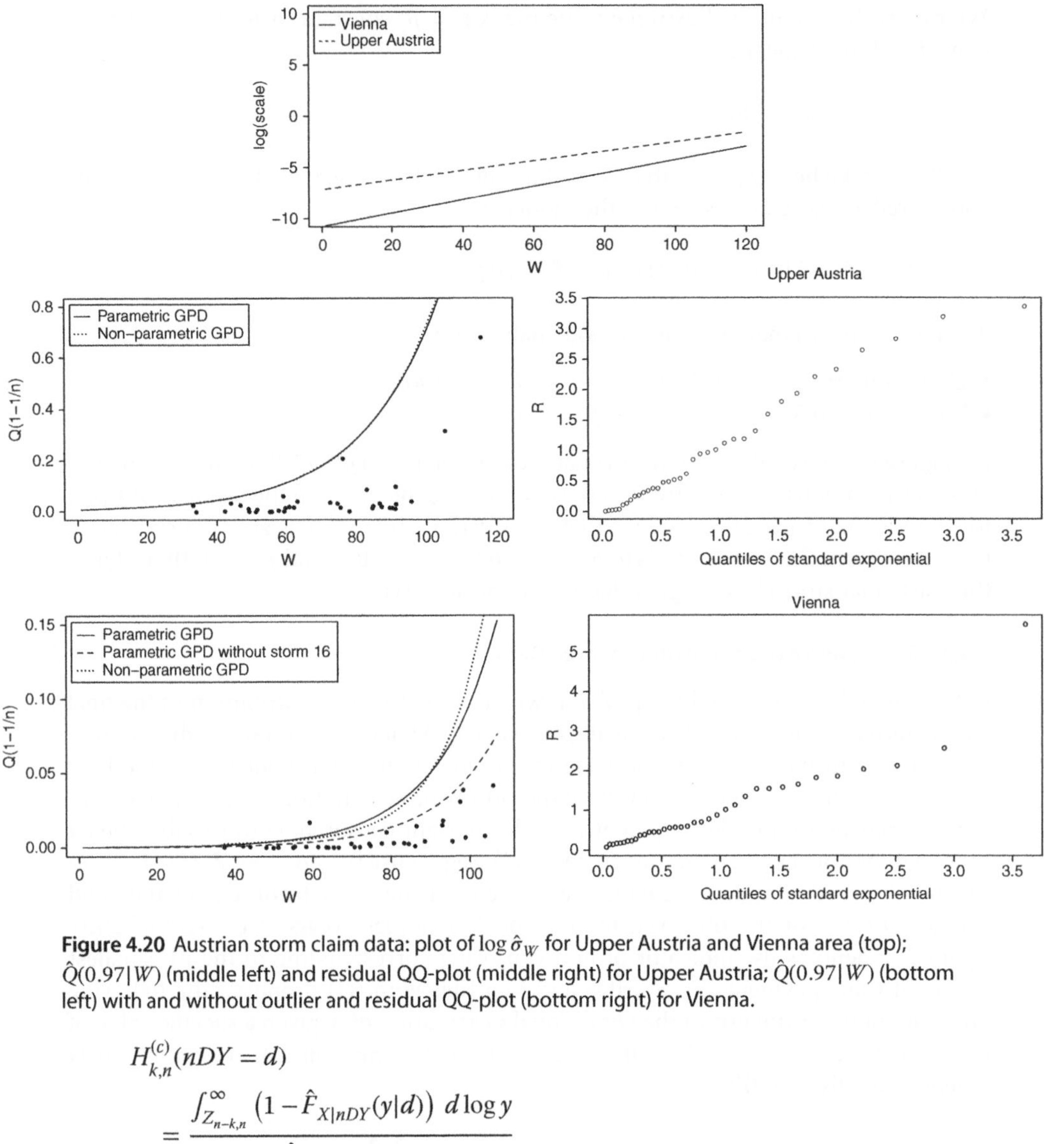

Figure 4.20 Austrian storm claim data: plot of $\log \hat{\sigma}_W$ for Upper Austria and Vienna area (top); $\hat{Q}(0.97|W)$ (middle left) and residual QQ-plot (middle right) for Upper Austria; $\hat{Q}(0.97|W)$ (bottom left) with and without outlier and residual QQ-plot (bottom right) for Vienna.

$$
\begin{aligned}
H_{k,n}^{(c)}&(nDY = d) \\
&= \frac{\int_{Z_{n-k,n}}^{\infty} \left(1-\hat{F}_{X|nDY}(y|d)\right) d\log y}{1-\hat{F}_{X|nDY}(Z_{n-k,n}|d)} \\
&= \frac{\sum_{j=1}^{k}\left(\prod_{i=1}^{n-j}\left[\left(1-\frac{W_{i,n}(d;h)}{1-\sum_{l=1}^{i-1}W_{l,n}(d;h)}\right)^{\Delta_{i,n}}\right]\log\frac{Z_{n-j+1,n}}{Z_{n-j,n}}\right)}{\prod_{i=1}^{n-k}\left[\left(1-\frac{W_{i,n}(d;h)}{1-\sum_{l=1}^{i-1}W_{l,n}(d;h)}\right)^{\Delta_{i,n}}\right]}
\end{aligned}
$$

Pareto QQ-plots adapted for censoring per chosen d value can then be defined as

$$
\left(-\log\left(1-\hat{F}_{X|nDY}(Z_{n-j+1,n}|d)\right), \log Z_{n-j+1,n}\right), \; j = 1, \ldots, n.
$$

Case study: MTPL data for Company A. In the MTPL application of Company A, the change of the X distribution given the number of development period becomes visible from the time plot of the claims in Figure 4.21 as a function of nDY_e. We applied the conditional Hill estimator adapted for censoring $H_{k,n}^{(c)}(nDY = d)$ with $d = 3, 8, 13$,

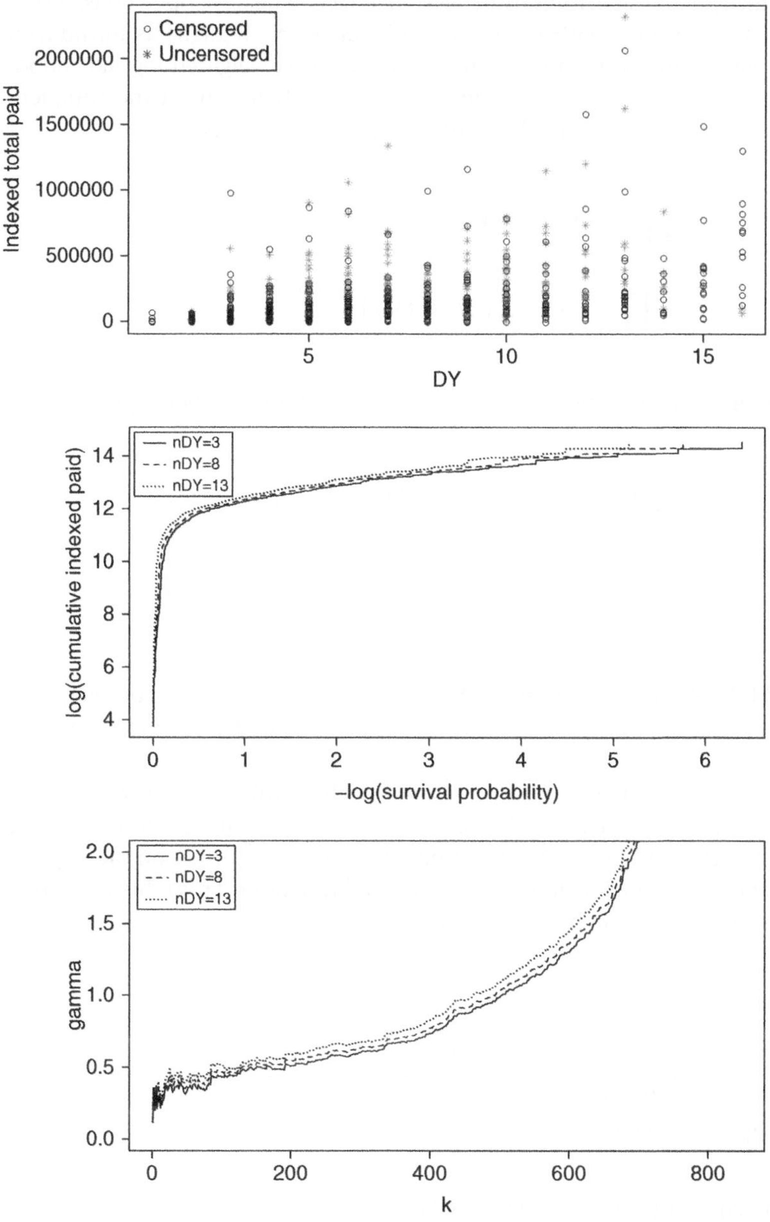

Figure 4.21 MTPL data for Company A: time plots of cumulative payments Z_i as a function of nDY_e (top); Pareto QQ-plots (middle) and Hill estimates (bottom) adapted for right censoring at development years $nDY = 3, 8, 13$.

$h = 15$, and the bi-weight kernel $K(u) = \frac{15}{16}(1 - u^2)^2 1_{\{|u|\le 1\}}$. Note that the Pareto QQ-plots show a heavier tail with increasing nDY, which is confirmed by the plots of $H^{(c)}_{k,n}(nDY = d)$. □

In order to derive a full model for the complete payments X as a function of $nDY = d$ a local version of the splicing algorithm from Section 4.3.2 can be developed considering random right censoring on X with the kernel weights $W_i = W_i(d; h)$ as introduced above. The EM algorithm can then be applied using a kernel weighted log-likelihood, comparable with the approach from Section 4.4.1. For instance, given the complete version of the data, the complete likelihood function is then given by

$$\mathcal{L}_{\text{complete}}(\boldsymbol{\Theta}; \mathcal{Y}) = \prod_{i=1}^{n} \left(\pi f_1(X_i; t^l, t, \boldsymbol{\Theta}_1)\right)^{1_{\{X_i\le t\}} W_i} \times \prod_{i=1}^{n} \left((1-\pi) f_2(X_i; t, T, \boldsymbol{\Theta}_2)\right)^{1_{\{X_i> t\}} W_i}.$$

The corresponding complete data weighted log-likelihood function then equals

$$\ell_{\text{complete}}(\boldsymbol{\Theta}; \mathcal{Y}) = \sum_{i=1}^{n} 1_{\{X_i\le t\}} W_i \left(\log \pi + \log f_1(X_i; t^l, t, \boldsymbol{\Theta}_1)\right) + \sum_{i=1}^{n} 1_{\{X_i> t\}} W_i \Big(\log(1-\pi) + \log f_2(X_i; t, T, \boldsymbol{\Theta}_2)\Big).$$

Case study: MTPL data for Company A. The above method from Beirlant and Reynkens [104], when applied with $d = 3, 8, 13$ years, yields the estimates given in Table 4.1.

The estimates of γ approximately correspond to the Hill estimates $H^{(c)}_{k,n}(nDY = d)$ at $k = 100$ (see Figure 4.21, bottom right). In Figure 4.22 the PP plots use the fitted survival functions at $nDY = 3, 8$ and 13 against the fitted RTF scale at $nDY = 8$, on the original scale and using the $-\log$ transformation of the survival functions. From this graph one observes also the increase in tail heaviness as a function of d. □

Table 4.1

	$nDY = 3$	$nDY = 8$	$nDY = 13$
π	0.859	0.846	0.826
t	390 000	390 000	390 000
M	2	2	2
$\boldsymbol{\alpha}$	(0.235,0.765)	(0.213,0.787)	(0.173,0.827)
$\mathbf{r}$	(1,5)	(1,4)	(1,4)
λ^{-1}	39 693	50 954	51 966
γ	0.441	0.453	0.468

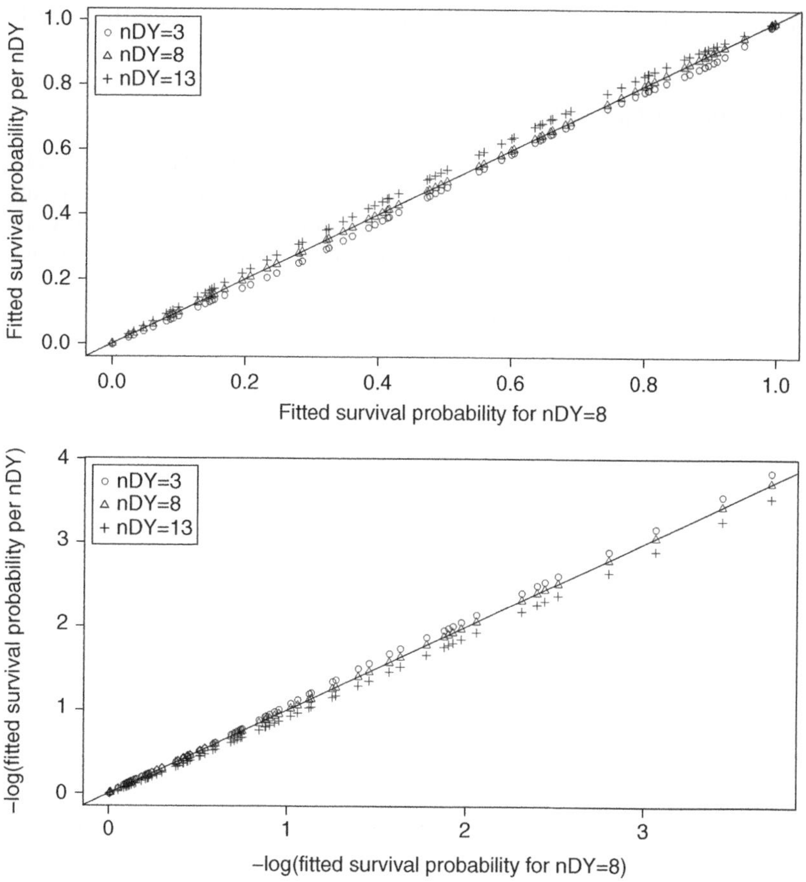

Figure 4.22 MTPL data for Company A: fit of splicing model at development years $nDY = 3, 8, 13$; PP plot of empirical survival function against splicing model RTF at $nDY = 8$ (top); idem with $-\log$ transformation (bottom).

4.5 Multivariate Analysis of Claim Distributions

Joint or multivariate estimation of claim distributions, for example originating from different lines of business which are possibly dependent, requires estimation of each component or marginal separately and of the dependence structure. The joint analysis of loss and allocated loss adjustment expenses (ALAE) forms another example in insurance. An early analysis of such a case is provided in Frees and Valdez [359]. A detailed EVA of such a data set using the concept of extremal dependence is found in Chapter 9 in Beirlant et al. [100].

We first model the multivariate tails using data that are large in at least one component, followed by a splicing exercise combining a tail and a modal fit. In a multivariate setting this program is of course much more complex in comparison with the univariate case. For the tail section we refer to the multivariate POT modelling using the multivariate generalized Pareto distribution, as introduced in Section 3.6. The joint modelling of "small" losses will be based on a multivariate generalization of the mixed Erlang distribution introduced by Lee and Lin [533]. Research in this matter has started only recently and here we only examine an ad hoc modelling for the Danish fire insurance data.

4.5.1 The Multivariate POT Approach

From (3.6.25) and (3.6.26) one observes the importance of estimating the stable tail dependence function l defined in Chapter 3. The estimation of the tail dependence can be performed non-parametrically or using parametric models. We refer to Kiriliouk et al. [488] for fitting parametric multivariate generalized Pareto models using censored likelihood methods.

A non-parametric estimator of an STDF is given by

$$\hat{l}_k(\mathbf{x}) = \frac{1}{k}\sum_{i=1}^{n} 1_{\{\text{there exists } j=1,\dots,d:\hat{F}_j(X_{i,j})>1-\frac{k}{n}x_j\}}, \tag{4.5.50}$$

with

$$\hat{F}_j(X_{i,j}) = \frac{R_{i,j}-0.5}{n} \text{ or } \frac{R_{i,j}}{n+1}$$

where $R_{i,j}$ denotes the rank of $X_{i,j}$ among $X_{1,j},\dots,X_{n,j}$:

$$R_{i,j} = \sum_{m=1}^{n} 1_{\{X_{m,j}\le X_{i,j}\}},\ j=1,\dots,d.$$

The estimator $\hat{l}_k$ is a direct empirical version of definition (3.6.22) of l with $u = n/k$.

A slightly different version is given by

$$\tilde{l}_k(\mathbf{x}) = \frac{1}{k}\sum_{i=1}^{n} 1_{\{X_{i,1}\ge X^{(1)}_{n-[kx_1]+1,n} \text{ or } \dots \text{ or } X_{i,d}\ge X^{(d)}_{n-[kx_d]+1,n}\}}, \tag{4.5.51}$$

where $X^{(j)}_{i,n}$ denotes the ith smallest observation of component j. Bias-reduced versions of these estimators were proposed in Fougères et al. [357] and Beirlant et al. [98]. In the bivariate case, $\hat{l}_k(1,1)$ or $\tilde{l}_k(1,1)$ then act as estimators of the extremal coefficient θ.

An estimator of the extremal dependence coefficient χ can be constructed on the basis of an estimator of $\chi(u)$ for $u \to 1$ using the estimator of $C(u,u)$

$$\hat{C}(u,u) = \frac{1}{n}\sum_{i=1}^{n} 1_{\{U_{i,1}<u,U_{i,2}<u\}}, \tag{4.5.52}$$

where $U_{i,j} = \hat{F}_j(X_{i,j})$ ($j = 1,2;\ i = 1,\dots,n$) with $\hat{F}_j$ denoting the empirical distribution function of the jth marginal and $X_{i,j}$ the ith observation of the jth component. Hence

$$\hat{\chi}(u) = 2 - \frac{\log \hat{C}(u,u)}{\log u}.$$

As an application note that an estimator of the parameter τ in the logistic dependence model can be obtained from $\chi(u) \to 2 - 2^{\tau}$ as $u \to 1$, from which

$$\hat{\tau}_u = \frac{\log\log \widehat{C}(u,u) - \log\log u}{\log 2}.$$

Setting

$$\tilde{Z}_{i,j} = \left(1 - \hat{F}_j(X_{i,j})\right)^{-1} = \left(1 - \frac{R_{i,j}}{n+1}\right)^{-1}, j = 1, 2,$$

the Hill estimator based on $\min(\tilde{Z}_{i,1}, \tilde{Z}_{i,2})$ $(i = 1, \ldots, n)$ leads to an estimator $\hat{\eta}_k$ of the coefficient of tail dependence η. Of course bias reduction techniques can be applied here too.

4.5.2 Multivariate Mixtures of Erlangs

Lee and Lin [533] defined a d-variate Erlang mixture where each mixture component is the joint distribution of d independent Erlang distributions with a common scale parameter $1/\lambda > 0$. The dependence structure is then captured by the combination of the positive integer shape parameters of the Erlangs in each dimension. We denote the positive integer shape parameters of the jointly independent Erlang distributions in a mixture component by the vector $\mathbf{r} = (r_1, \ldots, r_d)$ and the set of all shape vectors with non-zero weight by $\mathcal{R}$. The density of a d-variate Erlang mixture evaluated in $\mathbf{x} > \mathbf{0}$ can then be written as

$$f_{MME}(\mathbf{x}; \alpha, \mathbf{r}, \lambda) = \sum_{\mathbf{r}\in\mathcal{R}} \alpha_{\mathbf{r}} \Pi_{j=1}^{d} f_E(x_j, r_j, \lambda) = \sum_{\mathbf{r}\in\mathcal{R}} \alpha_{\mathbf{r}} \Pi_{j=1}^{d} \frac{\lambda^{r_j} x_j^{r_j-1} e^{-\lambda x_j}}{(r_j-1)!}. \tag{4.5.53}$$

Lee and Lin [533] showed that, given any density $f(\mathbf{x})$, the d-variate Erlang mixture

$$f_{MME}(\mathbf{x}; \lambda) = \sum_{r_1=1}^{\infty} \cdots \sum_{r_d=1}^{\infty} \alpha_{\mathbf{r}}(\lambda) \Pi_{j=1}^{d} f_E(x_j; r_j, \lambda)$$

with mixing weights

$$\alpha_{\mathbf{r}} = \int_{(r_1-1)/\lambda}^{r_1/\lambda} \cdots \int_{(r_d-1)/\lambda}^{r_d/\lambda} f(\mathbf{x}) d\mathbf{x}$$

satisfies $\lim_{\lambda\to\infty} F_{MME}(\mathbf{x}; \lambda) = F(\mathbf{x})$. The weights $\alpha_{\mathbf{r}}$ of the components in the mixture are defined by integrating the density over the corresponding d-dimensional rectangle of the grid formed by the shape parameters multiplied with the common scale. When the value of λ increases, this grid becomes more refined and the sequence of Erlang mixtures converges to the underlying distribution function.

Verbelen et al. [757] provided a flexible fitting procedure for multivariate mixed Erlangs (MMEs), which iteratively uses the EM algorithm, by introducing a computationally efficient initialization and adjustment strategy for the shape parameter vectors. Randomly censored and fixed truncated data can also be dealt with.

Case study: Danish fire insurance data. Here we consider a bivariate splicing model for the components building and contents, conditional on (building, contents) $\mathbf{t}^l = (1, 1)$. We first fitted a bivariate GPD based on a logistic extreme value distribution based on excesses over the threshold vector $\mathbf{t} = (7.32, 10.27)$ corresponding to $k = 35$ in the univariate extreme value plots. Univariate EVA leads to γ values around 0.5 for the building component and around 0.6 for the contents component. Fitting the GPD to each component leads to initial σ estimates. The parameter τ in the logistic dependence model can be estimated through estimating $\theta = l(1,1) = 2^\tau$ or by estimating $\chi(u) \to 2 - 2^\tau$ as $u \to 1$. These estimates are plotted in Figure 4.23 (middle) leading to $\hat{\tau} = \log \hat{l}_{35}(1,1)/\log 2 = 0.83$ or $\frac{\log\log \hat{C}(u,u) - \log\log u}{\log 2} = 0.81$ taking $u = 0.5$ from which $\hat{C}(0.5, 0.5) = 0.25$. We further consider this second estimate.

Concerning the tail dependence coefficient η, the level $\hat{\eta} = 0.65$ is dominating, while at the smallest k values the estimates increase systematically with decreasing k. As the $\hat{\chi}$ plot appears to indicate asymptotic dependence corresponding with η equal to 1, one has to be cautious interpreting the $\hat{\eta}$ plot which indeed ultimately for the smallest k tends to values around 1.

The bivariate (c.d.f) function of a splicing model with a bivariate mixed Erlang and a bivariate GPD is now given by

$$F(\mathbf{x}) = \begin{cases} 0, & \text{if } \mathbf{x} \leq \mathbf{t}^l, \\ \pi F_{MME}(\mathbf{x}), & \text{if } \mathbf{t}^l \leq \mathbf{x} \leq \mathbf{t}, \\ \pi F_{MME}(\mathbf{x}) + (1-\pi) F_{MGPD}(\mathbf{x}), & \text{if } \mathbf{x} \nleq \mathbf{t}, \end{cases}$$

with F_{MGPD} denoting the distribution function of the bivariate GPD as given in (3.6.25).

A bivariate mixed Erlang distribution was fitted along the method provided in Verbelen et al. [757] conditioned on $[1, 7.32] \times [1, 10.27]$, leading to $\mathbf{r}$ vectors (1,1) and (3,8) and α weights 0.92 and 0.08, and $1/\lambda = 1.49$. The proportion for the bivariate mixed Erlang fit is $\pi = 0.794$. The (c.d.f) corresponding to f_1 is then given by

$$F_1(\mathbf{x}; \mathbf{t}^l, \mathbf{t}, \boldsymbol{\Theta}) = \begin{cases} 0 & \text{if } \mathbf{x} \ngeq \mathbf{t}^l \\ \dfrac{\sum_{\mathbf{r} \in R} \alpha_{\mathbf{r}} \prod_{j=1}^d F_E(\min\{x_j, t_j\}; r_j, \lambda)}{\sum_{\mathbf{r} \in R} \alpha_{\mathbf{r}} \prod_{j=1}^d (F_E(t_j; r_j, \lambda) - F_E(t_j^l; r_j, \lambda))} & \text{if } \mathbf{x} \geq \mathbf{t}^l \text{ and } \mathbf{x} \ngeq \mathbf{t} \\ 1 & \text{if } \mathbf{x} \geq \mathbf{t}. \end{cases}$$

The bivariate distribution function of the fitted bivariate GPD is given by

$$F_2(x, y; \mathbf{t}, \boldsymbol{\gamma}, \boldsymbol{\sigma}, \tau)$$
$$= 2^{-1/\tau} \left\{ \left(\left(1 + \gamma_1 \frac{\min\{x - t_1, 0\}}{\sigma_1}\right)_+^{-\frac{\tau}{\gamma_1}} + \left(1 + \gamma_2 \frac{\min\{y - t_2, 0\}}{\sigma_2}\right)_+^{-\frac{\tau}{\gamma_2}} \right)^{1/\tau} \right.$$

$$-\left(\left(1+\gamma_1\frac{x-t_1}{\sigma_1}\right)_+^{-\frac{\tau}{\gamma_1}}+\left(1+\gamma_2\frac{y-t_2}{\sigma_2}\right)_+^{-\frac{\tau}{\gamma_2}}\right)^{1/\tau}\Bigg\}.$$

In order to guarantee that the marginal distributions have support on $(1, \infty)$ one has to impose the constraints $1 + \frac{\gamma_1}{\sigma_1}(1 - t_1) = 0$ and $1 + \frac{\gamma_2}{\sigma_2}(1 - t_2) = 0$, which then lead to the parameter values $(\gamma_1 = 0.57, \sigma_1 = 3.57)$ and $(\gamma_2 = 0.65, \sigma_2 = 6.05)$.

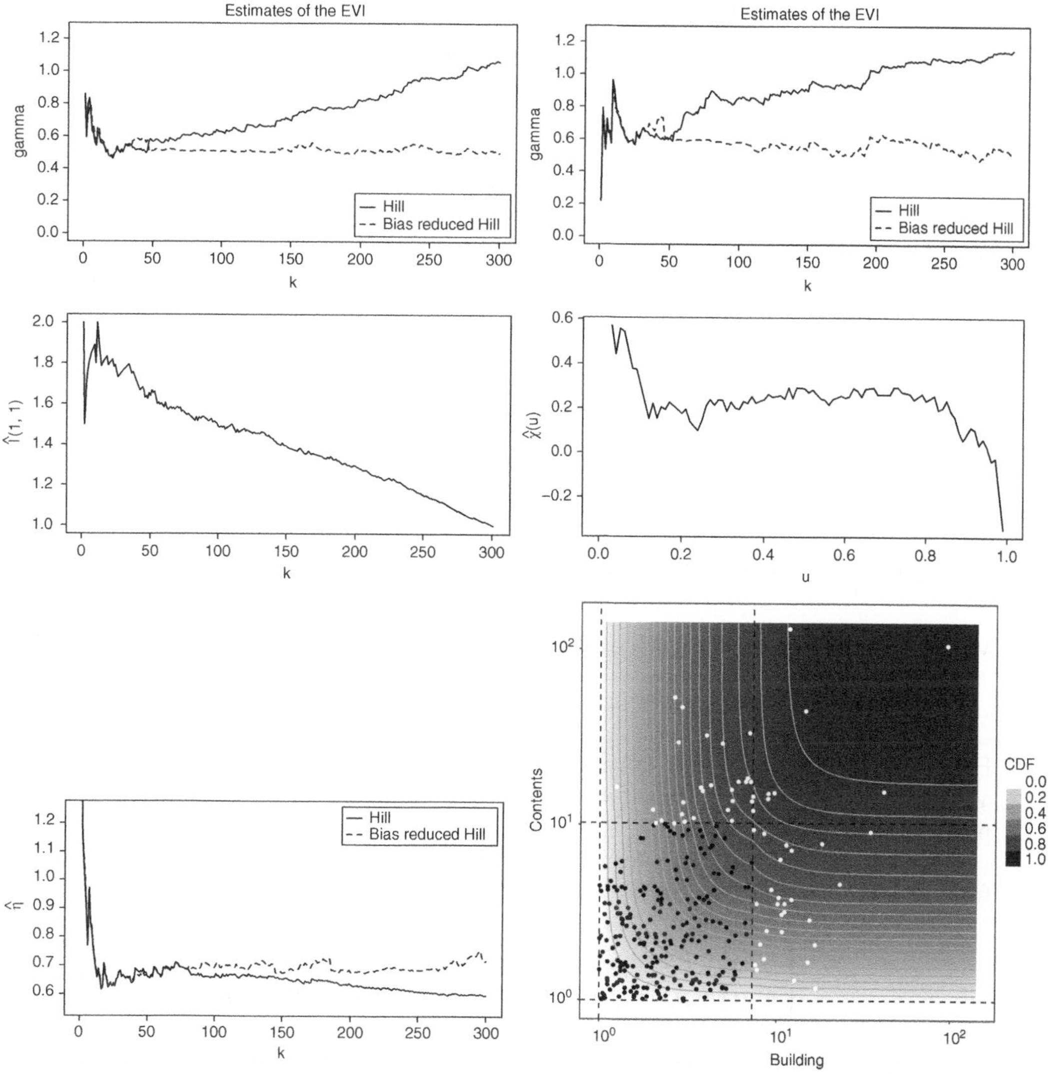

Figure 4.23 Danish fire insurance data, building and contents: Hill and bias-reduced Hill plots, building (top left) and contents (top right), plot of $\hat{l}_k(1,1)$ against k (middle left) and $\hat{C}(u,u)$ against $u \in (0,1)$ (middle right), plot of $\hat{\eta}_k$ (bottom left) and cumulative distribution function of the fitted bivariate splicing model (bottom right).

4.6 Estimation of Other Tail Characteristics

In Section 4.2.1.2 using EVA we discussed the estimation of an extreme quantile or a *VaR*

$$\mathrm{VaR}_{1-p}(X) = Q(1-p) = \inf\{x|F(x) \geq 1-p\}$$

in detail. Another popular tail characteristic is the *conditional tail expectation* $\mathrm{CTE}_{1-p}(X)$ defined by

$$\begin{aligned}\mathrm{CTE}_{1-p}(X) &= \mathbb{E}(X|X > Q(1-p)) \\ &= Q(1-p) + \mathbb{E}\,(X - Q(1-p)|X > Q(1-p)) \\ &= \mathrm{VaR}_{1-p}(X) + e(Q(1-p))\end{aligned}$$

when $\mathbb{E}(X) < \infty$, where e denotes the mean excess function defined in Section 3.4. If X is a continuous random variable, the CTE equals the Tail-VaR and the expected shortfall (ES) (cf. Section 7.2.2, where the role of these quantities for determining the solvency capital is discussed).

For an unlimited XL treaty with retention u, recall from Chapter 2 that the expected reinsured amount $\mathbb{E}(R)$ of a single claim X is given by

$$\Pi(u) := \mathbb{E}(\tilde{X}(u,\infty)) = \int_u^{\infty} (1 - F(z))dz = e(u)\overline{F}(u),$$

which is also referred to as the *pure premium* for R (see Chapter 7 for details). One immediately observes

$$\mathrm{CTE}_{1-p}(X) = \mathrm{VaR}_{1-p}(X) + \frac{\Pi(\mathrm{VaR}_{1-p}(X))}{p}.$$

With a finite layer size v in the XL treaty, the pure premium becomes

$$\mathbb{E}(\tilde{X}(u,v)) = \mathbb{E}\,(\min\{\max(X-u,0),v\}) = \Pi(u) - \Pi(u+v).$$

Hence the estimation of $\mathrm{VaR}_{1-p}(X)$ and $\Pi(u)$ at small and intermediate values of p, and at high and intermediate values of u is an important building block in measuring and managing risk.

When estimating $\mathrm{VaR}_{1-p}(X)$ for a two-component spliced distribution, we have from (4.3.43)

$$Q(1-p) = \begin{cases} Q_1((1-p)/\pi) & \text{if } 0 \leq p \leq \pi, \\ Q_2((1-p-\pi)/(1-\pi)) = Q_2\left(1 - \frac{p}{1-\pi}\right) & \text{if } \pi < p \leq 1, \end{cases} \tag{4.6.54}$$

where Q_1 denotes the quantile function of the ME component and Q_2 of the tail component. Q_1 can be obtained numerically. When the tail component is given by a simple Pareto distribution we have

$$Q_2(1-u) = t\, u^{-\gamma}, \; 0 < u < 1,$$

and hence with $t = X_{n-k,n}$ and $1 - \pi = (k+1)/(n+1)$, (4.6.54) yields $\hat{q}^{+}_{k,p}$ from (4.2.12) when $\pi < p \leq 1$. Using an upper-truncated Pareto or a generalized Pareto tail fit, one can use $\hat{q}^{T}_{k,p}$ or $\hat{q}^{ML}_{k,p}$, respectively, for $\pi < p \leq 1$.

When estimating $\Pi(u)$ we again identify two cases: $u \leq t = x_{n-k,n}$ and $u > t = x_{n-k,n}$, in which case the EVA modelling can be used.

When $u > t$, then from (4.3.43)

$$\begin{aligned}\Pi(u) &= \int_u^\infty \left\{1 - (\pi + (1-\pi)F_2(z))\right\} dz \\ &= (1-\pi)\int_u^\infty (1 - F_2(z)) \\ &=: (1-\pi)\Pi_2(u)\end{aligned}$$

where $\Pi_2(u)$ is given by the following expressions for the different possible EVA tail fits with EVI estimate smaller than 1:

- *Truncated Pareto fit*:

$$\begin{aligned}\hat{\Pi}^{\mathrm{TPa}}_{2,k}(u) &= \int_u^T \frac{\left(\frac{z}{t}\right)^{-1/\hat{\gamma}} - \left(\frac{T}{t}\right)^{-1/\hat{\gamma}}}{1-(T/t)^{-1/\hat{\gamma}}} dz \\ &= \frac{(u-T)\left(\frac{T}{t}\right)^{-1/\hat{\gamma}} + \left(u^{1-1/\hat{\gamma}} - T^{1-1/\hat{\gamma}}\right)\frac{t^{1/\hat{\gamma}}}{-1+1/\hat{\gamma}}}{1-\left(\frac{T}{t}\right)^{-1/\hat{\gamma}}};\end{aligned}$$

- *EPD fit*: using the notation from (4.2.18)

$$\begin{aligned}\hat{\Pi}^{\mathrm{EPD}}_{2,k}(u) &= \int_u^\infty \bar{G}_{\hat{\gamma},\hat{\delta},\hat{\tau}}(z/t) dz \\ &\approx t^{-1/\hat{\gamma}}\left\{\left(1-\frac{\hat{\delta}}{\hat{\gamma}}\right)\left(\frac{1}{\hat{\gamma}}-1\right)^{-1} u^{1-\hat{\gamma}^{-1}} + \frac{\hat{\delta}}{\hat{\gamma}t^{\hat{\tau}}}\left(\frac{1}{\hat{\gamma}}-\hat{\tau}-1\right)^{-1} u^{1+\hat{\tau}-\hat{\gamma}^{-1}}\right\};\end{aligned}$$

- *Generalized Pareto fit*:

$$\begin{aligned}\hat{\Pi}^{\mathrm{GPD}}_{2,k}(u) &= \int_u^\infty \left(1+\frac{\hat{\gamma}}{\hat{\sigma}}(z-t)\right)^{-1/\hat{\gamma}} dz \\ &= \frac{\hat{\sigma}}{1-\hat{\gamma}}\left(1+\frac{\hat{\gamma}}{\hat{\sigma}}(u-t)\right)^{1-1/\hat{\gamma}}.\end{aligned}$$

When $u < t$, we have from (4.3.43) that

$$\begin{aligned}
\Pi(u) &= \int_u^t \left(1 - \pi F_1(z)\right) dz + \int_t^{+\infty} \left(1 - (\pi + (1-\pi)F_2(z))\right) dz \\
&= (t-u) - \pi \int_u^t F_1(z) dz + (1-\pi) \int_t^{+\infty} (1 - F_2(z)) dz \\
&= (t-u) - (t-u)\pi + \pi \int_u^t (1 - F_1(z)) dz + (1-\pi)\Pi_2(t) \\
&= (1-\pi)(t-u) + \pi \Pi_1(u) + (1-\pi)\Pi_2(t).
\end{aligned}$$

Note that $\Pi(u) = 0$ for $u \geq T$ and $\Pi(u) = \Pi(t^l) + (t^l - u)$ for $u \leq t^l$. For the mixed Erlang distribution we get

$$\begin{aligned}
\Pi_1(u) &= \int_u^t \left(1 - \frac{F_1^*(z) - F_1^*(t^l)}{F_1^*(t) - F_1^*(t^l)}\right) dz \\
&= \frac{\left(F_1^*(t) - 1\right)(t-u) + (\Pi_1^*(u) - \Pi_1^*(t))}{F_1^*(t) - F_1^*(t^l)},
\end{aligned}$$

with

$$F_1^*(x) = \sum_{j=1}^{M} \alpha_j \left(1 - \sum_{n=0}^{r_j - 1} e^{-\lambda x} \frac{(\lambda x)^n}{n!}\right)$$

and, assuming that $r_n = n, n = 1, \ldots, M$,

$$\Pi_1^*(u) = \frac{1}{\lambda} e^{-\lambda u} \sum_{n=0}^{M-1} \sum_{k=n}^{M-1} \left(\sum_{j=k+1}^{M} \alpha_j\right) \frac{(\lambda u)^n}{n!}.$$

Case study: MTPL data for Company A. Based on the splicing model for the data of Company A within the interval censored framework, using an unbounded Pareto, the fit of which is shown in Figure 4.16, we calculate the XL pure premium $\Pi(u)$ as a function of u in Figure 4.24. We also add an estimate for $\Pi(u)$ when taking $u_i = \infty$, that is, considering only the lower bounds for the censored claims. The resulting value is significantly higher, which is consistent with the high estimates of the extreme value index as indicated in Figure 4.19. In order to compare with the classical approach using a statistical model for the ultimate estimates of the open claims, we also provide a comparison with the results based on the splicing model from Figure 4.11. This "classical" pure premium is also uniformly higher than the one obtained using interval censoring. □

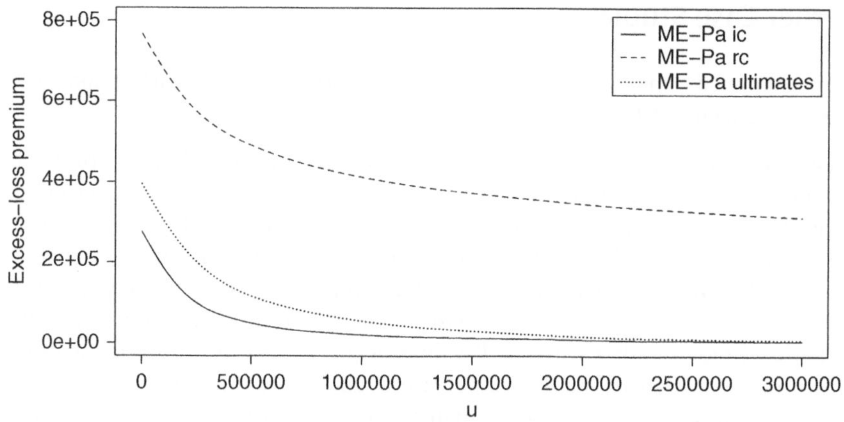

Figure 4.24 MTPL data for Company A: XL pure premium $\Pi(u)$ based on ME-Pa fit taking interval censoring into account. Comparison with the result when the upper bounds are ignored (right censoring) and when the premium is based on the ultimates.

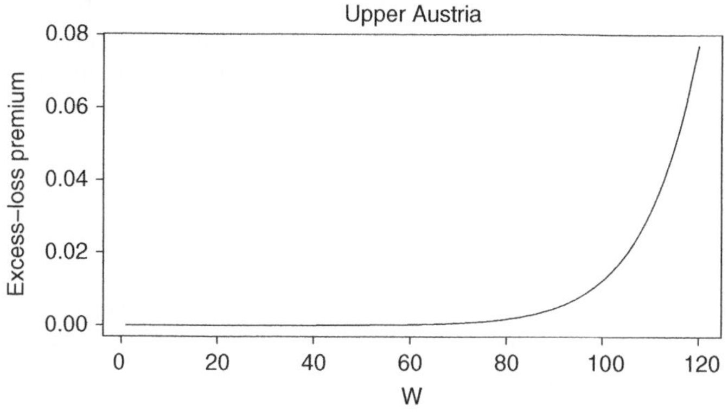

Figure 4.25 Austrian storm claim data: XL pure premium $\Pi(\exp(0.001w))$ for Upper Austria based based on a GPD regression fit with the wind index W as covariate.

Of course the estimation of $\Pi(u)$ can be extended to a regression context. For instance, when $u = u(x)$ is larger than a threshold function $\mu(x)$ of a one-dimensional covariate x, and using the GPD modelling approach, one obtains for $\hat{\gamma}(x) < 1$

$$\begin{aligned}\hat{\Pi}(u(x)) &= \hat{\bar{F}}_{Z|x}(\mu(x)) \int_{u(x)}^{\infty} \left(1 + \frac{\hat{\gamma}(x)}{\hat{\sigma}(x)}(c - \mu(x))\right)^{-1/\hat{\gamma}(x)} dc \\ &= \hat{\bar{F}}_{Z|x}(\mu(x)) \frac{\hat{\sigma}(x)}{1 - \hat{\gamma}(x)} \left(1 + \frac{\hat{\gamma}(x)}{\hat{\sigma}(x)}(u(x) - \mu(x))\right)^{1-1/\hat{\gamma}(x)}.\end{aligned}$$

The result of this procedure based on the GPD regression fit for the storm claim data of Upper Austria, with GPD$(0.445, e^{-7.2+0.046w}, 0)$, is shown in Figure 4.25.

4.7 Further Case Studies

We end this chapter by analysing the case studies on flood risk and earthquake risk which were introduced in Chapter 1.

- **Flood risk.** Here we model the aggregate annual loss data introduced in Section 1.3.4 (given as a percentage of the building value) for Germany and the UK. All presented derivative plots for Germany in Figure 4.27 based on the Pareto, log-normal, and Weibull QQ-plots ultimately are decreasing, while for the UK in Figure 4.26 the decrease in the Weibull derivative plot is small and this plot is closest to being constant when $\log x > -10$. The systematic decrease in the different estimators of γ with increasing threshold, together with the P-values of the T_B test for upper-truncation does indicate some evidence for a truncated Pareto tail. Indeed, for both countries the truncated Pareto model fits well. The estimates $\hat{T}$ of the right truncation point T are situated around 0.25 for the UK and 0.35 for Germany. However, for the UK data, a Weibull fit provides a valid alternative.
- **Earthquake risk.** We consider recent *magnitude data* of the 200 largest earthquakes in the Groningen area (the Netherlands) which are caused by gas extraction. In Figure 4.28 (top left), we present the exponential QQ-plot. A linear pattern is visible for a large section of the magnitudes data, while some concave curvature appears at the largest values. Along the Gutenberg–Richter (1956) law the magnitudes of independent earthquakes are drawn from a doubly truncated exponential distribution

$$P(M > m) = \frac{e^{-\lambda m} - e^{-\lambda T_M}}{e^{-\lambda m_0} - e^{-\lambda T_M}}, \quad m_0 < m < T_M.$$

  Kijko and Singh [487] provide a review of the vast literature on estimating the maximum possible magnitude T_M. The energy E released by earthquakes, expressed in megaJoules, relates to the magnitude M by

$$M = \log_{10}(E/2)/1.5 + 1.$$

  When transforming the magnitude data back to the energy scale, the Gutenberg–Richter model predicts a truncated Pareto tail. In Figure 4.28, plotting the Hill estimates we observe a systematic decrease with decreasing k, while the moment and ML-GPD estimators tend to -1 near $k = 1$. The estimates of $\hat{\gamma}_k^T$ stay rather stable at a level $\hat{\gamma} = 2$. The P-values of the T_B test for upper-truncation are boundary significant at significance level 0.05 for $k \in (30, 70)$. The amount of truncation is estimated around $\hat{D}_T \in (0.01, 0.02)$. The goodness of fit of the truncated Pareto fit is illustrated on the Pareto QQ-plot of the energy data where the truncated Pareto-model is fitted based on the top 50 values. The maximum magnitude $\hat{T}_M = \log_{10}\left(\hat{T}_E/2\right)/1.5 + 1$ is then estimated at 3.75 for the Groningen area.

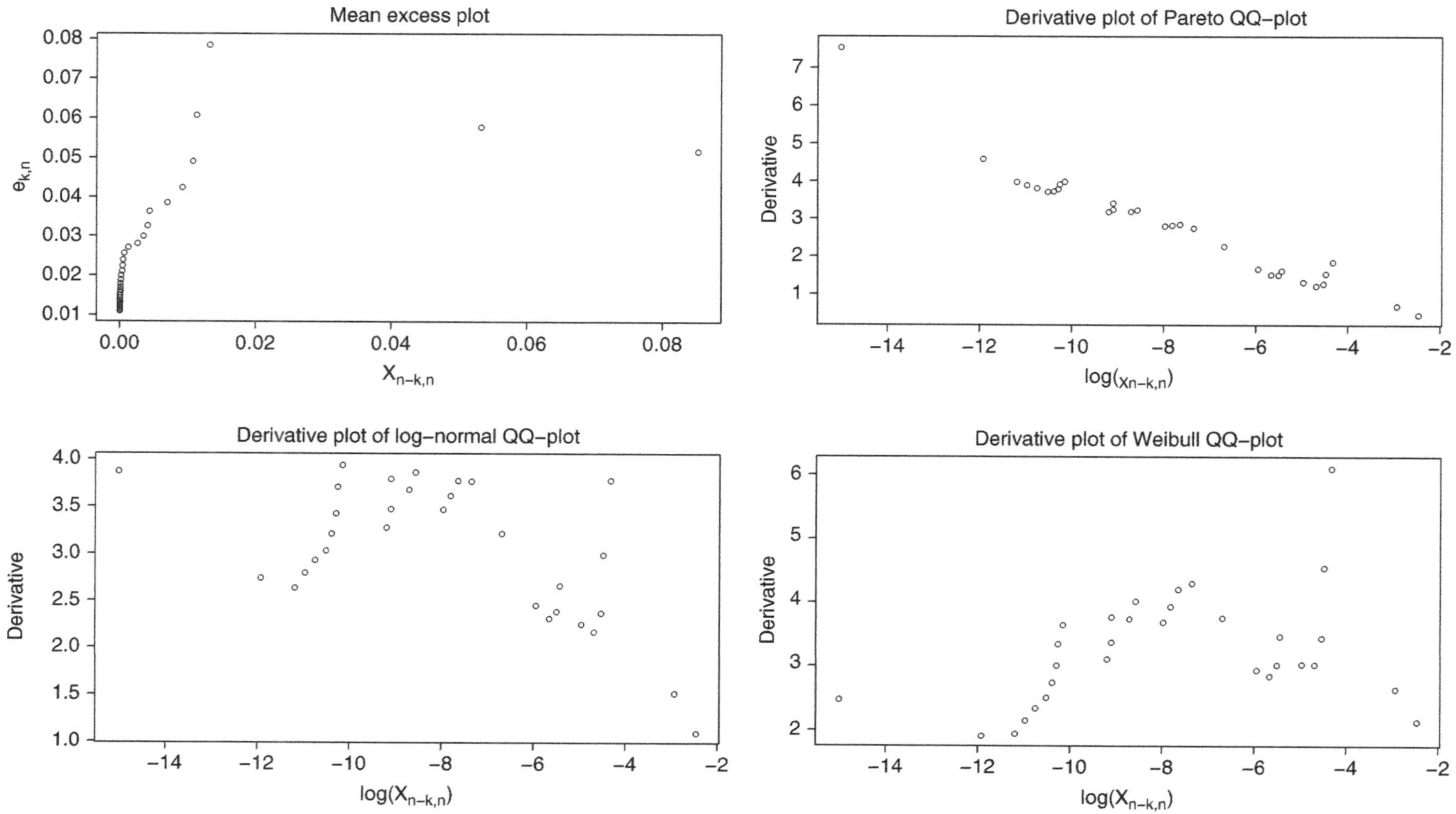

Figure 4.26 UK flood loss data: mean excess plot $(x_{n-k,n}, e_{k,n})$ (top left); Hill plot $(\log x_{n-k,n}, H_{k,n})$ (top right); log-normal derivative plot $(\log x_{n-k,n}, H_{k,n}/N_{k,n})$ (second line left); Weibull derivative plot $(\log x_{n-k,n}, H_{k,n}/W_{k,n})$ (second line right); γ estimates (third line left); P-values of T_B test (third line right); endpoint estimates $\hat{T}_k$ (bottom left); Pareto QQ-plot with truncated Pareto fit (full line) and Pareto fit (dashed line) (bottom right).

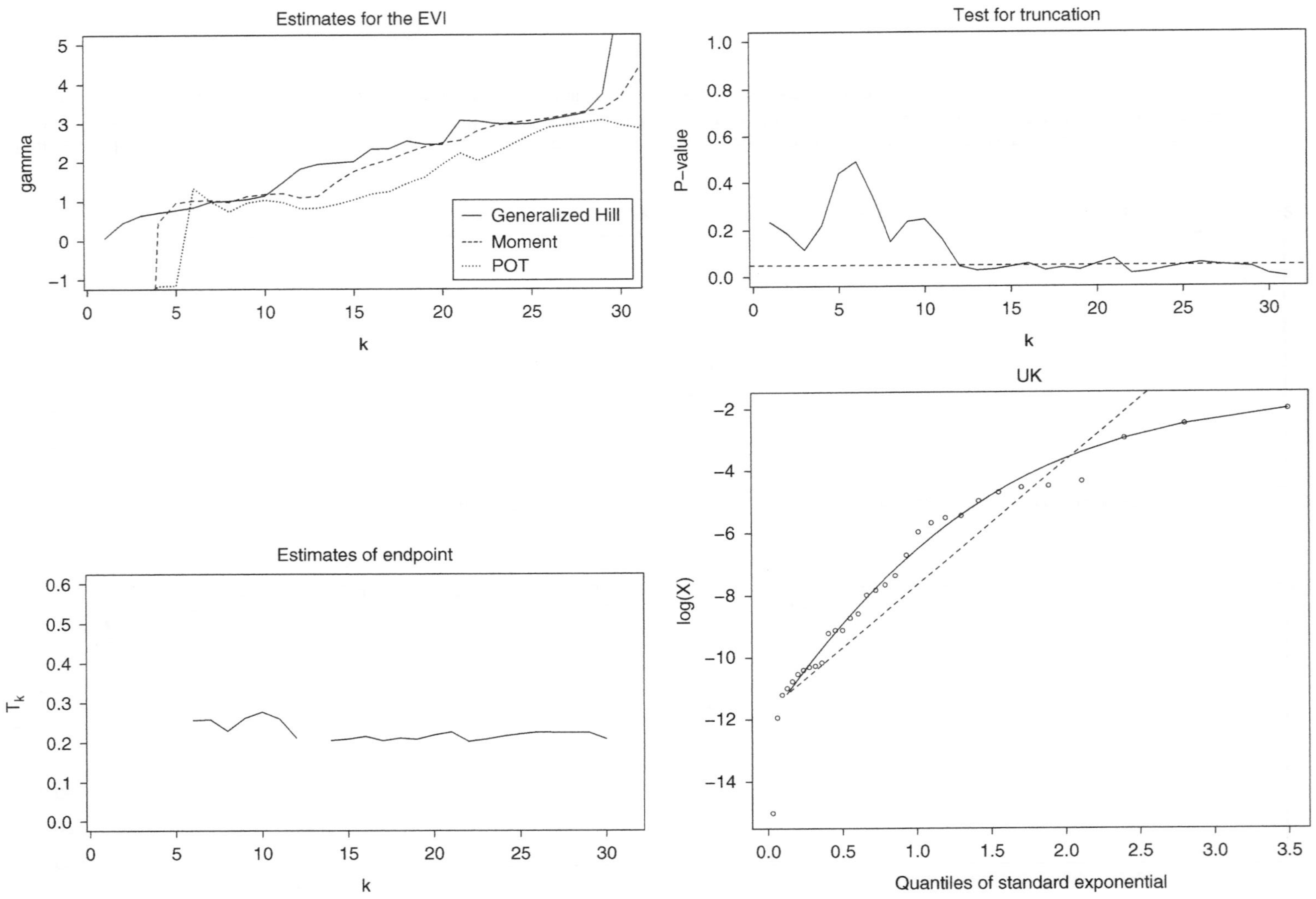

Figure 4.26 (Continued)

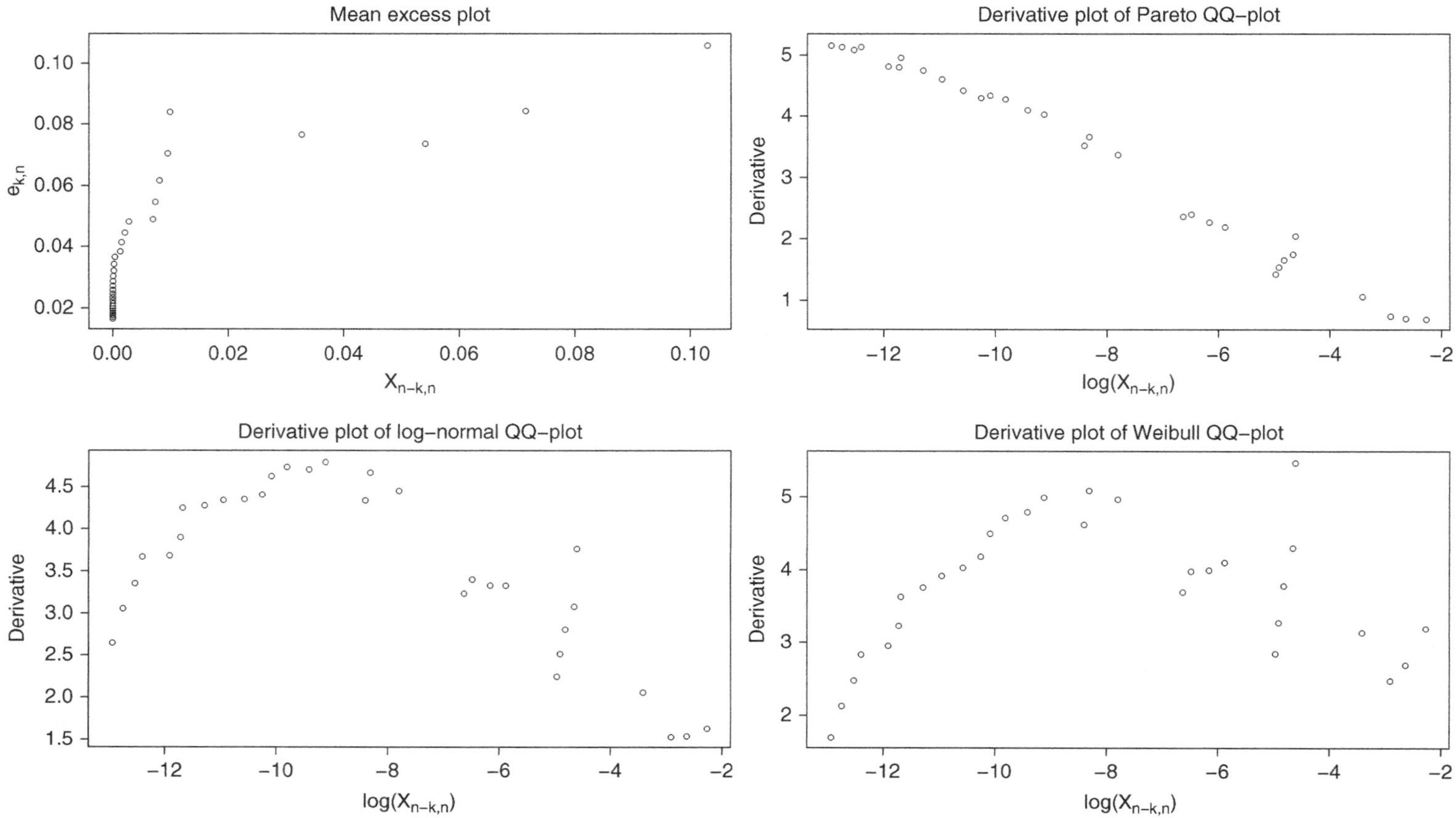

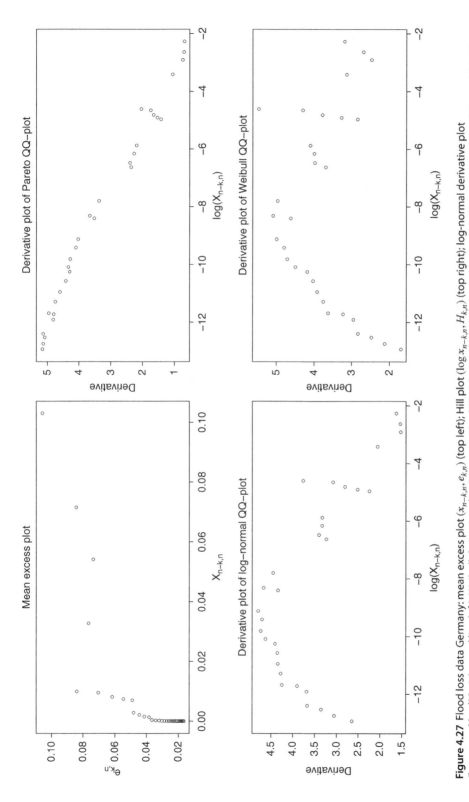

Figure 4.27 Flood loss data Germany: mean excess plot $(x_{n-k,n}, e_{k,n})$ (top left); Hill plot $(\log x_{n-k,n}, H_{k,n})$ (top right); log-normal derivative plot $(\log x_{n-k,n}, H_{k,n}/N_{k,n})$ (second line left); Weibull derivative plot $(\log x_{n-k,n}, H_{k,n}/W_{k,n})$ (second line right); γ estimates (third line left); P-values of T_B test (third line right); endpoint estimates $\hat{T}_k$ (bottom left); Pareto QQ-plot with truncated Pareto fit (full line), and Pareto fit (dashed line) (bottom right).

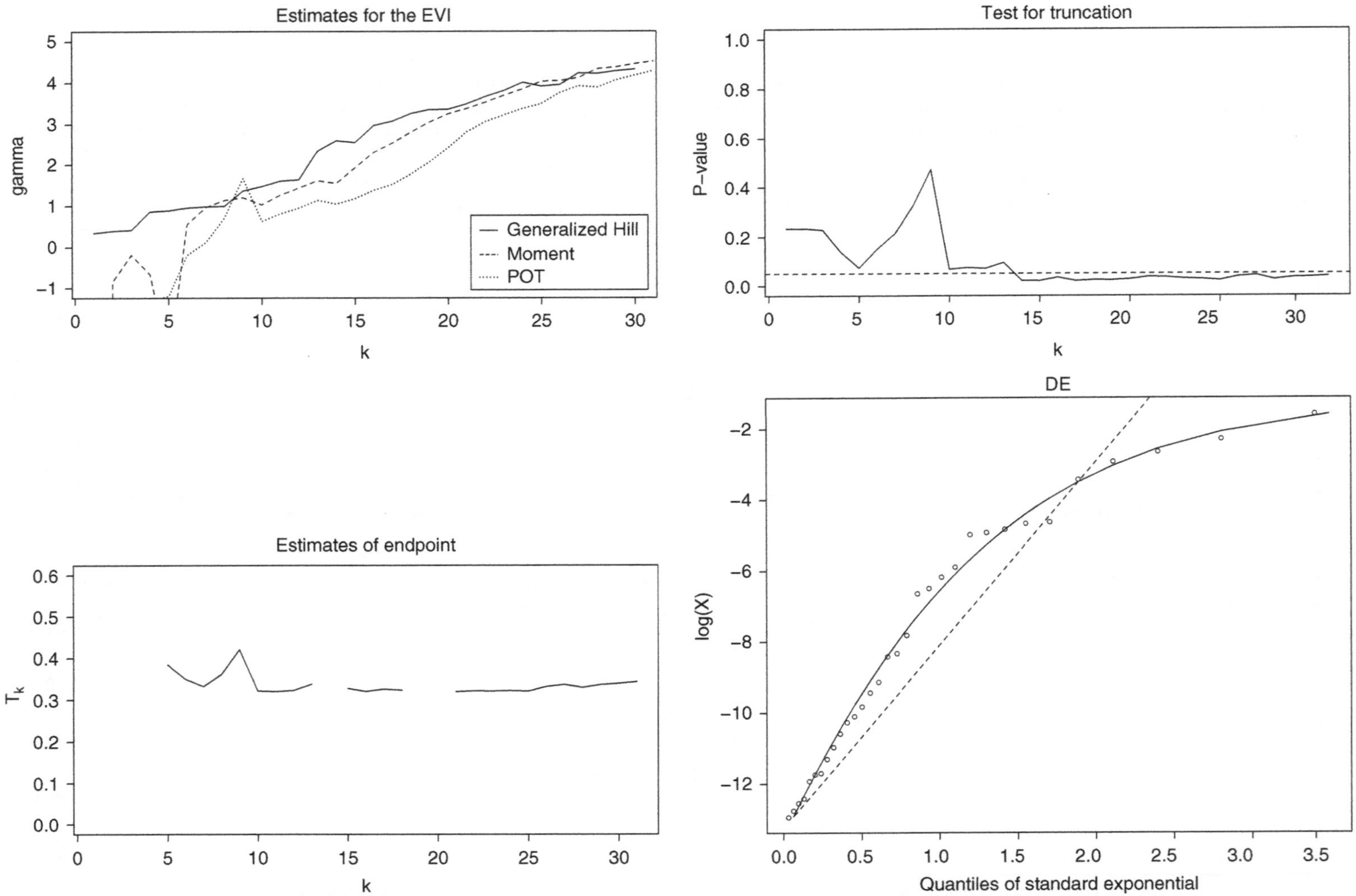

Figure 4.27 (Continued)

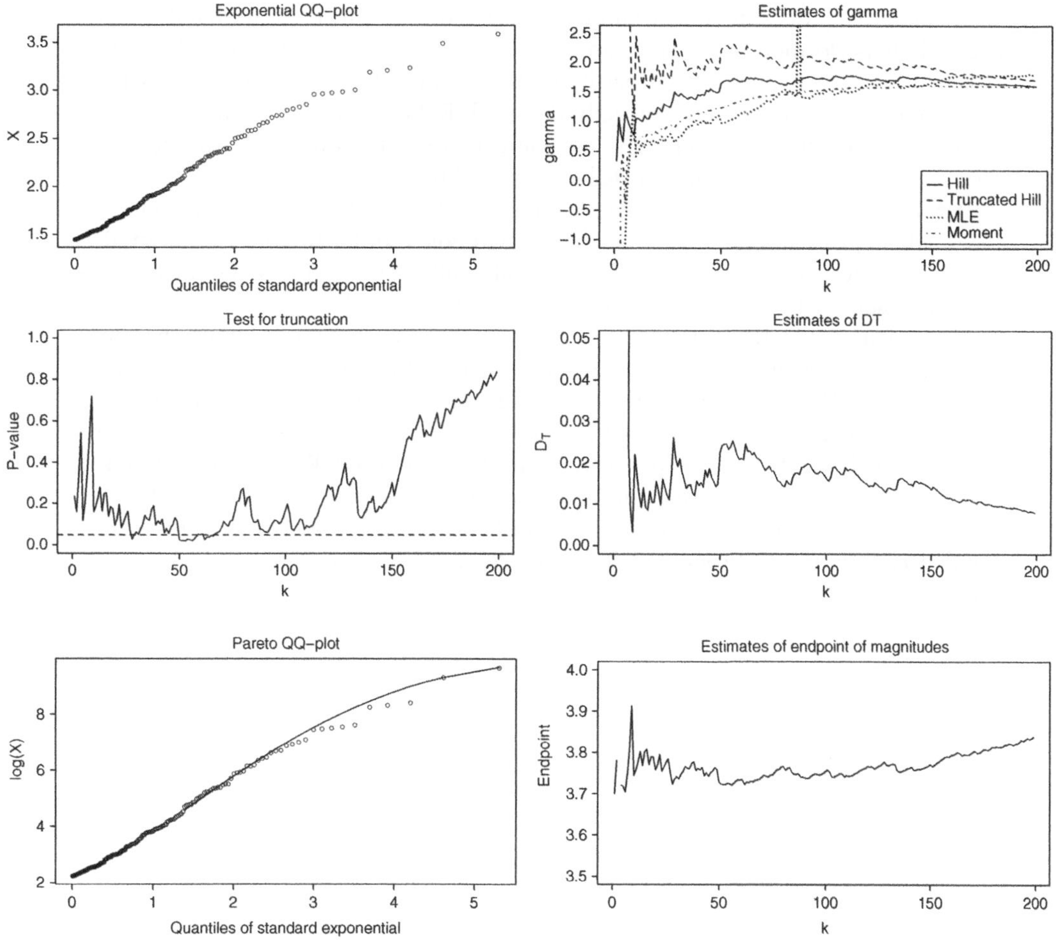

Figure 4.28 Earthquake magnitude data from Groningen area. Exponential QQ-plot based on magnitudes (top left); estimates of γ (top right); T_B P-value plot (middle left); $\hat{D}_T$ estimates (middle right); Pareto QQ-plot of energy values with truncated Pareto fit (bottom right); estimates of maximum magnitude $\hat{T}_M$ (bottom right).

4.8 Notes and Bibliography

In case of the Gumbel domain of attraction with $\gamma = 0$, EVA based on fitting a generalized Pareto distribution to POT values is known to exhibit slow convergence rates in many cases. To this end more specific models have been proposed, for example El Methni et al. [320] and De Valk and Cai [262].

In the last few decades some papers have appeared concerning robust estimation methods. Robust methods can improve the quality of extreme value data analysis by providing information on influential observations, deviating substructures and possible mis-specification of a model while guaranteeing good statistical properties over a whole set of underlying distributions around the assumed one. On the other hand an EVA precisely is performed to consider and emphasize the role of extremes. Hence in a risk management context it can hardly be the purpose to delete the most extreme observations when they were correctly reported. Robust and non-robust estimators then yield different scenarios for risk assessment should be compared. An interesting discussion on this can be found in Dell'Aquila and Embrechts [270].

EVA is an active field of research. A notable recent contribution is Naveau et al. [587], which gives an alternative to splicing methods in order to produce full models in a hydrological context. In De Valk [261] and Guillou et al. [415], some further new modelling approaches in the multivariate case are introduced.

A Bayesian approach to estimate the total cost of claims in XL reinsurance has been covered in Hesselager [441]. For the estimation of the Pareto index in XL treaties, see Reiss et al. [644]. Leadbetter [529] studied the connection between tail inference and high-level exceedance modelling, which is relevant for the XL case. Examples of early statistical analyses of large fire losses are Ramachandran [639], Ramlau-Hansen [640], and Corradin et al. [227]. Resnick [646] also studied the Danish fire insurance data set. Data for coverages of homes are analyzed in Grace et al. [404]. For glass losses, see Ramlau-Hansen [640]. Property reinsurance for the USA is covered in Gogol [394], for example.

5

Models for Claim Counts

Models and statistics for the number of claims can be built on counts in either discrete or continuous time. In practice, actuaries often look at the aggregate number of claims that have occurred within a fixed time interval (e.g., one calendar year) rather than studying the full stochastic process developing over time.

Whereas in real-life applications one only obtains information in discrete time, for instance with information about the date of the claim events, continuous time modelling can be a more flexible and tractable tool that also allows the implementation of stylized features (such as lack of memory) in a transparent and elegant way. From the continuous time stochastic process approach one can then go back to implications for the discrete time intervals. For this reason, in this chapter we start with the general mathematical treatment of claim number modelling in continuous time. Section 5.4 will then treat the discrete case, and Section 5.5 deals with statistical procedures to fit such counting models to data for both approaches. Finally, in Section 5.6.1 we will discuss how to move from models for a number of claims for the insurer to the ones for the reinsurer.

Assume that claims happen at specific time points $\{T_1, T_2, \dots, T_n, \dots\}$ that form an increasing sequence of necessarily dependent random variables. The claim counting process $\{N(t), t \geq 0\}$ can then be defined in terms of the claim instants by

$$\{N(t) = n\} = \{T_n \leq t < T_{n+1}\} .$$

5.1 General Treatment

A stochastic process $N = \{N(t); t \geq 0\}$ that counts the number of claims up to time t has to be a *counting process*. This means that we require the process to satisfy the following four and evident conditions:

(i) $N(t) \geq 0$
(ii) $N(t)$ is integer valued
(iii) If $s < t$, then $N(s) \leq N(t)$
(iv) for $s < t$, $N(t) - N(s)$ equals the number of claims that have occurred in the time interval $(s, t]$.

Reinsurance: Actuarial and Statistical Aspects, First Edition.
Hansjörg Albrecher, Jan Beirlant and Jozef L. Teugels.

We note that the sample paths of a counting process are non-decreasing and right-continuous. In many cases one assumes that the jumps are only of size one so that multiple claims at the same time instance are excluded.

5.1.1 Main Properties of the Claim Number Process

The distribution of $N(t)$, the number of claims up to time t, can be given in explicit form

$$p_n(t) = \mathbb{P}(N(t) = n), \; n \in \mathbb{N}.$$

It can also be introduced via the *(probability) generating function*

$$Q_t(z) := \mathbb{E}(z^{N(t)}) = \sum_{n=0}^{\infty} p_n(t) z^n, \tag{5.1.1}$$

which is at least defined for $|z| \leq 1$, but often the radious of convergence is larger. Apart from the expression $p_n(t)$ we will introduce a (c.d.f) of $N(t)$, that is, the probability of at most r claims up to time t:

$$F_{N(t)}(r) := \mathbb{P}(N(t) \leq r) = \sum_{n=0}^{r} p_n(t). \tag{5.1.2}$$

Among the main characteristics of a claim number process we should mention a variety of possible moments and derived quantities. The function $Q_t(z)$ can be used to evaluate these consecutive moments of $N(t)$. Here are a few of the most important qualitative indices.

(i) For $|z| < 1$, the rth order derivative of Q_t with respect to z is given by

$$Q_t^{(r)}(z) = \sum_{k=r}^{\infty} \frac{k!}{(k-r)!} p_k(t) z^{k-r}.$$

In terms of expectations this reads as

$$Q_t^{(r)}(z) = r!\, \mathbb{E}\left\{ \binom{N(t)}{r} z^{N(t)-r} \right\}. \tag{5.1.3}$$

The *factorial moments* of $N(t)$ can be obtained from this by letting $z \uparrow 1$:

$$Q_t^{(r)}(1) = \mathbb{E}\left([N(t)][N(t)-1]...[N(t)-r+1]\right) = r!\, \mathbb{E}\binom{N(t)}{r}. \tag{5.1.4}$$

In many cases $Q_t(z)$ will be analytic at $z = R > 1$. Then for $0 \leq u < R-1$

$$Q_t(1+u) = \sum_{r=0}^{\infty} \frac{u^r}{r!} Q_t^{(r)}(1) = \sum_{r=0}^{\infty} u^r \mathbb{E}\binom{N(t)}{r}. \tag{5.1.5}$$

(ii) From (5.1.4) one easily finds the successive moments of $N(t)$. Most importantly we begin with the *mean number of claims*, which is obtained from the first derivative of $Q_t(z)$ at 1:

$$\mathbb{E}(N(t)) = Q_t'(1). \tag{5.1.6}$$

(iii) The *variance of the number of claims* is obtained from (5.1.4) with $r = 1$ and $r = 2$ and gives

$$\mathrm{Var}(N(t)) = Q_t''(1) + Q_t'(1) - \{Q_t'(1)\}^2. \tag{5.1.7}$$

(iv) The *index of dispersion* is defined by

$$I_{N(t)} = \frac{\mathrm{Var}(N(t))}{\mathbb{E}(N(t))}. \tag{5.1.8}$$

This quantity is popular among actuaries. As we will see soon, it is equal to 1 for the Poisson process. As such, the value of the index of dispersion makes it possible to call a claim number process *overdispersed* (with respect to the Poisson case) if its index is greater than 1. The concept *underdispersed* is defined similarly. The index of dispersion has been introduced to indicate that claim numbers in a portfolio show a greater volatility than one could expect from a Poisson process; in particular overdispersion appears regularly in actuarial portfolios.

In the following we collect a broad set of models. Some of them – like the Poisson and the negative binomial model – are extremely popular while others are modifications intended to allow better data fitting. Several examples are based on an underlying probabilistic structure that has a natural actuarial context.

5.2 The Poisson Process and its Extensions

We start with some of the most popular examples of claim number processes, all based on Poisson processes. Standard references where more mathematical details and derivations can be found are Mikosch [577], Rolski et al. [652], and Grandell [406].

5.2.1 The Homogeneous Poisson Process

A homogeneous Poisson process $\tilde{N} = \tilde{N}_\lambda = \{\tilde{N}(t); t \geq 0\}$ with intensity (or rate) $\lambda > 0$ is a counting process that satisfies the following properties:

(i) *Start at 0: $\tilde{N}(0) = 0$ almost surely.*
(ii) *Independent increments: for any $0 \leq t_0 < t_1 < \ldots < t_n < \infty$, $n \in \mathbb{N}$, the increments $\Delta\tilde{N}_{t_i} := \tilde{N}(t_i) - \tilde{N}(t_{i-1})$ are mutually independent for $i = 1, \ldots, n$.*
(iii) *Poisson increments: for any $0 \leq s < t < \infty$*

$$\tilde{N}(t) - \tilde{N}(s) \sim \textit{Poisson}(\lambda(t-s)).$$

Without any doubt the *homogeneous Poisson process* $\tilde{N}(t)$ is the most popular among all claim number processes in the actuarial literature. Because of its benchmark character we deal with it first. We then offer a number of generalizations in which the homogeneous Poisson process acts as the main building block.

From the definition it follows that the increments are independent and also stationary, that is

$$\tilde{N}(t) - \tilde{N}(s) =_d \tilde{N}(t-s), \; 0 \leq s < t < \infty,$$

with

$$p_n(t) = e^{-\lambda t}\frac{(\lambda t)^n}{n!}, \qquad n \in \mathbb{N}.$$

Consequently

$$F_{\tilde{N}(t)}(r) = \frac{1}{r!}\int_{\lambda t}^{\infty} e^{-w} w^r dw,$$

while also $Q_t(z) = \exp\{-\lambda t(1-z)\}$ so that

$$Q_t^{(r)}(z) = e^{-\lambda t(1-z)}(\lambda t)^r.$$

It further follows that

$$\mathbb{E}(\tilde{N}(t)) = \lambda t \;\; ; \;\; \mathrm{Var}(\tilde{N}(t)) = \lambda t.$$

Properties of the homogeneous Poisson process $\tilde{N}(t)$ and jump arrival times

$$T_j := \inf\{t > 0 : \tilde{N}(t) \geq j\}, \; j \in \mathbb{N}$$

are as follows:

1. *The inter-arrival times* $W_j := T_j - T_{j-1}$, $j \in \mathbb{N}$ where $T_0 := 0$, are i.i.d. exponential random variables with mean $1/\lambda$:

$$F_{W_j}(x) = 1 - \exp(-\lambda x), \; j \in \mathbb{N}, \; x \geq 0.$$

2. *Order statistics property*: the conditional distribution of $(T_1, \ldots, T_n)$ given $\{\tilde{N}(t) = n\}$ for some $n \in \mathbb{N}$ equals the distribution of the order statistics $U_{1,n} \leq U_{2,n} \leq \ldots \leq U_{n,n}$ of independent uniform $(0, t)$ distributed random variables (e.g., see Sato [666]).

3. *Memoryless property*: it follows from the first property that the distribution of the time until the next arrival is independent of the time t we have already been waiting for that arrival:

$$\mathbb{P}(W_j > t + y | W_j > t) = \mathbb{P}(W_j > y) \quad \text{for all } y, t \geq 0.$$

This property gives the homogeneous Poisson process a special role among all claim number processes and may be seen as one of the main reasons for its popularity from a modelling perspective.

4. *Jump sizes*: as $t \downarrow 0$

$$\mathbb{P}(\tilde{N}(t) = k) = \begin{cases} 1 - \lambda t + o(t), & \text{if } k = 0, \\ \lambda t + o(t), & \text{if } k = 1, \\ o(t), & \text{otherwise,} \end{cases}$$

where the notation $f(t) = o(t)$ means $\lim_{t \to 0} f(t)/t = 0$. Hence, at any point in time, no more than one claim can occur with positive probability.

Finally, note that the homogeneous Poisson process can be constructed based on the sequence of i.i.d. exponential random variables (with rate $\lambda > 0$) $\{W_j, j \geq 1\}$, that is,

$$\tilde{N}(t) = \sum_{j=1}^{\infty} 1_{\{W_1 + \ldots + W_j \leq t\}}.$$

5.2.2 Inhomogeneous Poisson Processes

A more general definition of the Poisson process goes as follows:

A general Poisson process is a stochastic process $N = N_\mu = \{N_\mu(t); t \geq 0\}$ that satisfies the following properties:

(i) Càdlàg paths: the paths of N are almost surely càdlàg functions, that is right-continuous with existing left limits.

(ii) Start at 0: $N_\mu(0) = 0$ almost surely.

(iii) Independent increments: for any $0 \leq t_0 < t_1 < \ldots < t_n < \infty$, $n \in \mathbb{N}$, the increments $\Delta N_{t_i} := N_\mu(t_i) - N_\mu(t_{i-1})$ are mutually independent for $i = 1, \ldots, n$.

(iv) Poisson increments: for a càdlàg function $\mu : [0, \infty) \to [0, \infty)$ with $\mu(t) < \infty$ for all $t \geq 0$, the increments have the following distribution:

$$N_\mu(t) - N_\mu(s) \sim \mathit{Poisson}(\mu(t) - \mu(s)), \quad 0 \leq s < t < \infty.$$

The non-decreasing function μ is called the mean-value function of N_μ.

It is natural to call μ the mean-value function as it describes the expectation of the process increments:

$$\mathbb{E}(N_\mu(t) - N_\mu(s)) = \mu(t) - \mu(s), \quad 0 \leq s < t < \infty.$$

The homogeneous Poisson process described in the preceding subsection is the special case of a linear mean-value function,

$$\mu(t) = \lambda t, \quad t \geq 0,$$

for some intensity $\lambda > 0$.

An inhomogeneous Poisson process can be defined through a deterministic time-change of a homogeneous process.

1. Let $\tilde{N}_1(t)$ be a homogeneous Poisson process with intensity $\lambda = 1$ and let $\mu(.)$ be a valid mean-value function, then the process defined by $\{\tilde{N}_1(\mu(t)),\ t \geq 0\}$ is an inhomogeneous Poisson process with mean-value function μ.
2. Conversely, every inhomogeneous Poisson process N has a representation as a time-changed Poisson process, that is, $\{N_\mu(\mu^{-1}(t)),\ t \geq 0\}$ is a standard homogeneous Poisson process.

In other words, the intensity changing over time can equivalently be interpreted as going through time with constant intensity but varying speed. For this reason μ is also referred to as *operational time*: whereas time runs linearly for a homogeneous Poisson process, it can be seen to speed up or slow down according to μ for an inhomogeneous Poisson process.

If μ is continuous and strictly increasing with $\mu(t) \to \infty$, then the inverse function μ^{-1} exists and the inhomogeneous Poisson process can be converted back to a homogeneous one with intensity 1 by a time change using μ^{-1}.

In many applications it is even assumed that μ is absolutely continuous, that there exists a non-negative function $\lambda(.)$ such that

$$\mu(t) = \int_0^t \lambda(s)ds, \quad t \geq 0,$$

where the function $\lambda(.)$ is called the *intensity function*. Note that for a homogeneous Poisson process with intensity $\lambda > 0$ we have $\lambda(t) \equiv \lambda,\ t \geq 0$.

5.2.3 Mixed Poisson Processes

A far-reaching generalization of the ordinary Poisson process is obtained when the parameter λ is replaced by a random variable Λ with *mixing* or *structure distribution function* F_Λ. This increases the flexibility of the model due to additional parameters while keeping some of the main properties of the Poisson case. A common interpretation of this model extension is given in terms of a counting process which consists of two or more different sub-processes that individually behave as a Poisson process with a specific intensity value (e.g., with a heterogeneous group of policyholders each producing claims according to a simple Poisson process but with different intensities). Note, however, that here for each sample path of $N(t)$ the realization of Λ is chosen once at the beginning (i.e., in the above interpretation all counts then come from one of these sub-processes).

A mixed Poisson process can be represented as

$$N_\Lambda(t) = \tilde{N}(\Lambda t),\ t \geq 0,$$

where $\tilde{N}(t)$ is a homogeneous Poisson process and Λ is a positive random variable, and

$$p_n(t) = \int_0^\infty e^{-\lambda t}\frac{(\lambda t)^n}{n!}dF_\Lambda(\lambda), \tag{5.2.9}$$

while the generating function is given by

$$Q_t(z) = \int_0^\infty e^{-\lambda t(1-z)}\, dF_\Lambda(\lambda). \tag{5.2.10}$$

We evaluate the characteristics of the mixed Poisson process. It follows from the above equations that

$$Q_t^{(r)}(z) = \int_0^\infty e^{-\lambda t(1-z)}(\lambda t)^r dF_\Lambda(\lambda).$$

The latter relations immediately yield a linearly increasing expected growth

$$\mathbb{E}(N_\Lambda(t)) = \int_0^\infty \lambda t dF_\Lambda(\lambda) = t\mathbb{E}(\Lambda),$$

if $\mathbb{E}(\Lambda)$ exists. Similarly,

$$\mathrm{Var}(N_\Lambda(t)) = t\mathbb{E}(\Lambda) + t^2\mathrm{Var}(\Lambda),$$

and more generally

$$\mathbb{E}\binom{N_\Lambda(t)}{r} = \frac{t^r}{r!}\mathbb{E}(\Lambda^r)$$

(again given that the respective moments of Λ exist). The expression for the variance shows that among all mixed Poisson processes the Poisson process has the smallest variance. Actually the variance will be linear if and only if Λ is degenerate and this happens exactly for the Poisson case. The index of dispersion here equals

$$I_{N_\Lambda(t)} = 1 + t\,\frac{\mathrm{Var}(\Lambda)}{\mathbb{E}(\Lambda)},$$

which is constant in time for the homogeneous Poisson process. Since this index is always greater than 1, mixed Poisson processes are natural candidates to model overdispersed claim number processes.

Another way of highlighting the role of the random variable Λ is the observation that for $s \geq 0$,

$$\mathbb{E}\left\{e^{-s\frac{N_\Lambda(t)}{t}}\right\} = \int_0^\infty e^{-\lambda t(1-e^{-s/t})}\, dF_\Lambda(\lambda),$$

which for $t \uparrow \infty$ tends to the Laplace transform of $\hat{F}_\Lambda(s) = \mathbb{E}(e^{-s\Lambda})$ of F_Λ. This in turn implies that

$$\frac{N_\Lambda(t)}{t} \to_d \Lambda, \quad t \uparrow \infty. \tag{5.2.11}$$

Again, conditional on $N_\Lambda(t) = n$, the occurrence times of the n claim events are uniformly distributed on $(0, t)$, that is, mixed Poisson processes also satisfy the order statistics property. It also essentially characterizes the mixed Poisson processes (see [492], page 160). However, for a non-degenerate mixing variable, a mixed Poisson process no longer has independent increments (but conditionally independent ones).

The mixed Poisson process $N_\Lambda = \{N_\Lambda(t)\ t \geq 0\}$ was introduced to actuaries by Dubourdieu [310]. For a very thorough treatment of mixed Poisson processes and their stochastic properties, see Grandell [406]. We give a few special cases that have found their way into the actuarial literature.

Examples

(a) A mild extension of the classical Poisson process leads to *discrete mixtures of Poisson processes*. Here the structure distribution F_Λ has at most countably many different points of increase $\{\lambda_i, i \in \mathbb{N}\}$ with $\lambda_i > 0$, where $a_i := F_\Lambda(\lambda_i) - F_\Lambda(\lambda_i-) > 0$ and $\sum_{i=0}^{\infty} a_i = 1$. Then

$$Q_t^{(r)}(z) = \sum_{i=0}^{\infty} a_i e^{-\lambda_i t(1-z)} (\lambda_i t)^r .$$

We give a set of examples.

(i) For the special case of two points of increase, see Seal [691]. Here

$$p_n(t) = p \frac{(\lambda_1 t)^n}{n!} e^{-\lambda_1 t} + q \frac{(\lambda_2 t)^n}{n!} e^{-\lambda_2 t}$$

where $p + q = 1, 0 < p < 1$.

In order to illustrate the difference between the homogeneous Poisson process and a mixed Poisson process of this kind, in Figure 5.1 we show simulated paths from both models with the same expected number of claims per time unit. The mixture process shows two clusters of Poisson paths with different slopes.

(ii) If we take a binomial distribution as mixing law then $\lambda_i = i$ while $a_i = \binom{M}{i} p^i q^{M-i}, i = 0, 1, \ldots, M$. Then the generating function is

$$Q_t(z) = (p\, e^{-t(1-z)} + q)^M ,$$

which yields $\mathbb{E}(N(t)) = tpN$ and $\mathrm{Var}(N(t)) = t^2 pqN + tpN$.

(iii) If the mixing is according to a Poisson distribution, that is, $\lambda_i = i$ and $a_i = e^{-\mu}\frac{\mu^i}{i!}$, $i \geq 0$, then

$$Q_t(z) = e^{-\mu}(e^{\mu e^{-t(1-z)}} - 1).$$

This mixed Poisson process is then of the *Neyman-type A*. Here $\mathbb{E}(N(t)) = \mu t$ and $\mathrm{Var}(N(t)) = \mu t(1 + t)$.

(b) Another very popular special case of the mixed Poisson process is the *Pólya* (also *Pascal*) *process,* obtained by choosing a gamma distribution for the random variable Λ, that is,

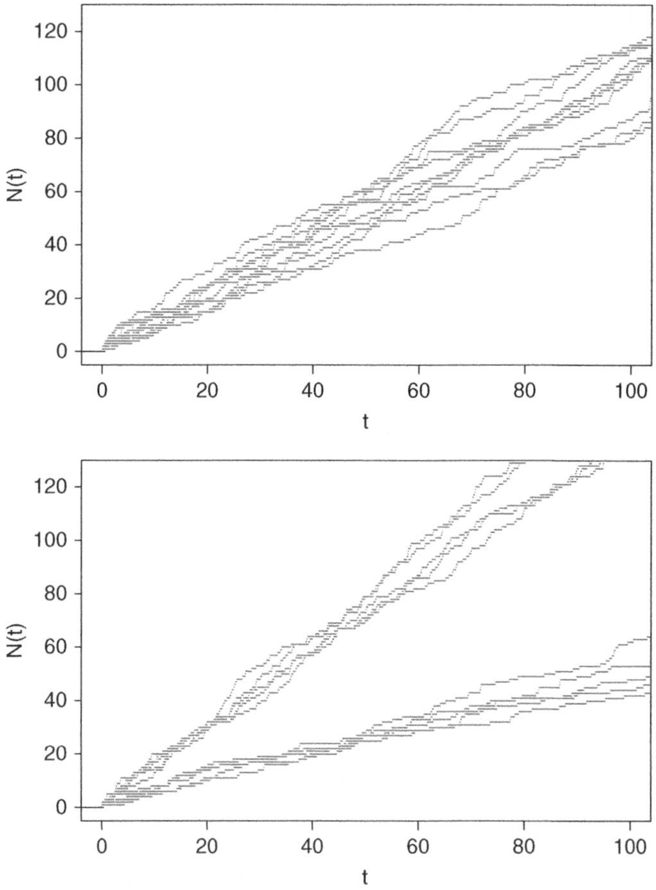

Figure 5.1 10 simulated Poisson paths on $[0, 100]$: homogeneous Poisson $\tilde{N}_1$ (top) and discrete mixed Poisson with $\lambda_1 = 0.5, \lambda_2 = 1.5$, and $p = 0.5$ (bottom).

$$dF_\Lambda(\lambda) = \frac{b^\alpha}{\Gamma(\alpha)} e^{-b\lambda} \lambda^{\alpha-1} d\lambda.$$

By a simple calculation we find the distribution of the number of claims up to time t to be

$$p_n(t) = \binom{\alpha + n - 1}{n} \left(\frac{b}{t+b}\right)^\alpha \left(\frac{t}{t+b}\right)^n, \tag{5.2.12}$$

which is a negative binomial distribution. In this form the result is due to Thyrion [744]. For some decades the negative binomial distribution has found applications in actuarial contexts. Thanks to the fact that the distribution contains two parameters, it naturally allows greater flexibility for data fitting than the single parameter Poisson distribution.

Here are the other characteristics of the process. First of all,

$$F_{N(t)}(r-1) = \frac{\Gamma(r+\alpha)}{\Gamma(\alpha)} \int_0^{t/b} \frac{u^{r-1}}{(1+u)^{r+\alpha}} du.$$

Then the probability generating function is given by $Q_t(z) = \left(\frac{b}{b+t(1-z)}\right)^{\alpha}$ from which

$$Q_t^{(r)}(z) = \frac{\Gamma(\alpha+r)}{\Gamma(\alpha)} \frac{b^{\alpha} t^r}{(b+t(1-z))^{\alpha+r}}$$

follows. The first few moments are given by

$$\mathbb{E}(N(t)) = \alpha \, \frac{t}{b} \; ; \; \mathrm{Var}(N(t)) = \alpha \, \frac{t}{b} \, \left(1 + \frac{t}{b}\right).$$

The Pólya process has been the prototype model for counting processes that show overdispersion, a property shared by many actual insurance data. Another reason why the Pólya process is popular is that one of its limiting forms is the Poisson process and so mentally also the latter seems close to the Pascal case. Indeed, if $\alpha \to \infty, b \to \infty$ while $\frac{\alpha}{b} \to \lambda$, then the Pólya process turns into the Poisson process.

(c) In [700], Sichel introduced a distribution that can be obtained from a Poisson distribution by mixing it with a general inverse Gaussian distribution of the form

$$h(\lambda) = \frac{dF_{\Lambda}(\lambda)}{d\lambda} = \frac{\mu^{-\eta} \lambda^{\eta-1}}{2K_{\eta}\left(\frac{\mu}{\beta}\right)} \exp\left\{-\frac{\lambda^2+\mu^2}{2\beta\lambda}\right\}$$

where the three parameters β, η and μ are non-negative. The function K_{θ} is the *modified Bessel function of the third kind* (or *MacDonald function*). An integral property of the Bessel function yields the following expression for the generating function

$$Q_t(z) = (1+2\beta t(1-z))^{-\eta/2} \frac{K_{\eta}\left(\frac{\mu}{\beta}\sqrt{1+2\beta t(1-z)}\right)}{K_{\eta}\left(\frac{\mu}{\beta}\right)}.$$

The outcome is called the *Sichel process.* From properties of the Bessel function one can derive an expression for the factorial moments in that

$$Q_t^{(r)}(1) = (\mu t)^r \frac{K_{\eta+r}\left(\frac{\mu}{\beta}\right)}{K_{\eta}\left(\frac{\mu}{\beta}\right)}.$$

The case $\eta = -1/2$ is particularly interesting since then the general inverse Gaussian distribution simplifies to the classical inverse Gaussian distribution. The resulting mixed process is called the *Poisson-inverse Gaussian process.* Wilmot [783] illustrates the usefulness of the Poisson-inverse Gaussian process as an alternative to the negative binomial distribution. In particular the distribution has been fitted to automobile frequency data. The structure function has a density given by

$$h(\lambda) = \frac{\mu}{\sqrt{2\pi b\lambda^3}} e^{-\frac{(\lambda-\mu)^2}{2b\lambda}},$$

while the generating function is given by the expression

$$Q_t(z) = \exp\left\{-\frac{\mu}{b}\left(\sqrt{1+2bt(1-z)}-1\right)\right\}.$$

(d) We shortly mention a few other choices that have been applied in specific portfolios.

(i) The *log-zero Poisson process* was, for example, used by Douglas [306] with

$$Q_t(z) = 1 - \theta - c\theta \log(1 - pe^{t(z-1)}).$$

(ii) The *Poisson-beta processes* take a bounded or unbounded beta-density for h. For the general case where

$$h(\lambda) = \frac{(b-a)^{-(p+q)}}{B(p,q)}(\lambda - a)^{p-1}(b-\lambda)^{q-1}, a < \lambda < b$$

see, for instance, Albrecht [31]. Special cases are given in Kupper [516] and Willmot [782].

(iii) Similarly the *Poisson-gamma processes* depart from a gamma density. Ruohonen [660] advocates the *Delaporte distribution*, which is obtained as a mixed Poisson with as structure density a generalized gamma of the form

$$h(\lambda) = \frac{\beta^\alpha}{\Gamma(\alpha)}(\lambda-\gamma)^{\alpha-1}\exp\{-(\lambda-\gamma)\beta\}, \ \lambda > \gamma.$$

See Delaporte [269], Willmot et al. [789], and Schröter [684].

(iv) Kupper [516] uses the structure density

$$h(\lambda) = \frac{\beta^\alpha}{\Gamma(\alpha)}\lambda^{-(\alpha+1)}e^{-\frac{\beta}{\lambda}},$$

whose resulting probabilities can be expressed in terms of Bessel functions.

5.2.4 Doubly Stochastic Poisson Processes

Whereas in mixed Poisson processes the constant (static) intensity of a homogeneous Poisson process is randomized, in a doubly stochastic Poisson process the entire mean-value function of an inhomogeneous Poisson process is randomized.

Generalities on doubly stochastic Poisson processes. Let $\tilde{N} = \tilde{N}_\lambda = \{\tilde{N}(t); t \geq 0\}$ denote the homogeneous Poisson process with intensity λ. Independently, let

$M = \{M(t); t \geq 0\}$ be a stochastic process that has almost surely non-decreasing càdlàg paths with $M(0) = 0$ and $M(t) < \infty$ for $t < \infty$. Then the process $\{N(t) = \tilde{N}_\lambda(M(t)); t \geq 0\}$ is called a *doubly stochastic Poisson process directed by M*. Doubly stochastic Poisson processes are also called *Cox processes* based on the paper [229] where this concept appeared first in a general form (the first treatment of a specific form of such a process for actuarial purposes goes back even further, to Ammeter [36], see the Notes at the end of the chapter).

Hereafter, unless otherwise stated, we assume the intensity λ of the underlying homogeneous Poisson process to be 1.

If there exists a non-negative process $\Lambda = \{\Lambda(t); t \geq 0\}$ such that M has the representation

$$M(t) =_d \int_0^t \Lambda(s)ds, \; t \geq 0,$$

then Λ is called the intensity process.

If $M(t) = \Lambda(0) \cdot t$ for some non-negative random variable $\Lambda(0)$, then the transformed process is again a mixed Poisson process.

Given a sample path $\mu(t)$ of $M(t)$, it holds that for $s < t$

$$\begin{aligned}\mathbb{P}(N(t) - N(s) = k | M(t) = \mu(t), M(s) = \mu(s)) \\ = \frac{1}{k!}(\mu(t) - \mu(s))^k e^{-(\mu(t)-\mu(s))}.\end{aligned}$$

Note that we do not need information concerning the full path of Λ between s and t, but rather the values at s and t. This is not the case if we write this in terms of the intensity process $\{\Lambda(t); t \geq 0\}$:

$$\begin{aligned}\mathbb{P}(N(t) - N(s) = k | \Lambda(u) = \lambda(u), \; s \leq u \leq t) \\ = \frac{1}{k!}\left(\int_s^t \lambda(u)du\right)^k e^{-\int_s^t \lambda(u)du}.\end{aligned}$$

From the conditional distribution of $N(t)$ given $\{\Lambda(u), \; 0 \leq u \leq t\}$ we obtain

$$Q_t(z) = \mathbb{E}_\Lambda\left(e^{-(1-z)\int_0^t \Lambda(u)du}\right).$$

For the first two moments of doubly stochastic Poisson processes, note that from the conditional distribution of $N(t) - N(s)$ given $\{\Lambda(u), \; s \leq u \leq t\}$ we find

$$\mathbb{E}(N(t) - N(s) | \Lambda(u) = \lambda(u), \; s \leq u \leq t) = \int_s^t \lambda(u)du,$$

from which

$$\mathbb{E}(N(t) - N(s)) = \int_s^t \mathbb{E}(\Lambda(u))du,$$

while from

$$\mathrm{Var}(N(t) - N(s) | \Lambda(u) = \lambda(u), \; s \leq u \leq t) = \int_s^t \lambda(u)du,$$

we obtain

$$\mathrm{Var}(N(t) - N(s)) = \int_s^t \mathbb{E}(\Lambda(u))du + \mathrm{Var}\left(\int_s^t \Lambda(u)du\right),$$

from which the overdispersion for doubly stochastic Poisson processes follows.

A probabilistic construction of the claim arrival process of a doubly stochastic Poisson process can be based on the one-to-one correspondence between the claim arrival times and claim number process

$$N(t) = \sum_{j=1}^{\infty} 1_{\{T_j \le t\}}, \; t \ge 0. \tag{5.2.13}$$

If W_j ($j \in \mathbb{N}$) denote independent standard exponentially distributed random variables, then

$$T_j = \inf\{t > 0 \, : \, W_1 + \ldots + W_j \le M(t)\}, \; j \in \mathbb{N}, \tag{5.2.14}$$

$$N(t) = \sum_{j=1}^{\infty} 1_{\{W_1 + \ldots + W_j \le M(t)\}}, \; t \ge 0.$$

For a fixed time horizon $T > 0$, a sample path $\{\hat{N}(t); t \in [0, T]\}$ of $N(t)$ can now be generated using the following steps for a given stochastic model for $M(t)$:

1. Simulate a path $\{\hat{M}(t); t \in [0, T]\}$ from $M(t)$ according to some suitably fine discretization grid.
2. Repeatedly draw independent standard exponentially distributed random variables until their sum exceeds the level $\hat{M}(T)$, and compute the claim arrival times according to (5.2.14).
3. Determine the sample path $\{\hat{N}(t); t \in [0, T]\}$ from the sampled claim arrival times using (5.2.13).

While this algorithm is constructed directly on the very nature of a doubly stochastic Poisson process, more efficient algorithms are available. See Korn et al. [497] for an application of the acceptance/rejection method in this context.

Doubly stochastic Poisson processes directed by Lévy subordinators. In [695], Selch introduced Cox processes $\tilde{N}_\lambda(M(t))$ directed by a Lévy subordinator M. Lévy processes are characterized by their independent and stationary increments. In comparison with linear functions which have constant increments, Lévy processes have i.i.d. increments. Homogeneous Poisson processes hence are an example of Lévy processes. In the following we only state some basic facts and properties that we need for later purposes. For more mathematical details on Lévy processes, see, for example, Sato [666], Schoutens [683], and Applebaum [44].

A stochastic process $X = \{X(t); t \ge 0\}$ is called a Lévy process if it has the following properties:

(i) Start at 0: $X(0) = 0$ almost surely.

(ii) Independent increments: for any $0 \le t_0 < t_1 < \ldots < t_n < \infty$, $n \in \mathbb{N}$, the increments $\Delta X_{t_i} := X(t_i) - X(t_{i-1})$ are mutually independent for $i = 1, \ldots, n$.

(iii) Stationary increments: for any $0 < s < t$ *the increments satisfy* $X(t) - X(s) =_d X(t-s)$.
(iv) Stochastic continuity: for all $t \geq 0$ *and* $\epsilon > 0$ *one has* $\lim_{s\to t} \mathbb{P}(|X(t)-X(s)| > \epsilon) = 0$.

For every Lévy process X one can construct a version that has càdlàg paths almost surely, and we can hence assume the càdlàg property (cf. [666]). A Lévy process with almost surely non-decreasing paths is called a Lévy subordinator, denoted here by M.

Due to the stochastic continuity of the process, the jumps of a Lévy process cannot appear at any fixed point in time with a positive probability. Furthermore, due to the càdlàg property of the paths, only countably many jumps can occur in any finite time interval and the number of jumps with size larger than some arbitrary $\epsilon > 0$ is finite.

One then considers the *jump measure*

$$J(A) = |\{t \geq 0 : (t, dX(t-)) \in A\}|, \; A \in \mathcal{B}([0,\infty) \times \mathbb{R}_0)$$

where $dX(t-) = X(t) - \lim_{s\uparrow t} X(s)$, and the so-called *Lévy measure*

$$\nu_X(B) = \mathbb{E}\left(J([0,1] \times B\right), \; B \in \mathcal{B}(\mathbb{R}_0).$$

One typically imposes an integrability condition such as $\int_{|x|<\epsilon} |x|\nu_X(dx) < \infty$ for the small jumps to be well behaved. The famous *Lévy–Khintchine representation* completely characterizes the distribution of the Lévy process in terms of the Lévy measure and an extra drift parameter b_X under the above integrability condition:

Let M be a Lévy subordinator. The Laplace transform $\widehat{F}_{M(t)}$ *of* $M(t)$ *can be expressed in terms of the so-called Laplace exponent* $\Psi_M : [0,\infty) \to [0,\infty)$ *by*

$$\widehat{F}_{M(t)}(u) = \exp\left(-t\Psi_M(u)\right),$$

where the Laplace exponent is a Bernstein function derived from the characteristics of the subordinator as

$$\Psi_M(u) = u\, b_X + \int_0^\infty (1 - e^{-ux})\nu_M(dx), \; u \geq 0.$$

We give here some classical examples of Lévy subordinators.

1. The homogeneous Poisson process with intensity $\lambda > 0$ with Laplace exponent $\Psi_M(u) = \lambda(1 - \exp(-u))$ and Lévy characteristics $b_M = 0$ and $\nu_M = \lambda 1_{\{1\in B\}}$, $B \in \mathcal{B}((0,\infty))$.
2. The gamma (α, β) subordinator: with $M(t)$ being gamma distributed with density

$$f_{M(t)}(x) = \frac{\beta^{\alpha t}}{\Gamma(\alpha t)} x^{\alpha t-1} e^{-\beta x}, \; x > 0.$$

The Laplace exponent is

$$\Psi_M(u) = \alpha \log\left(1 + \frac{u}{\beta}\right), \quad u \geq 0,$$

while $\widehat{F}_{M(t)}(u) = \left(1 + \frac{u}{\beta}\right)^{-\alpha t}$, $b_M = 0$ and

$$\nu_M(dx) = \alpha \exp(-\beta x)\frac{1}{x}1_{(0,\infty)}(x)dx.$$

Moreover

$$\widehat{F}_{N(t)}(s) = \left(1 + \frac{\lambda}{\beta}(1 - e^{-s})\right)^{-\alpha t},$$

from which

$$\mathbb{E}(N(t)) = \frac{\alpha}{\beta}\lambda t.$$

3. The inverse Gaussian (β, η) subordinator: here $M(t)$ is assumed to follow the classical inverse Gaussian distribution as given in example (c) of a mixed Poisson process with $\mu = (\beta/\eta)t$ and $b = \eta^{-2}$ (cf. Section 5.2.3):

$$f_{M(t)}(x) = \frac{\beta t}{\sqrt{2\pi x^3}} \exp\left(-\frac{1}{2x}(\eta x - \beta t)^2\right), \quad x > 0.$$

The Laplace exponent is

$$\Psi_M(u) = \beta\left(\sqrt{2u + \eta^2} - \eta\right), \quad u \geq 0.$$

Then $\widehat{F}_{M(t)}(u) = \exp\left(-\beta t\left\{\sqrt{2u + \eta^2} - \eta\right\}\right)$, $b_M = 0$ and $\nu_M(dx) = \frac{1}{\sqrt{2\pi}}\beta x^{-3/2} \exp\left(-\frac{1}{2}\eta^2 x\right)1_{(0,\infty)}(x)dx$. Moreover

$$\widehat{F}_{N(t)}(s) = \exp\left(-\beta t\left\{\sqrt{2\lambda(1 - e^{-s}) + \eta^2} - \eta\right\}\right),$$

from which

$$\mathbb{E}(N(t)) = \frac{\beta}{\eta}\lambda t.$$

Due to the jumps of the subordinator, the basic Poisson model is here extended to allow for simultaneous claim arrivals. The Lévy subordinator M serves as the (random) operational time (also referred to as *stochastic clock*).

The discontinuity of paths of M entails that M is not differentiable. It follows that no random intensity Λ exists for these Cox processes. Also, the process cannot be converted back to the underlying independent Poisson process by a time change with the inverse of the directing process.

Moreover, Selch [695] showed that $\tilde{N}_\lambda(M(t))$ is a Poisson cluster process, which means that the cluster sizes are i.i.d. and their arrival times are determined by an independent Poisson process:

$$\{N(t)\,:\,t \geq 0\} =_d \left\{ \sum_{j=1}^{L(t)} Y_j \,:\, t \geq 0 \right\},$$

where $\{L(t)\,:\,t \geq 0\}$ denotes a Poisson process with intensity $\lambda_L = \Psi_M(\lambda)$, and where, independent of L, $Y_1, Y_2, \ldots$ are i.i.d. copies of a cluster size random variable Y with Laplace transform $\widehat{F}_Y(u) = 1 - \frac{\Psi_M(\lambda[1-\exp(-u)])}{\Psi_M(\lambda)}$. Hence the waiting times between clusters should be exponentially distributed for this model to be applicable.

It may be desirable that the operational time, although randomly running slower and faster, behaves on average like real time, that is,

$$\mathbb{E}(M(t)) = t, \text{ for all } t \geq 0,$$

which by the properties of Lévy processes is equivalent to

$$\mathbb{E}(M(1)) = 1. \tag{5.2.15}$$

Condition (5.2.15) is referred to as *time normalization.*

Note that the marginal distributions of $N(t)$ at a given time point t equal that of a mixed Poisson process. For the finite-dimensional distribution of N with time points $0 = t_0 < t_1 < \ldots < t_n$ and integers $0 = k_0 \leq k_1 \leq \ldots \leq k_m$, Selch [695] provides the expression

$$\mathbb{P}(N(t_1) = k_1, \ldots, N(t_n) = k_n) = (-\lambda)^{k_n} \Pi_{j=1}^{n} \frac{1}{(\Delta k_j)!} \widehat{F}^{(\Delta k_j)}_{M(\Delta t_j)}(\lambda) \tag{5.2.16}$$

with $\Delta k_j = k_j - k_{j-1}$ and $\Delta t_j = t_j - t_{j-1}$ $(j = 1, \ldots, n)$.

For the marginal distribution of $N(t)$ and a gamma (α, β) subordinator, this formula must reduce to a negative binomial distribution. Indeed, then the derivatives of the Laplace transform $\widehat{F}_{M(t)}(\lambda)$ are

$$\widehat{F}^{(k)}_{M(t)}(\lambda) = (-1)^k (\alpha t + k - 1) \ldots (\alpha t) \left(\frac{1}{\beta + \lambda} \right)^k \left(\frac{\beta}{\beta + \lambda} \right)^{\alpha t}$$

so that

$$\mathbb{P}(N(t) = k) = \frac{(-\lambda)^k}{k!}\hat{F}^{(k)}_{M(t)}(\lambda) = \binom{\alpha t + k - 1}{k}\left(\frac{\beta}{\lambda+\beta}\right)^{\alpha t}\left(\frac{\lambda}{\lambda+\beta}\right)^{k}$$

which corresponds to (5.2.12).

We summarize the four models based on Poisson processes (PP) in Table 5.1.

To illustrate the differences between the inhomogeneous Poisson and doubly stochastic Poisson counting process models, Figure 5.2 depicts a simulated path from each as well as a sample path of a homogeneous Poisson process with the same number of expected claims. We also give the expected value curve together with bands that are four standard deviations wide in each case. While the inhomogeneous Poisson process follows a non-linear trend but shows Poisson jumps, the doubly stochastic Poisson example allows for larger jump sizes, but maintains the same expected claim number per time unit throughout the realization.

We emphasize that a main difference between a Pólya process (i.e., mixed Poisson process with gamma-distributed Λ) and a doubly stochastic Poisson process directed by a gamma subordinator is that in the latter case for each new time interval the realization of Λ is sampled anew, whereas for the Pólya process the same value is used throughout. From this point of view, the interpretation of heterogeneous policies to motivate this model is more intuitive for the doubly stochastic Poisson process, as at each time instant the next claim(s) may come from a different group of policies with another claim intensity. In particular, when considering model fitting to discrete claim counts in Section 5.4 under an i.i.d. assumption for the available realizations of claim numbers, the doubly stochastic Poisson process is arguably the more natural continuous-time version of this model.

Table 5.1

Homogeneous PP	**Inhomogeneous PP**
$\tilde{N}_\lambda(t)$	$N_\mu(t) = \tilde{N}_1(\mu(t))$
$\mu(t) = \lambda t,\ \lambda(u) \equiv \lambda$	$\mu(t) = \int_0^t \lambda(u)du$
$\mathbb{E}(\tilde{N}_\lambda(t)) = \mathrm{Var}(\tilde{N}_\lambda(t)) = \lambda t$	$\mathbb{E}(N_\mu(t)) = \mathrm{Var}(N_\mu(t)) = \mu(t)$
Mixed PP	**Doubly stochastic PP**
$N_\Lambda(t) = \tilde{N}_1(\Lambda t)$	$N_M(t) = \tilde{N}_1(M(t))$
$M(t) = \Lambda t$	$M(t) = \int_0^t \Lambda(u)du$
$\mathbb{E}(N_\Lambda(t)) = t\mathbb{E}(\Lambda)$	$\mathbb{E}(N_M(t)) = \int_0^t \mathbb{E}(\Lambda(u))du$
$\mathrm{Var}(N_\Lambda(t)) = t\mathbb{E}(\Lambda) + t^2\mathrm{Var}(\Lambda)$	$\mathrm{Var}(N_M(t)) = \int_0^t \mathbb{E}(\Lambda(u))du + \mathrm{Var}\left(\int_0^t \Lambda(u)du\right)$

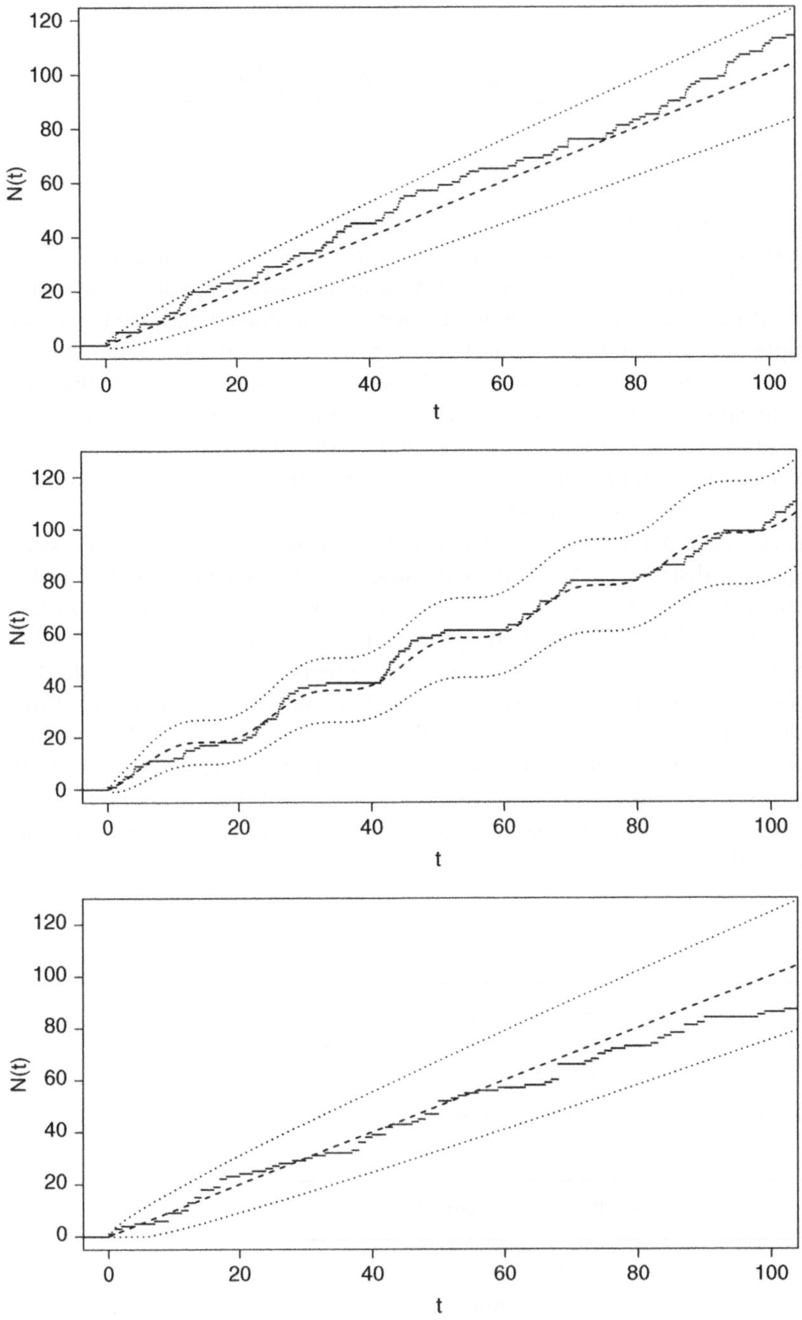

Figure 5.2 Simulated path on $[0, 100]$ with mean value and mean value +/- 2 standard deviation curves: homogeneous Poisson $\tilde{N}_1$ (top); inhomogeneous Poisson with $\lambda(u) = 1 + \sin(\pi u/10)$ (middle); doubly stochastic Poisson directed by a gamma subordinator ($\alpha = 2, \beta = 2$) (bottom).

Multivariate extensions. The general framework of Cox processes allows several multivariate extensions. *Multivariate Cox processes* driven by shot-noise processes are treated in Scherer et al. [670]. Another natural way to introduce dependence between claim numbers of different lines of business or categories is to start with a priori independent Poisson processes for each, but then direct them all by the same Lévy subordinator. This causes simultaneous claim occurrences in the different business lines in an intuitive way (cf. Selch [695]). To this end let $N_j = \{N_j(t);\, t \geq 0\}$ count the claims arriving in each portfolio $j = 1, \ldots, d$ up to time $t \geq 0$. In a traditional modelling approach claim arrivals in each portfolio j are assumed to follow independent homogeneous Poisson processes $\tilde{N}_j = \{\tilde{N}_{j,\lambda_j}(t);\, t \geq 0\}$ with individual intensities $\lambda_j > 0$. Let $\tilde{\mathbf{N}} = (\tilde{N}_{1,\lambda_1}, \ldots, \tilde{N}_{d,\lambda_d})$ be the corresponding multivariate standard Poisson process with independent marginals with intensity vector $\lambda = (\lambda_1, \ldots, \lambda_d)$. The multivariate Cox process $\mathbf{N} = (N_1, \ldots, N_d)$ is then defined through

$$\begin{aligned} \mathbf{N} &= \tilde{\mathbf{N}}_\lambda(M(\cdot)) \\ &= \left\{ \left(\tilde{N}_{1,\lambda_1}(M(t)), \ldots, \tilde{N}_{d,\lambda_d}(M(t)) \right) ;\, t \geq 0 \right\}. \end{aligned} \tag{5.2.17}$$

This process naturally generates dependence between the d portfolios, as the stochastic factor M similarly affects all components. Furthermore, while the underlying Poisson process yields only jumps of size one, the subordinator and therefore time can now jump, leading to simultaneous claim arrivals within and between the individual portfolios in the time-changed process.

Expression (5.2.16) for the finite-dimensional distributions of $\mathbf{N}$ with time points $0 = t_0 < t_1 < \ldots < t_n$ and integer vectors $\mathbf{0} = \mathbf{k}_0 \leq \mathbf{k}_1 \leq \ldots \leq \mathbf{k}_n$ with $\mathbf{k}_i = (k_i^{(1)}, \ldots, k_i^{(d)})$ now generalizes to (cf. [695]):

$$\mathbb{P}(\mathbf{N}(t_1) = \mathbf{k}_1, \ldots, \mathbf{N}(t_n) = \mathbf{k}_n) = (-\lambda)^{\mathbf{k}_n} \Pi_{i=1}^n \frac{1}{(\Delta \mathbf{k}_i)!} \hat{F}_{M(\Delta t_i)}^{(|\Delta \mathbf{k}_i|)}(|\lambda|) \tag{5.2.18}$$

with $\Delta \mathbf{k}_i = (k_i^{(j)} - k_{i-1}^{(j)};\, j = 1, \ldots, d)$, $\Delta t_i = t_i - t_{i-1}$, $i = 1, \ldots, n$, where we use the notation $|\mathbf{x}| = |(x_1, \ldots, x_d)| = |x_1| + \ldots + |x_d|$, $\mathbf{x}^{\mathbf{k}} = x_1^{k_1} \ldots x_d^{k_d}$ and $\mathbf{k}! = k_1! \ldots k_d!$.

5.3 Other Claim Number Processes

In this section we deal with a number of examples of claim number processes which are further extensions of the Poisson model in various directions.

5.3.1 The Nearly Mixed Poisson Model

A further generalization of the mixed Poisson model, hence of the Poisson process, is obtained by the assumption that the counting process satisfies the following condition (cf. (5.2.11)): there exists a non-negative random variable Λ such that as $t \uparrow \infty$,

$$\frac{N(t)}{t} \to_d \Lambda.$$

A counting process that satisfies this condition will be called *nearly mixed Poisson.* We mention that this definition lacks a dynamic background, while the mixed Poisson process can be constructed using precise dependence rules between the successive variables $\{T_i, i \geq 1\}$. Nevertheless, the nearly mixed Poisson process encompasses quite a number of other models.

The convergence in distribution is equivalent to pointwise convergence of the corresponding Laplace transforms. Hence for all $\theta \geq 0$ one has that

$$\mathbb{E}\{e^{-\theta \frac{N(t)}{t}}\} \to \mathbb{E}\{e^{-\theta\Lambda}\} = \widehat{F}_\Lambda(\theta) \tag{5.3.19}$$

where $\widehat{F}_\Lambda(\theta)$ is the Laplace transform of a distribution F_Λ that we keep calling the *mixing distribution.* By the continuity of the limit $\widehat{F}_\Lambda$ we see that (5.3.19) is equivalent to the condition that for all values of $\theta \geq 0$,

$$\mathbb{E}\left(\left\{1-\frac{\theta}{t}\right\}^{N(t)}\right) \to \widehat{F}_\Lambda(\theta).$$

By the definition of the generating function

$$Q_t\left(1-\frac{w}{t}\right) \to \widehat{F}_\Lambda(w)$$

for all $w \geq 0$. The main difference to the mixed Poisson model is that for the latter there is equality in the above relation while for the nearly mixed Poisson case, the generating function attains the quantity $\widehat{F}_\Lambda(w)$ only in the limit. Clearly, conditioning on the random variable Λ we see that, for any $u \in \mathbb{R}$,

$$\mathbb{E}\left\{\exp\, iu\frac{N(t)-t\Lambda}{\sqrt{t}}\right\} = \int_0^\infty \mathbb{E}\left\{\exp\left(iu\frac{N(t)-t\lambda}{\sqrt{t}}|\Lambda=\lambda\right)\right\} dF_\Lambda(\lambda).$$

By the central limit theorem for the Poisson process one arrives at

$$\mathbb{P}\left\{\frac{N(t)-t\Lambda}{\sqrt{t}} \leq x\right\} \to \int_0^\infty \Phi\left(\frac{x}{\sqrt{\lambda}}\right) dF_\Lambda(\lambda),$$

where Φ is the standard normal distribution function. Also the *infinitely divisible processes* and *renewal processes* discussed below are examples of nearly mixed Poisson processes.

5.3.2 Infinitely Divisible Processes

The Poisson process directed by a Lévy subordinator discussed in Section 5.2.4 is itself a Lévy process (in fact, another Lévy subordinator), and as such has i.i.d. increments. The general class of counting processes with i.i.d. increments is of interest. For this to hold the distribution at each time point t has to be *infinitely divisible* (i.e., is the distribution of a sum of n i.i.d. random variables for any $n \in \mathbb{N}$), but from Sato [666, Cor. 27.5] it follows that the class of discrete infinitely divisible distributions coincides with the one of compound Poisson distributions with integer-valued jump size distribution. The probability generating function of an *infinitely divisible counting process* hence can be expressed as

$$Q_t(z) = e^{-\lambda t(1-g(z))}$$

where $g(z) := \mathbb{E}(z^G)$ is the probability generating function of a discrete random variable G on the strictly positive integers. In the special case where $g(z) = z$ one gets back to the classical *Poisson process*. For other choices of $g(z)$, the counting process should be called a *(discrete) compound Poisson process*. Note that instead of "stretching" the time axis as in the subordination approach one here specifies explicitly a discrete distribution for the number of additional claims at each Poisson instant. In fact, all processes that can be constructed this way correspond to a Lévy subordinator (e.g., see [666, Thm. 24.11]), but the explicit alternative construction in terms of the random variable G is quite intuitive.

The probabilities are given in the form

$$p_n(t) = \sum_{k=0}^{\infty} e^{-\lambda t} \frac{(\lambda t)^k}{k!} g_n^{*k}$$

where $\{g_n^{*k}; n = 1, 2, \ldots\}$ is the k-fold convolution of the distribution $\{g_n\}$ with probability generating function $g(z)$. Needless to say the explicit evaluation of the above probabilities is mostly impossible because of the complex form of the convolutions.

A general form for $Q_t^{(r)}(z)$ can be given by using Faa di Bruno's formula

$$Q_t^{(r)}(z) = Q_t(z) \sum_{i=0}^{r} (\lambda t)^i \sum_{\{a_j\}} \frac{r!}{a_1! \ldots a_r!} \prod_{k=1}^{r} \left\{ \frac{g^{(k)}(z)}{k!} \right\}^{a_k}$$

where the inside sum runs over all integers $a_j \geq 0$ for which $a_1 + a_2 + \cdots + a_r = i$ and simultaneously $a_1 + 2a_2 + \cdots + ra_r = r$. From the first two r values one obtains the expressions

$$\mathbb{E}(N(t)) = \lambda t\, \mathbb{E}(G), \mathrm{Var}(N(t)) = \lambda t\, \mathbb{E}(G^2).$$

Kupper [518] gives some nice applications of such a construction of infinitely divisible processes to model claim counts, where the Poisson process refers to the number of accidents and the number of casualties in accident i is given by the random variable G_i.

Any *infinitely divisible process* provides an example of a nearly mixed Poisson process as long as $\mathbb{E}(G) < \infty$. Here $\widehat{F}_\Lambda(w) = \exp\{-\lambda w \mathbb{E}(G)\}$ so that the limiting mixing distribution is again degenerate but now at the point $\lambda \mathbb{E}(G)$.

Here are a few more explicit cases that have appeared in the actuarial literature.

(i) As a first particular case one finds a *generalized Poisson–Pascal process* introduced by Kestemont and Paris [486], where $g(z) = (1 - \beta(z-1))^{-\alpha}$ is the generating function of a negative binomial random variable (this is not to be confounded with the gamma-subordinated Poisson process, where the overall number of claims up to time t is negative binomially distributed).

(ii) If $g(z) = (e^{\theta z} - 1)(e^{\theta} - 1)^{-1}$ then G is a truncated Poisson and the corresponding claim process is of *Neyman-type A*.

(iii) If $g(z) = ((1-\theta)z)(1-\theta z)^{-1}$ with $0 < \theta < 1$, then G is a truncated geometric. The generating function is

$$Q_t(z) = \exp\left\{-\lambda t\left(1 - \frac{(1-\theta)z}{1-\theta z}\right)\right\}.$$

The probabilities can be evaluated explicitly in terms of Laguerre polynomials

$$p_n(t) = \mathbb{P}(N(t) = n) = \begin{cases} e^{-\lambda t} & \text{if } n = 0, \\ e^{-\lambda t}\,\theta^n L_n^{-1}\left(-\frac{\lambda t(1-\theta)}{\theta}\right) & \text{if } n \geq 1. \end{cases}$$

The latter counting process is called the *Pólya–Aeppli process*.

5.3.3 The Renewal Model

If the inter-claim times $\{W_i = T_i - T_{i-1}; i = 1, 2, \dots\}$ (with $T_0 = 0$) are i.i.d. (non-negative) random variables with cumulative distribution function (c.d.f.) F_W, then the corresponding claim number process $N(t)$ is called a *renewal process*. This process was introduced in risk theory by Sparre Andersen [707], so that often this model is referred to as the *Sparre Andersen model*. Since

$$\{N(t) = n\} = \{T_n \leq t < T_{n+1}\}$$

one has that

$$p_n(t) = F_W^{*(n)}(t) - F_W^{*(n+1)}(t).$$

In general, successive convolutions of a distribution cannot be evaluated explicitly. The only simple exception is provided by the exponential case $F_W(t) = 1 - e^{-\lambda t}$, for which $N(t)$ is a Poisson process. The Poisson process is actually the only mixed Poisson process that is simultaneously a renewal process.

The generating function is also cumbersome, since

$$Q_t(z) = 1 + \left(1 - \frac{1}{z}\right)\sum_{n=1}^{\infty} F_W^{*(n)}(t)z^n. \tag{5.3.20}$$

From (5.3.20), (5.1.4) and $\mu = \mathbb{E}(W)$ it follows that

$$\mathbb{E}(N(t)) = Q_t'(1) = \sum_{n=1}^{\infty} F_W^{*(n)}(t) =: U(t) \sim \frac{t}{\mu}$$

where $U(t)$ is the classical *renewal function generated by* W. The evaluation of the variance is already somewhat tedious but leads to

$$\text{Var}(N(t)) = 2\,U * U(t) + U(t) - U^2(t) \sim \frac{\sigma^2}{2\mu^3}\,t$$

where $\sigma^2 = \text{Var}(W)$.

Any *renewal process* generated by a distribution with a finite mean is a nearly mixed Poisson. This follows from the so-called weak law of large numbers from renewal theory: if the mean μ of the interclaim time distribution is finite, then $N(t)/t \to_d \Lambda$ where Λ is degenerate at the point $1/\mu$.

The renewal model is an extension of the Poisson process that allows for an elegant mathematical treatment. We refer to Feller [345] and Rolski et al. [652] for detailed accounts. At the same time, its practical importance is limited, as the implicitly introduced particular memory pattern between claim times is typically hard to interpret or justify in the insurance context.

5.3.4 Markov Models

Another model for the claim number process can be provided by a *pure (non-homogeneous) birth process*. Here the probabilities $\{p_n(t); n \geq 0, t \geq 0\}$ satisfy a system of Kolmogorov difference-differential equations, well-known from the theory of continuous-time Markov chains. Then

$$\frac{dp_n(t)}{dt} = -\lambda_n(t)p_n(t) + \lambda_{n-1}(t)p_{n-1}(t), \; n \geq 1, \tag{5.3.21}$$

where we take $\lambda_{-1}(t) = 0$. The transition probabilities can sometimes be explicitly evaluated.

(i) The first such example is the Poisson process where $\lambda_n(t) = \lambda$.
(ii) The mixed Poisson process also has a representation of the above form, with

$$\lambda_n(t) = \frac{\int_0^\infty \lambda^{n+1} e^{-\lambda t}\, dF_\Lambda(\lambda)}{\int_0^\infty \lambda^n e^{-\lambda t}\, dF_\Lambda(\lambda)}.$$

(iii) Another one is provided by the *Yule process* where $\lambda_n(t) = n\lambda$. It is then easy to show that for $n \geq 0$,

$$p_n(t) = e^{-\lambda t}(1 - e^{-\lambda t})^n,$$

so that the generating function is given by

$$Q_t(z) = e^{-\lambda t}(1 - z(1 - e^{-\lambda t}))^{-1}.$$

For details and more examples, see [492, Ch. 7].

5.4 Discrete Claim Counts

In this section we consider models when fixing the time unit under consideration (which typically will be one year). If the portfolio (and risk exposure) remains unchanged over the years, this number is often assumed to be an i.i.d. random variable over the years (otherwise, it is common to apply volume corrections to justify the assumption). Clearly, sampling the counting process of Sections 5.2 and 5.3 at equidistant time instants gives candidates for such discrete claim counts, and whenever the continuous-time process is of Lévy type then the resulting discrete counts will inherit the i.i.d. property. Whereas this discussed approach gives the general picture, one can also start directly by fitting a discrete distribution to (annual) aggregate claim counts (one may also argue that a certain discretization is in any case natural, since exact claim times will often not be available). The Poisson and negative binomial distribution then play a special role, not only because of their interpretations originating from the continuous-time view, but also due to the simplicity under which one can perform resulting aggregate claim calculations, as we will see in Chapter 6.

So let us now take the (also traditional) viewpoint of modelling the discrete claim counts directly. In this section we then omit to refer to time as a parameter.

A very popular set of claim number distributions is the $(a, b, 1)$ *class*.[1]. It is based on a simple recursion for $p_n := \mathbb{P}(N = n)$, namely

$$p_n = \left(a + \frac{b}{n}\right) p_{n-1}, \quad n \geq 2. \tag{5.4.22}$$

Note that the recursion does not specify the quantities p_1 and p_0. Nevertheless the requirement $\sum_{n=0}^{\infty} p_n = 1$ eliminates one of the latter two parameters, so that the $(a, b, 1)$ class may be used for data-fitting using the three parameters a, b and $\rho := p_0$.

One can rewrite $Q(z) := \sum_{n=0}^{\infty} p_n z^n$ in the form $Q(z) = \rho + (1-\rho)R(z)$ where $\rho \in [0, 1]$, and $R(z)$ is again a generating function of a discrete probability distribution $\{r_n; n \geq 1\}$. The effect of the parameter ρ is to introduce an eventual extra weight at the point 0. This is often done by *truncation* in the sense that $\rho = \mathbb{P}(N = 0)$ while $R(z)$ is the generating function of the probabilities $\mathbb{P}(N = k|N > 0)$. If we shift the distributions one integer to the left and take $\rho = 0$ then we of course get the classical unshifted distributions.

Relation (5.4.22) can be solved in a variety of ways. We use generating functions $Q(z) = \mathbb{E}(z^N)$. Then (5.4.22) turns into a first-order differential equation

$$(1 - az)\, Q'(z) = (a + b)\, Q(z) + p_1 - (a + b)\rho \tag{5.4.23}$$

where $\rho = p_0 = \mathbb{P}(N = 0)$. We have to solve this equation with the side condition $Q(1) = 1$ while we also know that $Q(0) = \rho$. These two conditions suffice to determine the constant of integration as well as the unknown quantity p_1.

- Case 1: $a = 0$

  This case is rather easy and quickly leads to the expression

$$Q(z) = \rho + (1 - \rho)\frac{e^{bz} - 1}{e^{b} - 1}.$$

  From this it follows that

$$p_n = \frac{1 - \rho}{1 - e^{-b}} \frac{b^n}{n!} e^{-b}, \quad n \geq 1,$$

1 We take this nomenclature from Klugman et al. [491]. It is also often referred to as the Panjer class or Sundt–Jewell class (cf. [721]), as these scholars introduced it to actuarial circles. Note, however, that the class had already been studied in detail in 1969 by Johnson and Kotz [468].

which is a *shifted Poisson distribution*. To illustrate this, take $r_n = e^{-b}\frac{b^n}{n!}$ then $p_n = \frac{r_n}{1-r_0}$. Put another way, if $\rho = e^{-b}$ then we fall back on the classical Poisson model with parameter b.

- Case 2: $a \neq 0$.
  It seems advantageous to introduce an auxiliary quantity $\Delta := 1 + b/a$. The solution to the differential equation (5.4.23) is slightly more complicated.
  - Subcase (a): $\Delta \neq 0$
    Then the generating function is given by

$$Q(z) = \rho + (1-\rho)\frac{(1-az)^{-\Delta} - 1}{(1-a)^{-\Delta} - 1}. \tag{5.4.24}$$

Here

$$R(z) = \frac{(1-az)^{-\Delta} - 1}{(1-a)^{-\Delta} - 1}.$$

This expression is akin to the probability generating function for a negative binomial distribution with parameters Δ and a. Hence

$$r_n = \binom{\Delta + n - 1}{n} a^n (1-a)^{\Delta}$$

and with $r_0 = (1-a)^{\Delta}$ we see that the negative binomial distribution is a member of the $(a, b, 1)$ class.

  - Subcase (b): $\Delta = 0$
    Now the solution is

$$Q(z) = \rho + (1-\rho)\frac{\log(1-az)}{\log(1-a)} \tag{5.4.25}$$

which easily leads to the *logarithmic distribution*

$$p_n = \frac{1-\rho}{-\log(1-a)}\frac{a^n}{n}, \qquad n \geq 1. \tag{5.4.26}$$

Here a shift is unnecessary since the logarithmic distribution is concentrated on the positive integers.

We consider two special cases.

(i) If $\Delta = -m$, a negative integer, then we run into a distribution that is related to the *binomial distribution*. Indeed, choosing

$$r_n = \binom{m}{n} c^n (1-c)^{m-n},$$

then with $r_0 = (1-c)^m$ and $c = \frac{a}{a-1}$ we see that the binomial distribution is also a member of the $(a, b, 1)$ class.

(ii) If $\Delta = -\theta \in (-1, 0)$ a similar calculation relates to the *Engen distribution* introduced by Willmot in [784]

$$r_n := r_n(\theta, a) = \frac{\theta}{(1-a)^\theta - 1} \frac{a^n \Gamma(n-\theta)}{n!\Gamma(1-\theta)}, \quad n \geq 1$$

which is again a distribution concentrated on the positive integers.

5.5 Statistics of Claim Counts

The statistical toolbox for modelling claim counts can be divided into two cases: whether one only knows the number of claims in a given fixed time window, say one year, or whether the specific arrival times or dates of the claims are known. Clearly, the statistical information is much richer in the second case and procedures for fitting Poisson, homogeneous or inhomogeneous, or even Cox processes, can, for instance, be based on waiting times between subsequent claims.

5.5.1 Modelling Yearly Claim Counts

Statistical analysis of claim count data is typically performed by generalizing the normal theory using the *exponential family of distributions*, which comprises popular models for claim counts such as the Poisson, binomial and other models, and which can be extended to important cases such as the negative binomial distribution.

The exponential family of distributions is defined through the probability density functions f for which

$$\log f(y; \theta, \phi) = \frac{y\theta - b(\theta)}{a(\phi)} + c(y; \phi)$$

for parametrization (θ, ϕ) with θ the parameter of interest and ϕ a nuisance parameter, and some functions b only depending on θ, and a and c only depending on ϕ.

Typical examples are

(i) the Poisson distribution with

$$\log f(y; \lambda) = y \log \lambda - \lambda - \log y!\,,$$

for which $\theta = \log \lambda$, $a(\phi) = \phi = 1$, $b(\theta) = \exp(\theta) = \lambda$, and $c(y; \phi) = -\log y!$

(ii) the binomial model with

$$\log f(y; p) = y \log \frac{p}{1-p} + n \log(1-p) + \log \binom{n}{y},$$

for which $\theta = \log \frac{p}{1-p}$, $a(\phi) = \phi = 1$, $b(\theta) = -n\log(1-p) = n\log(1+e^\theta)$, and $c(y; \phi) = \log \binom{n}{y}$

(iii) the negative binomial model as introduced in (5.2.12) with $t = 1$ satisfies

$$\log f(y; b, \alpha) = -y \log(1+b) + \alpha \log \frac{b}{1+b} + \log \frac{\Gamma(\alpha+y)}{\Gamma(\alpha) y!},$$

so that $\theta = -\log(1+b)$, $a(\phi) = \phi = 1$, $b(\theta) = -\alpha \log \frac{b}{1+b} = -\alpha \log(1-e^\theta)$, and $c(y;\phi) = \log \frac{\Gamma(\alpha+y)}{\Gamma(\alpha)y!}$. Hence the negative binomial example only belongs to the exponential family for known α.

This family can be treated using classical likelihood theory stating that with

$$U := \frac{d \log f(Y;\theta)}{d\theta} \text{ and } U' := \frac{d^2 \log f(Y;\theta)}{d\theta^2}$$

one has

$$\mathbb{E}(U) = 0 \text{ and } \mathrm{Var}(U) = \mathbb{E}(U^2) = -\mathbb{E}(U'),$$

from which we obtain that

$$\mathbb{E}(Y) = b'(\theta) \text{ and } \mathrm{Var}(Y) = b''(\theta) a(\phi)$$

since $U = (Y - b(\theta))/a(\phi)$ and $U' = -b''(\theta)/a(\phi)$. Hence ϕ is a dispersion parameter, while $b'(\theta)$ gives the expected value of the variable Y:

(i) in the Poisson case we obtain the well-known results

$$\mu = \mathbb{E}(Y) = b'(\theta) = e^\theta = \lambda \text{ and } \mathrm{Var}(Y) = b''(\theta)a(\phi) = e^\theta = \lambda$$

(ii) in the binomial setting

$$\mu = \mathbb{E}(Y) = b'(\theta) = n\frac{e^\theta}{1+e^\theta} = np,$$
$$\mathrm{Var}(Y) = b''(\theta)a(\phi) = n\frac{e^\theta}{(1+e^\theta)^2} = np(1-p) < \mathbb{E}(Y)$$

(iii) in the negative binomial case

$$\mu = \mathbb{E}(Y) = b'(\theta) = \alpha\frac{e^\theta}{1-e^\theta} = \alpha/b,$$
$$\mathrm{Var}(Y) = b''(\theta)a(\phi) = \alpha\frac{e^\theta}{(1-e^\theta)^2} = \frac{\alpha}{b}\frac{1+b}{b} = \mathbb{E}(Y)\left(1+\frac{1}{b}\right)$$
$$> \mathbb{E}(Y).$$

These results confirm the underdispersion in case of the binomial model and the overdispersion in the negative binomial case as it was already found for all mixed Poisson models.

Now the expected value $\mu = \mathbb{E}(Y)$ can be modelled as a function of a covariate (vector), either in a *parametric* or *non-parametric way*. Here we consider the application with accident year or claim arrival year t as a covariate. While in practice the non-parametric approach often gives a start, we first discuss here the parametric modelling approach since the non-parametric solution uses the theory behind the parametric analysis.

- *Parametric modelling* happens through *generalized linear models* (GLM). A standard reference here is McCullagh and Nelder [566]. A GLM allows the transformation of the systematic part of a model without changing the distribution of the variable under consideration. The *link function* g connects the mean μ_i of the observed variables $Y_i,\ i = 1, \dots, n$, with the covariate. Assuming a simple linear predictor we then propose

$$g(\mu_i) = \beta_0 + \beta_1 t_i,\ i = 1, \dots, n,$$

where g is a smooth monotonic function and $t_1, \dots, t_n$ are the values of the covariate. For instance, when using the Poisson or negative binomial model for Y, $g = \log$ is a popular link function. The *linear predictors*

$$\eta_i = \beta_0 + \beta_1 t_i,\ i = 1, \dots, n,$$

are used in maximizing the log-likelihood based on the vector $\mathbf{y} = (y_1, \dots, y_n)$ of observations

$$\ell(\Theta; \mathbf{y}) = \sum_{i=1}^{n} \log f_Y(\theta_i, \phi, y_i) = \sum_{i=1}^{n} \left\{ \frac{y_i\theta_i - b(\theta_i)}{a(\phi)} - c(y_i, \phi) \right\},$$

where $\Theta = (\theta_1, \dots, \theta_n)$, and $\theta_i = (g \circ b')^{-1}(\eta_i)$ denotes the inverse function of the composition of g and b'. The functions a, b and c may vary with i, for instance to allow different binomial parameters $n = n_i$ for each observation of a binomial response or to allow for different α in the negative binomial model. Also, for practical work, it suffices to consider $a_i(\phi) = \phi/\omega_i$ where the weights ω_i are known constants. It will often be convenient to consider $\mathrm{Var}(Y)$ as a function of $\mathbb{E}(Y) = \mu$, and then we can define a function V such that $V(\mu) = b''(\theta)/\omega$ so that $\mathrm{Var}(Y) = V(\mu)\phi$.

Likelihood maximization then proceeds by partial differentiation of ℓ with respect to each β parameter:

$$\frac{\partial \ell}{\partial \beta_j} = \frac{1}{\phi} \sum_{i=1}^{n} \omega_i \left(y_i - b_i'(\theta_i)\right) \frac{\partial \theta_i}{\partial \beta_j}.$$

Moreover

$$\frac{\partial \theta_i}{\partial \beta_j} = \frac{\partial \theta_i}{\partial \mu_i}\frac{\partial \mu_i}{\partial \beta_j} = \frac{1}{b_i''(\theta_i)}\frac{\partial \mu_i}{\partial \beta_j},$$

where the last step follows from $\partial\mu/\partial\theta = b''(\theta)$. Hence

$$\frac{\partial \ell}{\partial \beta_j} = \frac{1}{\phi}\sum_{i=1}^{n} \frac{y_i - \mu_i}{b_i''(\theta_i)/\omega_i}\frac{\partial \mu_i}{\partial \beta_j},$$

and the equations to solve are

$$\sum_{i=1}^{n} \frac{y_i - \mu_i}{V(\mu_i)}\frac{\partial \mu_i}{\partial \beta_j} = 0,\ j = 0, 1, \ldots$$

These equations are in fact the solving equations of non-linear least squares regression analysis with objective function $\sum_{i=1}^{n}(y_i - \mu_i)^2/V(\mu_i)$, where μ_i depends non-linearly on the β parameters. This set of equations is then solved iteratively.

In many cases the nature of the response distribution is not known precisely, and it is only possible to specify the relationship between the variance and the mean of the responses, that is, the function V can be specified, but little more. Then, *quasi-likelihood* using the log quasi-likelihood

$$q(\mu_1, \ldots, \mu_n) = \sum_{i=1}^{n} \int_{y_i}^{\mu_i} \frac{y_i - z}{\phi V(z)} dz$$

leads to exactly the same equations as the full likelihood approach.

As a goodness-of-fit criterion the *scaled deviance*

$$D^*(\mathbf{y}, \Theta) = 2\left(\ell(\mathbf{y}; \mathbf{y}) - \ell(\hat{\Theta}; \mathbf{y})\right)$$

is proposed, which constitutes twice the difference between the maximum achievable log-likelihood obtained by setting $\hat{\mu}_i = y_i$, and that achieved by the model under investigation. Moreover it can be shown that for large enough sample size n

$$D^*(\mathbf{y}, \Theta) \approx_d \chi^2_{n-\dim(\beta)}, \tag{5.5.27}$$

where $\dim(\beta)$ denotes the number of regression parameters β. In our example $\dim(\beta_0, \beta_1) = 2$.

The scaled deviance based on the fitted model can be written as

$$D^*(\mathbf{y}, \hat{\Theta}) = \frac{2}{\phi}\sum_{i=1}^{n} w_i\left(y_i(\tilde{\theta}_i - \hat{\theta}_i) - b(\tilde{\theta}_i) + b(\hat{\theta}_i)\right) =: \frac{D(\mathbf{y}, \hat{\Theta})}{\phi} \tag{5.5.28}$$

with $\tilde{\theta}_i$ denoting the estimates of the parameters θ_i based on the full model with n parameters with maximum achievable likelihood $\ell(\mathbf{y}; \mathbf{y})$. Here $D(\mathbf{y}, \hat{\Theta})$ is known as the *deviance* for the current model.

Hence based on (5.5.27) and (5.5.28) we find an estimator for the parameter ϕ:

$$\hat{\phi} = \frac{D(\mathbf{y}, \hat{\Theta})}{n - \dim(\beta)}. \tag{5.5.29}$$

Another important goodness-of-fit measure for a given regression model is the *explained deviance*

$$1 - \frac{D(\mathbf{y}, \hat{\Theta})}{D(\text{no regression})}$$

where D(no regression) denotes the deviance for the model with only an intercept $\eta_i = \beta_0$ $(i = 1, \ldots, n)$. This generalizes the adjustment coefficient in classical linear regression and expresses which percentage of the total deviance before regression has been explained by using a regression model, which is the complement of how much unexplained deviance remains after using the postulated regression model.

Finally, when ϕ is unknown, nested GLMs with p_0 and p_1 $(p_0 < p_1)$ parameters (and corresponding deviances denoted by D_0 and D_1) can be compared with the generalized likelihood ratio test, rejecting [H_0: model with p_0 parameters holds] for large values of

$$F = \frac{(D_0 - D_1)/(p_1 - p_0)}{D_1/(n - p_1)}$$

with $F \approx_d F_{p_1 - p_0, n-p_1}$ under H_0.

(i) *Poisson model*: The deviance of the Poisson model is given by

$$D(\mathbf{y}, \hat{\Theta}) = 2 \sum_{i=1}^{n} \left(y_i \log \frac{y_i}{\hat{\mu}_i} - (y_i - \hat{\mu}_i) \right).$$

If a constant term is included in the model it can be shown that $\sum_i (y_i - \hat{\mu}_i) = 0$, and hence then

$$D(\mathbf{y}, \hat{\Theta}) = 2 \sum_{i=1}^{n} y_i \log \frac{y_i}{\hat{\mu}_i},$$

which for large μ can be approximated by

$$D(\mathbf{y}, \hat{\Theta}) \approx \sum_{i=1}^{n} (y_i - \hat{\mu}_i)^2 / \hat{\mu}_i,$$

which is the so-called *Pearson χ^2 statistic.*
Overdispersion can be deduced from fitting a Poisson model with which the ϕ parameter using (5.5.29) is larger than 1.

(ii) *Negative binomial model*: here the log-likelihood is given by

$$\ell = \sum_{i=1}^{n} \left(y_i \log \frac{\mu_i}{\mu_i + \alpha} - \alpha \log \left(1 + \frac{\mu_i}{\alpha} \right) + c(y_i) \right),$$

from which a maximum likelihood estimator of α can be traced back through a non-linear equation involving the digamma function. The deviance is then given by

$$D(\mathbf{y}, \hat{\Theta}) = 2\sum_{i=1}^{n}\left(y_i\left[\log\frac{y_i}{y_i+\hat{\alpha}} - \log\frac{\hat{\mu}_i}{\hat{\mu}_i+\hat{\alpha}}\right]\right.$$
$$\left. -\hat{\alpha}\left[\log\left(1+\frac{y_i}{\hat{\alpha}}\right) - \log\left(1+\frac{\hat{\mu}_i}{\hat{\alpha}}\right)\right]\right).$$

- *Non-parametric modelling* can be performed using *generalized additive modelling* (GAM) where the expected counts are modelled through a smooth non-parametric function $h : [0, T] \to \mathbb{R}$ not depending on specific parameters. Here the linear predictor predicts through some smooth monotonic function of the expected value of the response, and the response may follow any exponential family distribution, or simply have a mean-variance relationship permitting to use the quasi-likelihood approach. The resulting model is then of the kind

$$g(\mu_i) = h(t_i),\ i = 1, \ldots, n.$$

Given knot locations $s_1 < s_2 < \ldots < s_m$, one defines a basis $b_1(t) = 1, b_2(t) = t, b_{i+2}(t)$ $(i = 1, \ldots, m-2)$ where $b_{i+2}(t)$ denote cubic splines linked to the knots. Then, the function h at t_i is estimated through

$$\eta_i = \beta_1 + \beta_2 t_i + \beta_3 b_3(t_i) + \ldots + \beta_m b_m(t_i),\ i = 1, \ldots, n,$$

which yields a generalized linear model in the unknown parameters $\beta_1, \ldots, \beta_m$. This then leads to minimizing the penalized objective function using a smoothing parameter λ which controls the weight to be given to the smoothing:

$$\sum_{i=1}^{n}\frac{(y_i-\mu_i)^2}{V(\mu_i)} + \lambda\beta^t S\beta,$$

with $\beta = (\beta_1, \ldots, \beta_m)^t$ and S a symmetric $m \times m$ matrix only depending on the knots. Such analysis allows to validate a parametric proposal for the shape of a parametric regression function of λ or μ. We refer to Wood [795] for more details.

Case study: Dutch fire insurance data. The explained deviances for the Poisson GAM fit is 78.5% while for the negative binomial fit it is only 23%. Testing the null hypothesis of constant means μ_i cannot be rejected using the negative binomial model, but can be rejected under the Poisson model. Indeed, the horizontal line at the overall mean level jumps out of the Poisson confidence bound in 2005 (see Figure 5.3).

Case study: MTPL data for Company A. The explained deviances of the Poisson GAM fit is 52%, while for the negative binomial it is only 13.8%. Testing the null hypothesis of constant means μ_i over the different years cannot be rejected with a P-value of 0.26. Note that the horizontal line in Figure 5.4 at the overall mean level is

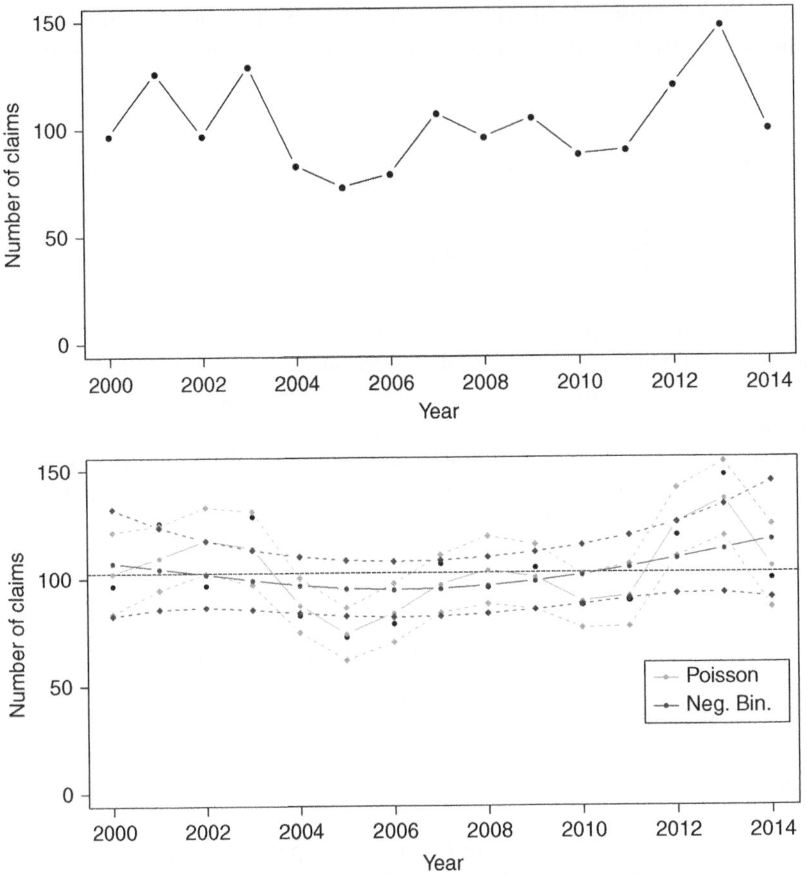

Figure 5.3 Dutch fire insurance data, claim counts: time plot of the yearly occurrences (top); non-homogeneous Poisson and negative binomial fits using GAM on μ_i, together with horizontal line from Poisson fit with no regression $\mu_i = \mu$ (bottom).

situated completely within the Poisson and negative binomial confidence bounds. In this case the premium volume (not disclosed here) was also available from 1998 on as an exposure measure ω_t. Using the exposure as an offset, this is, considering the GAM regression with response $\log(\mu_t/\omega_t)$, the explained deviance of the Poisson fit on this interval equals 94.5% (see Figure 5.4). Note that since the reinsurer only observes claims reaching the reporting threshold, the Poisson distributions for year i is thinned by a value $\mathbb{P}(T \leq v_i - 1)$, where T denotes the reporting delay and v_i is the number of years separating evaluation date and the claim occurrence year. Here we can refer to Figure 1.3 for estimates of these thinning probabilities. The delays being rather small in this case study there was hardly any influence on the resulting estimates.

Case study: MTPL data for Company B. As discussed in Chapter 1 (see Figure 1.2) the reporting thresholds for this company changed drastically from 1995 onwards, from which one can expect some significant changes in the mean number of claims reported to the reinsurer. The Poisson and negative binomial GAM fits hardly show any difference. The Poisson and negative binomial explained deviances are both about 80%.

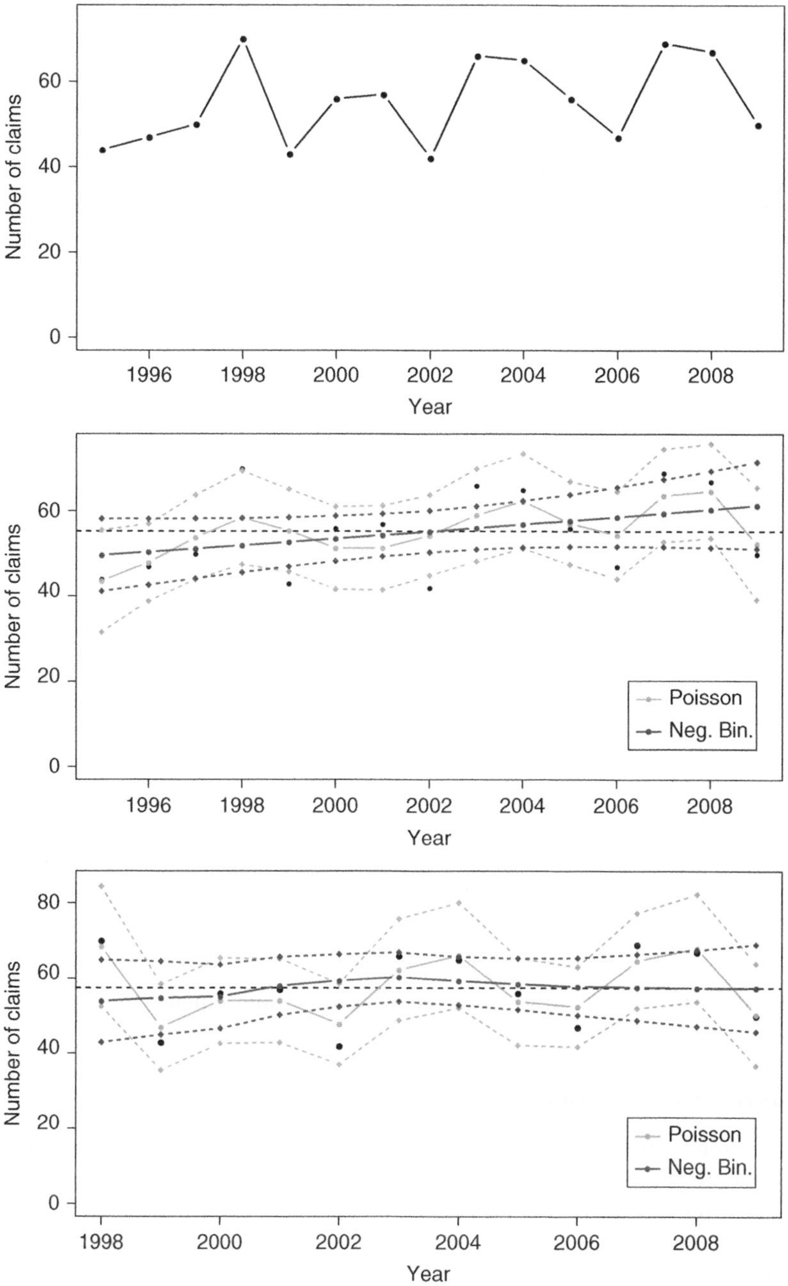

Figure 5.4 MTPL data for Company A, claim counts: time plot of the yearly occurrences (top); non-homogeneous Poisson and negative binomial fits using GAM on μ_i, together with horizontal line from Poisson fit with no regression $\mu_i = \mu$ (middle); Poisson and negative binomial fits with exposure as offset (bottom).

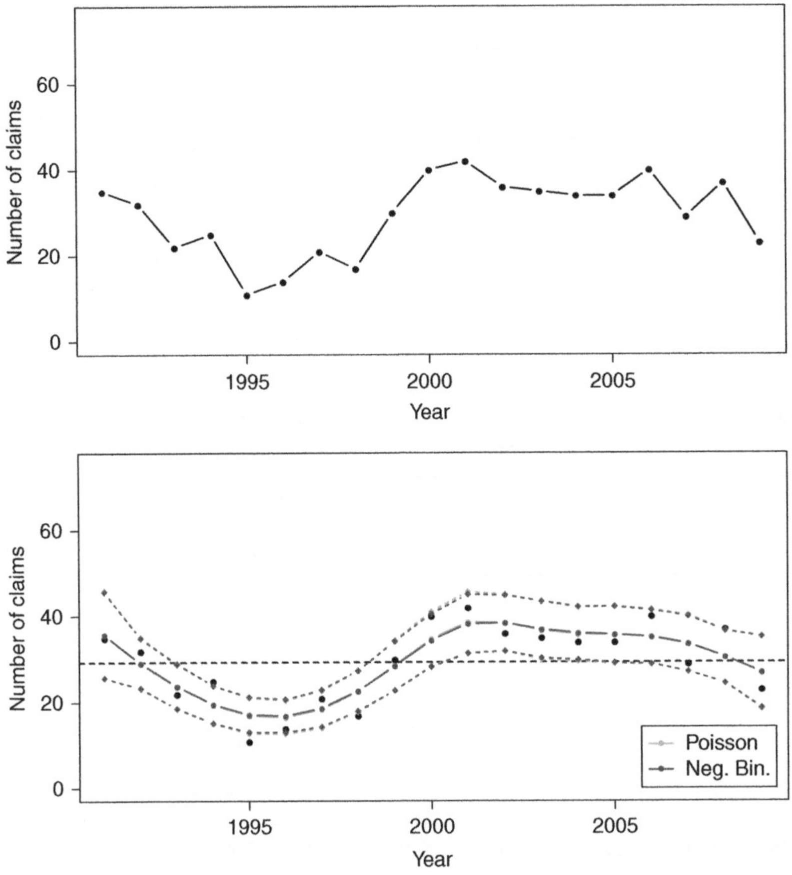

Figure 5.5 MTPL data for Company B, claim counts: time plot of the yearly occurrences (top); non-homogeneous Poisson and negative binomial fits using GAM on μ_i, together with horizontal line from Poisson fit with no regression $\mu_i = \mu$ (bottom).

The test for constant mean is strongly rejected with a P-value equal to 2.2×10^{-6} (see Figure 5.5).

5.5.2 Modelling the Claim Arrival Process

If one has more refined information on the times of claim occurrences, and hence the waiting times between these events, one can analyze the Poisson properties for these data based on the order statistics property and the independence and exponentiality of the waiting times W_i, $i = 1, \ldots, n$. To verify the order statistics property one plots the time points T_j against j:

$$(j, T_j),\ j = 1, \ldots, n. \tag{5.5.30}$$

This plot should closely follow the line $f(j) = \bar{w}j$, where $\bar{w}$ denotes the average waiting time, as an estimate of $1/\lambda$. This line hence represents how the arrival times increase with the claim number if the intensity were constant. Linearity corresponds to the uniform distribution of the arrival times over the interval $[0, T]$ where T represents the end of the inspected time interval. Deviation from this linear structure is then a first indication that the counting process is not a homogeneous Poisson process. A formal statistical test on uniformity can, for instance, be performed using the Kolmogorov–Smirnov test for uniformity comparing the empirical distribution function $\hat{\Gamma}_n(t) = n^{-1}\sum_{i=1}^{n} 1_{[0,t]}(T_i)$ against the c.d.f. t/T of the uniform $[0, T]$ distribution, rejecting uniformity for too large values of $D_{KS} = \sup_{t\in[0,T]} |\hat{\Gamma}_n(t) - t/T|$.

The exponentiality of the waiting times W_i can also be verified through the exponential QQ-plot of these inter-arrival times:

$$\left(-\log\left(1 - \frac{i}{n+1}\right); \log W_{i,n}\right), \quad i = 1, \ldots, n. \tag{5.5.31}$$

A formal test for exponentiality based on the exponential QQ-plot is given by the Shapiro–Wilk test based on the correlation coefficient r_Q of this QQ-plot and rejecting exponentiality for too low values of r_Q. Of course graphically one can also inspect for a constant derivative of mean excess plot based on the observed W values.

In order to see how the intensity λ varies over time one can use a moving average approach, plotting

$$1/\hat{\lambda}_i = \frac{1}{\min(n, i+m) - \max(1, i-m) + 1} \sum_{j=\max(1,i-m)}^{\min(n,i+m)} W_j. \tag{5.5.32}$$

against i ($i = 1, \ldots, n$). Of course under a simple Poisson counting process no trends should be visible in such plots. If deviations from a homogeneous Poisson behavior are detected in the plots, one can start evaluating the inhomogenous Poisson behavior and try to estimate the mean value function $\mu(t) = \int_0^t \lambda(u)du$ through (5.5.32). Then after a time change using the inverse function $\mu^{-1}(t)$ one can validate the homogeneous Poisson behavior of the time-changed process in order to inspect possible inhomogeneous Poisson behavior.

When large clusters of arrivals appear that cannot be predicted using an inhomogeneous Poisson behavior, a next option is to consider Cox process models, for instance directed by a Lévy subordinator. In this setting Zhang and Kou [806] proposed estimating $\mathbb{E}(\Lambda(t))$ using a kernel estimator based on the observed arrival times $\hat{T}_1 \leq \hat{T}_2 \leq \ldots \leq \hat{T}_n$ in the observed time window $[0, T]$

$$\hat{E}(\Lambda(t)) = \sum_{i=1}^{n} \frac{1}{h} K\left(\frac{1}{h}(\hat{T}_i - t)\right)$$

with bandwidth h.

For the estimation of a Cox process directed by a Lévy subordinator model such as the gamma or inverse Gaussian subordinators, we assume that the process N is observed up to a time horizon $T > 0$ on a discrete time grid $0 := t_0 < t_1 < \ldots < t_m := T$ with m monitoring points. Moreover we assume an equidistant grid with step size $h = T/m$, that is, $t_i = hi$ for $i = 0, 1, \ldots, m$. For instance the grid size could be a day or 1 week. We denote the observations of N, or the observed frequencies, at times t_i as

$$\hat{f}_i = \hat{N}(t_i),\ i = 0, 1, \ldots, m,$$

while $\hat{f}_0 = 0$. Following the Lévy property of such process N the increments $\Delta\hat{f}_i = \hat{f}_i - \hat{f}_{i-1}$ $(i = 1, \ldots, m)$ are i.i.d. observations of the infinitely divisible distribution $N(h)$. Using (5.2.16) we find that the log likelihood of the Cox process with the intensity parameter λ of the underlying Poisson process $\tilde{N}$ and the parameter vector Θ of the Lévy subordinator (i.e., $\Theta = (\alpha, \beta)$ in the gamma case, and $\Theta = (\beta, \eta)$ in the inverse Gaussian model) is given by

$$\begin{aligned} \ell\left(\lambda, \Theta | \Delta\hat{f}_i,\ i = 1, \ldots, m\right) &= \sum_{i=1}^{m} \log\left\{\frac{(-\lambda)^{\Delta\hat{f}_i}}{\Delta\hat{f}_i!}\hat{F}^{(\Delta\hat{f}_i)}_{M(h)}(\lambda, \Theta)\right\} \\ &= \hat{f}_m \log(\lambda) - \sum_{i=1}^{m}\log(\Delta\hat{f}_i!) + \sum_{i=1}^{m}\log\left((-1)^{\Delta\hat{f}_i}\hat{F}^{(\Delta\hat{f}_i)}_{M(h)}(\lambda, \Theta)\right). \end{aligned} \tag{5.5.33}$$

This results in optimizing

$$\hat{f}_m \log(\lambda) + \sum_{i=1}^{m}\log\left((-1)^{\Delta\hat{f}_i}\hat{F}^{(\Delta\hat{f}_i)}_{M(h)}(\lambda, \Theta)\right)$$

with respect to (λ, Θ).

Note that the intensity parameter λ can be selected upfront assuming time normalization as the sample mean of the process distribution

$$\hat{\lambda}_0 := \frac{1}{T}\hat{N}(T).$$

Case study: Dutch fire insurance data. The results of the moving average plots in Figure 5.6 with $m = 50$, with the yearly average waiting times added to it, clearly indicate non-constant intensity, both in the short term but also over the different years. Hence a homogeneous Poisson model is not appropriate for the entire time period.

Assume for the moment that the intensities were constant within each year. The plot (5.5.30) clearly shows a deviation from uniformity (Kolmogorov–Smirnov P-value 0.00016) (Figure 5.6, top right) and the piecewise linear mean value function μ with yearly levels equal to the estimated intensity in Figure 5.6 (top right) shows large fluctuations. Then the time-changed arrival times $\hat{\mu}(T_i)$ approximately will behave as arrival times of a standard homogeneous Poisson process. The plot of (5.5.31) for the time-changed waiting times $\hat{\mu}(T_i) - \hat{\mu}(T_{i-1})$ corresponds better to the requested linear pattern required for a homogeneous Poisson process, with Shapiro–Wilk P-value equal to 0.06. However, plotting the moving averages from (5.5.32) for the time-changed process still shows a considerable amount of volatility through time. In fact a seasonal effect is visible by plotting the boxplot of the time-changed waiting times per month computed over the available years, with shorter waiting times in winter and summer, hence higher intensity in winter and summer.

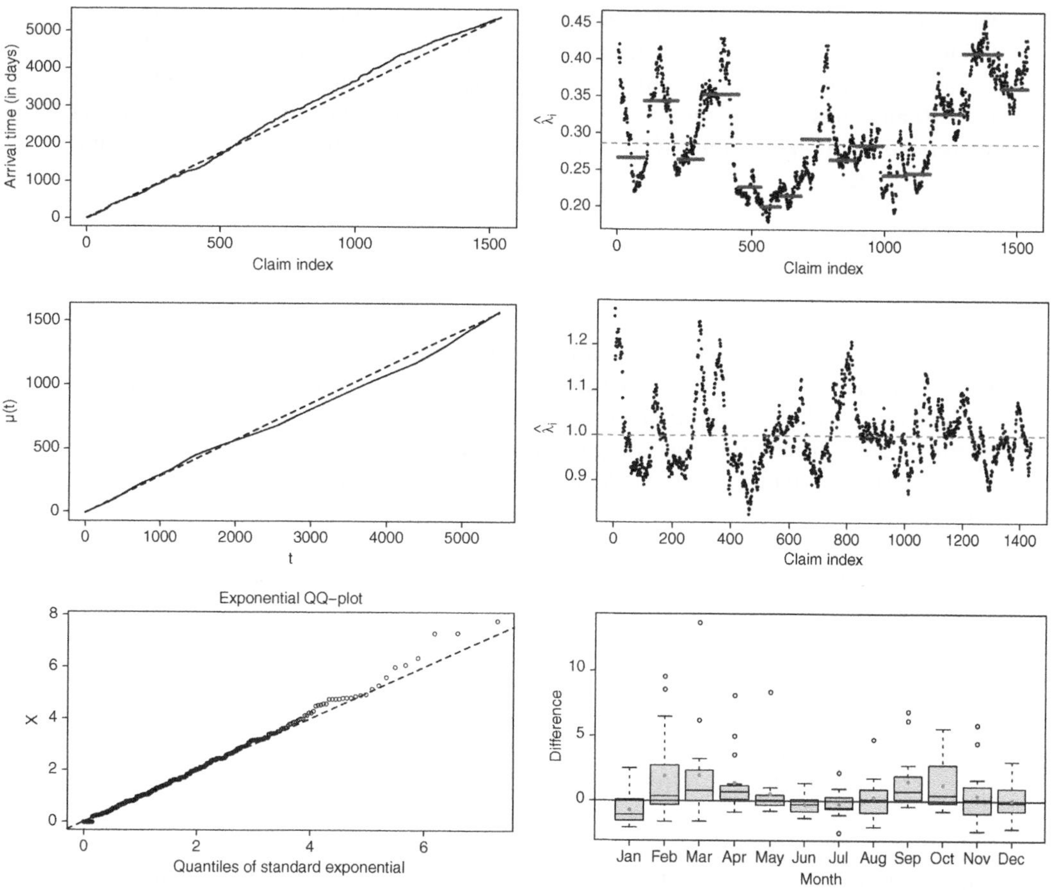

Figure 5.6 Dutch fire insurance data: distribution of time points (5.5.30) (top left); moving average estimates of intensity function (with $m = 50$) (top right); estimated time transformation (middle left); moving average estimates of intensity function after time transformation (middle right); exponential QQ-plot (5.5.31) based on μ-transformed waiting times (bottom left); boxplots of monthly mean waiting times after time transformation (bottom right).

To explore the seasonality effect further, a seasonal ARIMA model is fitted to the observed intensities (e.g., see Shumway [699] or Box and Jenkins [158]). Given that many days did not have a claim at all, the data were divided into one-week intervals. The average waiting time per week is then used as an estimate of $1/\lambda$ in that week.

A seasonal ARIMA process, or SARIMA$(p, d, q)(P, D, Q)_m$ process X_t, is an ARIMA (p, d, q) process where the residuals ϵ_t are ARIMA(P, D, Q) with respect to lag m:

$$(1 - \sum_{i=1}^{p} \phi_i L^i)(1 - L)^d X_t = (1 + \sum_{i=1}^{q} \theta_i L^i)\epsilon_t,$$

$$(1 - \sum_{i=1}^{P} \Phi_i L_m^i)(1 - L_m)^D X_t = (1 + \sum_{i=1}^{Q} \Theta_i L_m^i)\epsilon_t,$$

where $L_m X_t = X_{t-m}$, $L = L_1$, while ϕ_i, Φ_i are constants that determine the autocorrelation of the process with past values, and θ_i, Θ_i modulate how much past shocks have impacted on the current value. All ϵ_t are assumed to be i.i.d. normally distributed with mean 0 and constant variance σ^2. The seasonality of the data can be inspected using an *auto-correlation function* (ACF), which is defined for any process with mean μ_t and variance σ_t^2 as

$$R(s,t) = \frac{\mathbb{E}\left[(X_s - \mu_s)(X_t - \mu_t)\right]}{\sigma_s \sigma_t}.$$

In a seasonal process one expects significant auto-correlation with the data from the previous corresponding season. The *partial auto-correlation function* (PACF) is defined as the auto-correlation conditional on all values of the process between s and t, which eliminates the correlation originating from the values of X between s and t. Using the function AUTO.ARIMA in R, an ARIMA$(2,1,1)(0,0,1)_{52}$ model with mean 0 is selected given by

$$\lambda_t = 1.4241\lambda_{t-1} - 0.3275\lambda_{t-2} - 0.0966\lambda_{t-3} + \epsilon_t - 0.9848\epsilon_{t-1} + 0.2245\epsilon_{t-52} - 0.2211\epsilon_{t-53},$$

with $\epsilon_t \sim \mathcal{N}(0, 0.01648)$. When looking for a seasonal ARIMA with lag 26 the resulting model ARIMA$(2,1,1)(0,0,1)_{26}$ is

$$\begin{aligned} \lambda_t = {} & 1.4254\lambda_{t-1} - 0.3318\lambda_{t-2} - 0.0936\lambda_{t-3} + \epsilon_t - 0.9860\epsilon_{t-1} \\ & + 0.0543\epsilon_{t-26} - 0.0535\epsilon_{t-27} + 0.2261\epsilon_{t-52} - 0.2229\epsilon_{t-53}, \end{aligned} \tag{5.5.34}$$

with $\epsilon_t \sim \mathcal{N}(0, 0.01643)$. In Figure 5.7 we contrast the observed intensities with a simulated process from model (5.5.34), and add the ACF and PACF plots as well as a normal QQ-plot of the residuals of the fitted model. The observed data seem to show larger clustering. Process (5.5.34) also generates negative intensity values, which of course is not admissible. The normality of the residuals is also strongly rejected, with a correlation of 0.75 on the normal QQ-plot of the residuals. Several power transformations were therefore used, leading to an ARIMA$(2,1,1)(0,0,2)_{26}$ model based on the log-transformed intensities:

$$\begin{aligned} \log \lambda_t = {} & 1.7013 \log \lambda_{t-1} - 1.6929 \log \lambda_{t-2} - 0.0084 \log \lambda_{t-3} + \epsilon_t - 0.9873\epsilon_{t-1} \\ & + 0.0723\epsilon_{t-26} - 0.0713\epsilon_{t-27} + 0.1682\epsilon_{t-52} - 0.1661\epsilon_{t-53}, \end{aligned} \tag{5.5.35}$$

with $\epsilon_t \sim \mathcal{N}(0, 0.08385)$. Now the correlation coefficient of the normal QQ-plot of the residuals is 0.95, and the clusters are better approximated using (5.5.35), but still the clustering is underestimated.

Applying the likelihood method based on (5.5.33) using a daily observation grid over the entire observation period using a gamma and an inverse Gaussian model leads to the following parameters:

- gamma: $\hat{\lambda} = 0.1188$, $\hat{\alpha} = 5.1500$ and $\hat{\beta} = 2.1414$
- inverse Gaussian: $\hat{\lambda} = 0.1861$, $\hat{\beta} = 2.7018$ and $\hat{\eta} = 1.7598$.

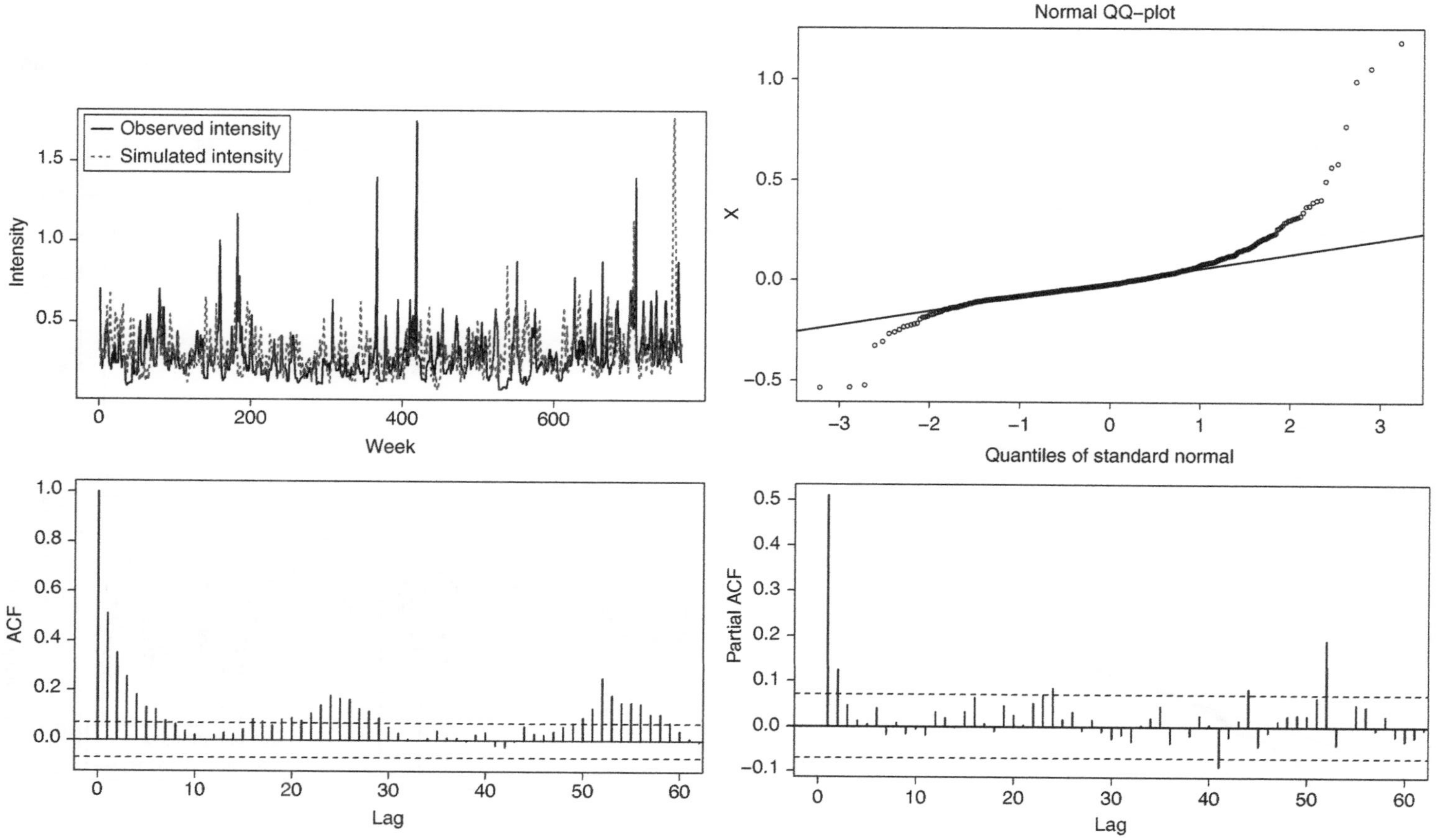

Figure 5.7 Dutch fire insurance data. Model (5.5.34) fit: observed and simulated intensities (top left); normal QQ-plot of residuals (top right); auto-correlation (second line left); partial auto-correlation (second line right). Model (5.5.35) fit: observed and simulated intensities (third line left); normal QQ-plot of residuals (third line right); auto-correlation (bottom left); partial auto-correlation (bottom right).

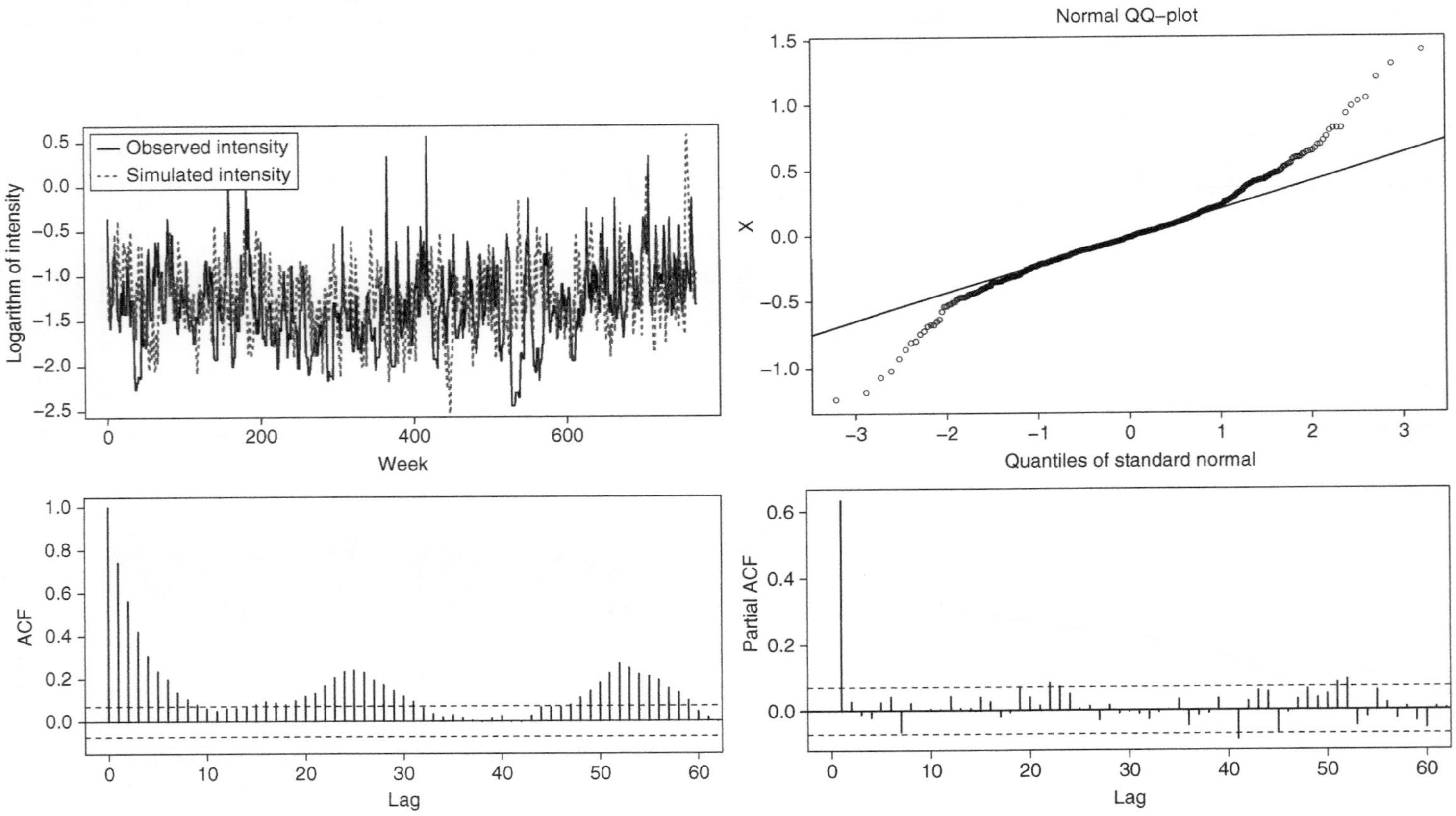

Figure 5.7 (Continued)

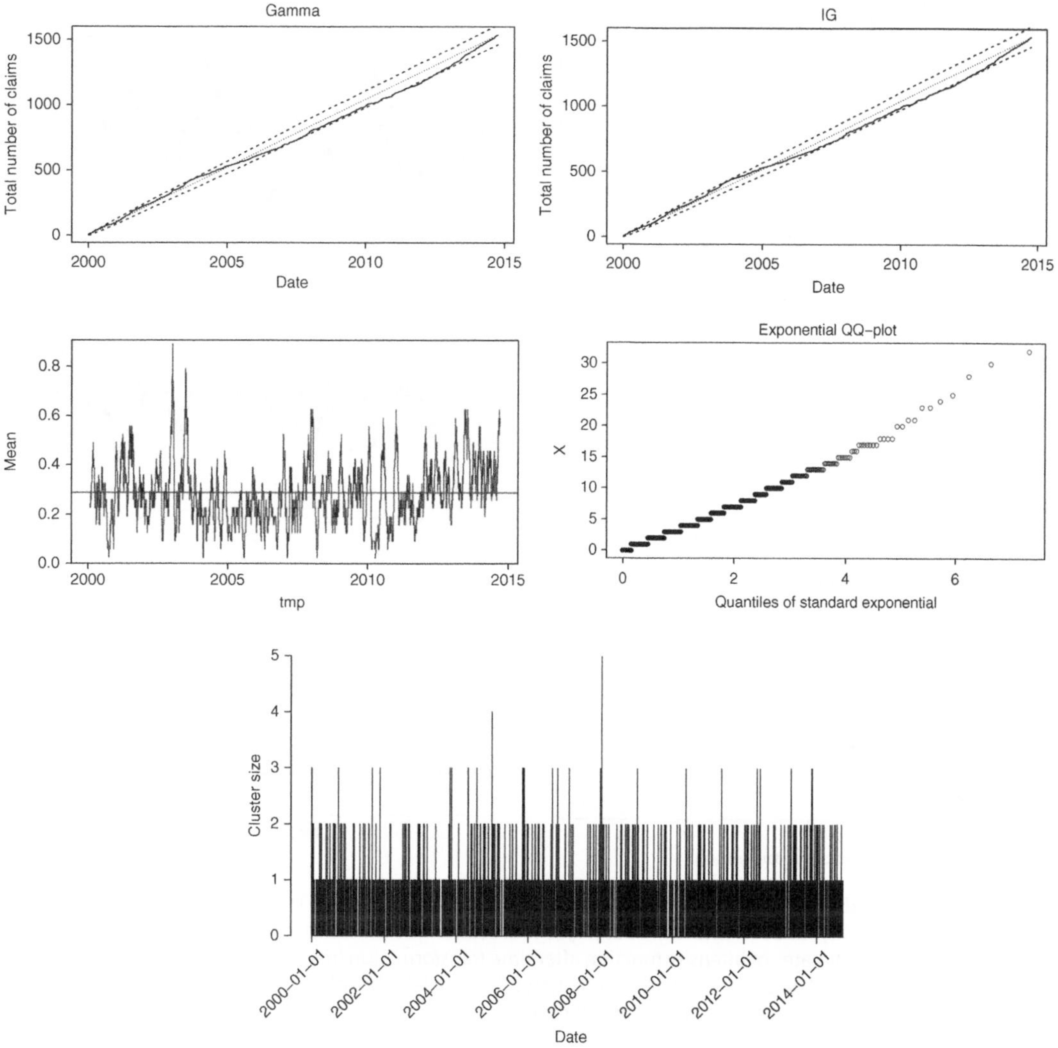

Figure 5.8 Dutch fire insurance data. Simulated confidence intervals and observed cumulative counting data for fitted Cox models directed by a gamma Lévy subordinator (top left) and an inverse Gaussian Lévy subordinator (top right); mean values of locally fitted Cox gamma models using 30 days moving windows (middle left) and exponential QQ-plot based on waiting times between clusters (middle right); cluster sizes of a simulated claim number process following the Cox model with the IG fitted parameters (bottom).

We simulated 100 sample paths from the fitted Cox models in order to construct confidence intervals at each time point. Both models do fit inside these bands, while the log-likelihood values in the fitted parameters are −3633.67 for the gamma model and −3634.07 for the inverse Gaussian model. In order to evaluate the fit of these models, the Cox models directed by the gamma model were applied locally using a moving window of 30 days. The recorded mean values of the estimated local processes are also given in Figure 5.8, in which no stochastic structure was detected. The exponentiality of

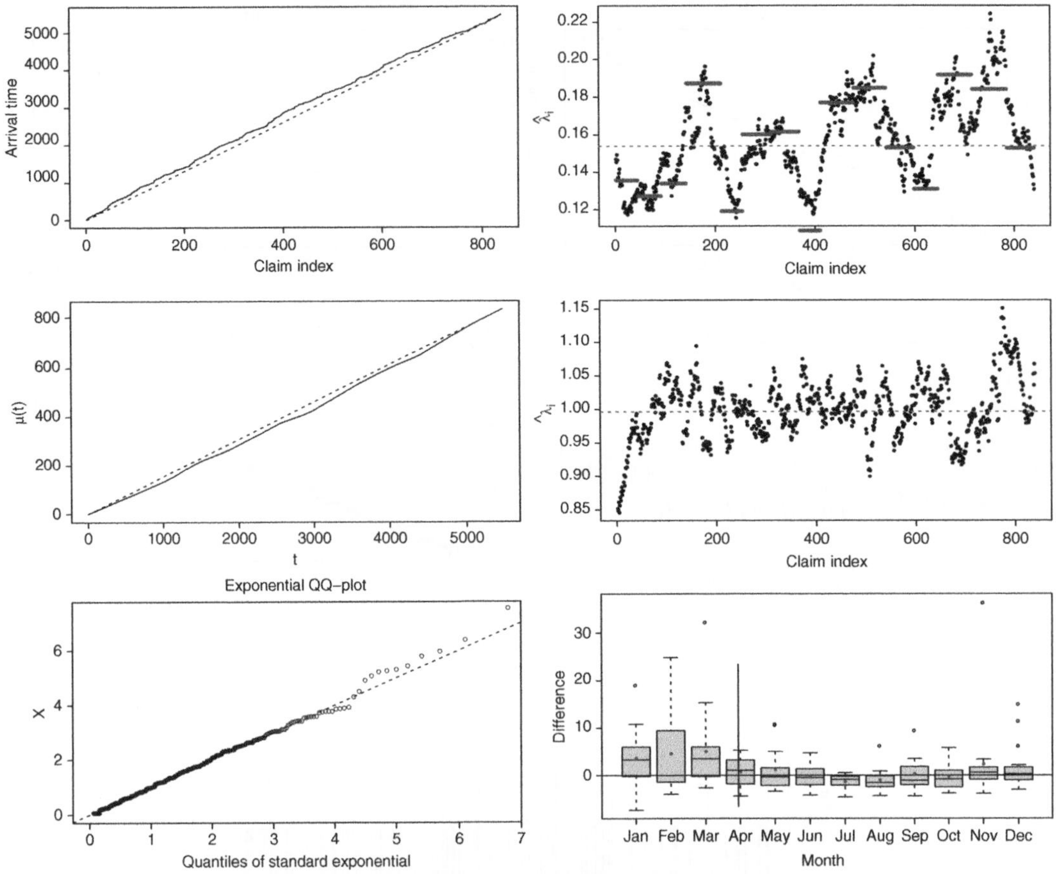

Figure 5.9 MTPL data for Company A: distribution of time points (5.5.30) (top left); moving average estimates of intensity function (with $m = 50$) (top right); estimated time transformation (middle left); moving average estimates of intensity function after time transformation (middle right); exponential QQ-plot (5.5.31) based on μ-transformed waiting times (bottom left); boxplots of monthly mean waiting times after time transformation (bottom right).

the waiting times between clusters also appears to be satisfied, and simulated processes based on the inverse Gaussian fit show similar cluster sizes, as illustrated in Figure 1.9.

Case study: MTPL data for Company A. We repeated the heterogeneous Poisson analysis with a time transformation for the MTPL occurrence data of Company A. The results are given in Figure 5.9. The Kolmogorov–Smirnov P-value equals 0.044, indicating that the homogeneous Poisson process model can be rejected here. The P-value of the Shapiro–Wilk test for exponentiality of the waiting times after time transformation equals 0.3735 so that a homogeneous Poisson process model cannot be rejected. However, some seasonal effects do appear but could not be proven to be significant, even with a time series analysis.

Applying the likelihood method based on (5.5.33) to using a daily observation grid over the entire observation period using a gamma and an inverse Gaussian model leads to the following parameters:

- gamma: $\hat{\lambda} = 0.0915$, $\hat{\alpha} = 3.8200$ and $\hat{\beta} = 2.2735$
- inverse Gaussian: $\hat{\lambda} = 0.1277$, $\hat{\beta} = 2.1592$ and $\hat{\eta} = 1.7932$.

Again we simulated 100 sample paths from the fitted Cox models in order to construct confidence intervals at each time point. Both models do fit inside these bands, while the log-likelihood values in the fitted parameters are almost equal: −2462.77 for the gamma model and −2462.73 for the inverse Gaussian model. In order to evaluate the fit of these models, the Cox models directed by the gamma model were applied locally using a moving window of 30 days. The recorded mean values of the estimated local processes are also given in Figure 5.10, in which no stochastic structure was detected. Also the exponentiality of the waiting times between clusters appears to be satisfied, and simulated processes based on the inverse Gaussian fit (see Figure 5.10) show similar cluster sizes as in Figure 1.8.

For the estimation of a multivariate Cox process directed by a Lévy subordinator we again assume that the multivariate process $\mathbf{N}$ is observed up to a time horizon $T > 0$ on a discrete time grid $0 := t_0 < t_1 < \dots < t_m := T$ with m monitoring equidistant grid points with step size $h = T/m$, that is, $t_i = hi$ for $i = 0, 1, \dots, m$. We denote the observations of $\mathbf{N}$, or the observed frequencies, at times t_i as

$$\hat{\mathbf{f}}_i = (\hat{f}_i^{(1)}, \dots, \hat{f}_i^{(d)}) = \hat{\mathbf{N}}(t_i),\ i = 0, 1, \dots, m,$$

while $\hat{\mathbf{f}}_0 = \mathbf{0}$. Following the Lévy property of such a process $\mathbf{N}$, the increments $\Delta\hat{\mathbf{f}}_j = \hat{\mathbf{f}}_j - \hat{\mathbf{f}}_{j-1}$ $(j = 1, \dots, m)$ are i.i.d. observations of the infinitely divisible distribution $\mathbf{N}(h)$. Using (5.2.18) optimizing the log likelihood of the Cox process with the intensity parameter vector λ of the underlying Poisson process $\tilde{\mathbf{N}}_\lambda$ and the parameter vector Θ of the Lévy subordinator (i.e., $\Theta = (\alpha, \beta)$ in the gamma case, and $\Theta = (\beta, \eta)$ in the inverse Gaussian model) results in optimizing

$$\sum_{j=1}^{d} \hat{f}_j^{(m)} \log(\lambda_j) + \sum_{i=1}^{m} \log\left((-1)^{|\Delta\hat{\mathbf{f}}_i|} \hat{F}_{M(h)}^{(|\Delta\hat{\mathbf{f}}_i|)}(\lambda, \Theta)\right) \tag{5.5.36}$$

with respect to λ, Θ.

Here, the use of time normalization means

$$\hat{\lambda}_0 := \frac{1}{T}\hat{\mathbf{N}}(T).$$

Case study: Danish fire insurance data. Selch [695] applied the likelihood method based on (5.5.36) to the complete Danish fire insurance data set using time normalization, which leads to $\alpha = \beta$ in the case of the gamma subordinator and $\beta = \eta$ for the inverse Gaussian subordinator (Table 5.2). Hence we only report $\hat{\alpha}$ for the gamma and $\hat{\beta}$ for the inverse Gaussian model.

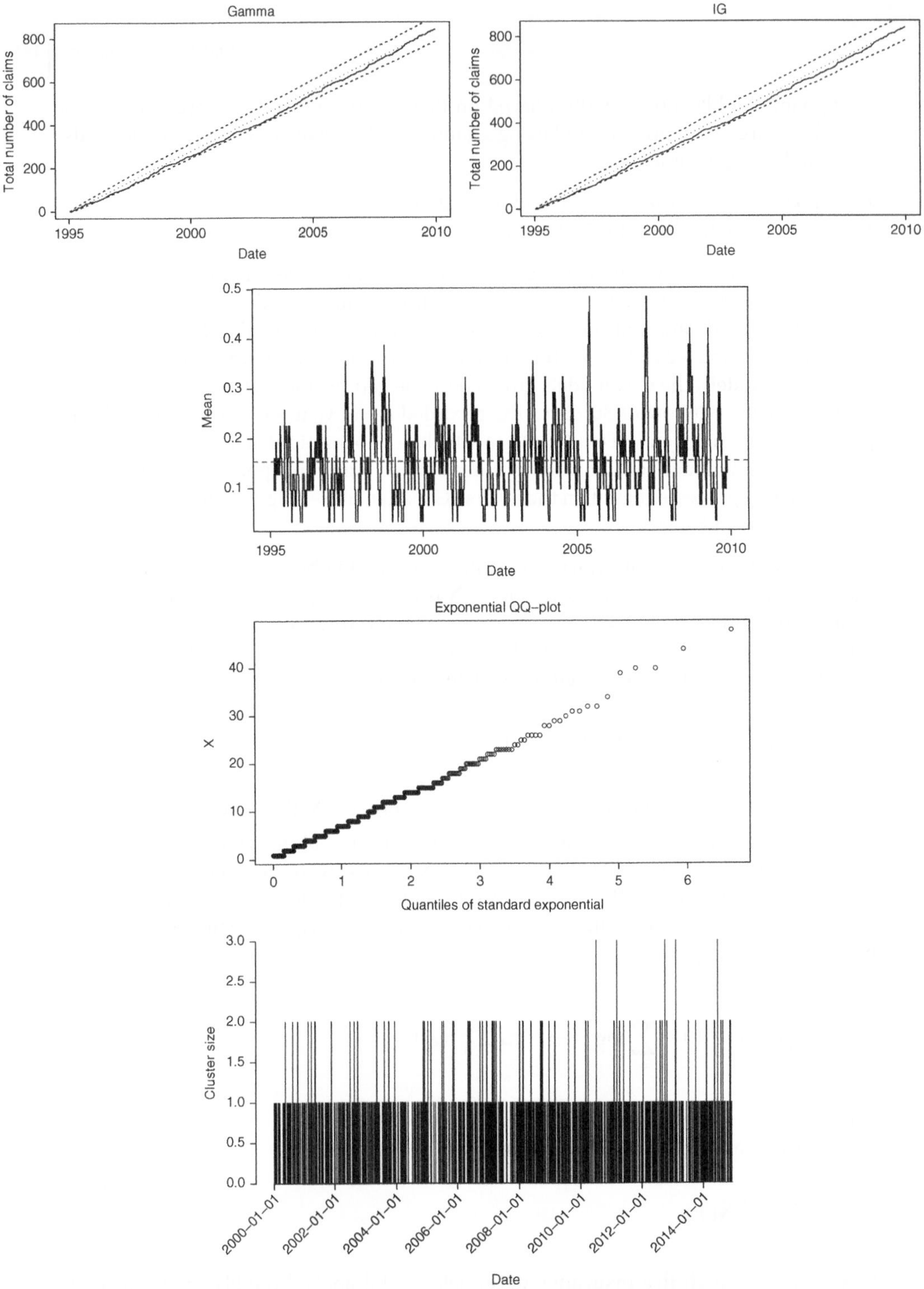

Figure 5.10 MTPL data for Company A: Simulated confidence intervals and observed cumulative counting data for fitted Cox models directed by a gamma Lévy subordinator (top left) and an inverse Gaussian Lévy subordinator (top right); mean values of locally fitted Cox gamma models using 30 days moving windows (second line) and exponential QQ-plot based on waiting times between clusters (third line); cluster sizes of a simulated claim number process following the Cox model with the IG fitted parameters (bottom).

Table 5.2

	λ_1	λ_2	λ_3	
inverse Gaussian	180.9	152.6	56.0	$\hat{\alpha} = 11.24$
gamma	180.9	152.6	56.0	$\hat{\beta} = 148.97$

It was observed that simulated cluster sizes are larger than the observed ones. Hence there is room for further model improvements in this case.

5.6 Claim Numbers under Reinsurance

In this section we look at the number of claims when reinsurance has been taken. If $\{N(t); t \geq 0\}$ denotes the number of original claims, then after reinsurance the insurer keeps $\{N_D(t); t \geq 0\}$ while the reinsurer gets $\{N_R(t); t \geq 0\}$. For all but XL reinsurance, the latter quantities are easily determined in terms of the former. We therefore deal with the XL case in a separate subsection.

1) In the case of *QS reinsurance* the reinsurer accepts a portion a of all first-line claims X_i, while the first-line insurer keeps $1 - a$ of these claims, but then

$$N_D(t) = N_R(t) = N(t) \,.$$

2) When there is a *SL reinsurance* contract with retention C, then the first-line insurer has to deal with all the incoming claims while for the reinsurer

$$N_R(t) = 1_{\left\{\sum_{i=1}^{N(t)} X_i > C\right\}} .$$

Indeed the reinsurer will only see one (aggregate) claim if the total claim amount in the original portfolio exceeds the threshold. However, in practice the reinsurer may also want to look into the settlement of (at least the larger) claims, leading to a big aggregate loss.

3) For *surplus reinsurance* the situation is somewhat more complex. The first-line insurer has to deal with all incoming claims since he can only shift part of an incoming claim to the reinsurer if the sum insured Q overshoots the retention line M. Therefore $N_D(t) = N(t)$. The reinsurer will only see the claims for which $Q > M$. This means that for the reinsurer the number of incoming claims equals the number of original claims for which $Q > M$. Correspondingly, if we have a distribution of Q available, $N_R(t)$ can be determined as in Section 5.6.1, with $s_M = \mathbb{P}(Q > M)$.

5.6.1 Number of Claims under Excess-loss Reinsurance

Under an XL contract, the first-line insurer has to pay the part of each claim that is below the retention, so that $N_D(t) = N(t)$. On the other hand, the reinsurer only has to pay for claims which touch the reinsured layer. Let $s_u = \mathbb{P}(X > u)$ be the probability that a claim exceeds the retention level u. Then

$$N_R(t) = \sharp\{k : 1 \le k \le N(t) | X_k > u\}$$

with

$$\tilde{p}_n(t) := \mathbb{P}(N_R(t) = n) = \sum_{k=n}^{\infty} p_k(t) \binom{k}{n} s_u^n (1 - s_u)^{k-n}, \quad n \in \mathbb{N}. \tag{5.6.37}$$

Here $p_k(t) = \mathbb{P}(N(t) = k)$ denotes the probability that there are exactly k claims occurring in the time interval from 0 to t.

An alternative expression can be obtained if we look at the generating function $\tilde{Q}_t(z) = \mathbb{E}(z^{N_R(t)})$. It is a straightforward calculation to relate it to the generating function $Q_t(z) = \mathbb{E}(z^{N(t)})$ of the original claim number:

$$\tilde{Q}_t(z) = Q_t((1 - s_u) + s_u z). \tag{5.6.38}$$

Written in this fashion, the variable $N_R(t)$ can be considered as a *thinned* version of the original $N(t)$.

From the above, one can quickly derive information on the moments of $N_R(t)$. For any non-negative integer k

$$\mathbb{E}\binom{N_R(t)}{k} = \frac{1}{k!}\tilde{Q}^{(k)}(1) = s_u^k\, \mathbb{E}\binom{N(t)}{k}.$$

In particular

$$\mathbb{E}(N_R(t)) = s_u\, \mathbb{E}(N(t)),$$

$$\mathrm{Var}(N_R(t)) = s_u^2\, \mathrm{Var}(N(t)) + s_u(1 - s_u)\, \mathbb{E}(N(t)).$$

For the dispersion one has

$$\tilde{I}(t) = s_u\, I(t) + (1 - s_u),$$

which indicates that the smaller s_u, the closer the dispersion gets to the value 1, which constitutes the Poisson case (this, by the way, is a neat way to intuitively see why the Poisson distribution is a natural model for rare events).

Let us illustrate the above procedure for a number of examples that have been introduced in the previous part.

(i) Our first example refers to the *mixed Poisson process*. If we look at the reinsured part then clearly,

$$\begin{aligned}\tilde{Q}_t(z) &= \int_0^{\infty} \exp\{-\lambda t(1 - ((1 - s_u) + s_u z))\} dF_\Lambda(\lambda) \\ &= \int_0^{\infty} \exp\{-\lambda t s_u(1 - z)\} dF_\Lambda(\lambda) = Q_{s_u t}(z).\end{aligned}$$

This simple relation shows that we only need to replace the time-variable t by $s_u t$ (i.e., a "slowing down" of the original claim number process). All other ingredients of the original mixed Poisson process remain the same. For example, generally

$$\mathbb{E}\binom{N_R(t)}{k} = \frac{(s_u t)^k}{k!}\mathbb{E}(\Lambda^k).$$

Let us illustrate the above with two special cases.

- For the *Pólya process* we have

$$\tilde{p}_n(t) = \binom{\alpha+n-1}{n}\left(\frac{b}{s_u t+b}\right)^{\alpha}\left(\frac{s_u t}{s_u t+b}\right)^{n}.$$

- For the *Sichel process* we obtain the generating function

$$\tilde{Q}_t(z) = [1+2\beta s_u t(1-z)]^{-\theta/2}\frac{K_\theta\left(\frac{\mu}{\beta}\sqrt{1+2\beta s_u t(1-z)}\right)}{K_\theta(\frac{\mu}{\beta})}.$$

(ii) For an *infinitely divisible process* we again look at what happens with the thinned process. From (5.6.38) we get

$$\tilde{Q}_t(z) = e^{-\lambda t(1-g((1-s_u)+s_u z))}$$

which can be rewritten in the form

$$\tilde{Q}_t(z) = e^{-\tilde{\lambda}t(1-\tilde{g}(z))}$$

where

$$\tilde{\lambda} = \lambda(1-g(1-s_u))$$

and

$$\tilde{g}(z) = \frac{g((1-s_u)+s_u z)-g(1-s_u)}{1-g(1-s_u)}.$$

It takes a little extra calculation to see that $\tilde{g}(z) = \mathbb{E}(z^{\tilde{G}})$ where

$$\tilde{g}_m := \mathbb{P}(\tilde{G}=m) = \frac{1}{1-g(1-s_u)}\sum_{n=m}^{\infty}\binom{n}{m}(1-s_u)^{n-m}s_u^m.$$

The latter formula can in itself be interpreted as a thinning of the original variable G. We therefore see that the thinned process remains infinitely divisible.

- For the *Neyman-type A-process*, the thinned process is of the same type but needs $\tilde{\theta} = s_u\theta$ as a new parameter.
- Similarly for the *Pólya–Aeppli process*, the thinned process is of the same type with $\tilde{\theta} = \frac{\theta s_u}{1-\theta(1-s_u)}$.

(iii) Next we deal with the *Sparre Andersen model.* In renewal theory thinning occurs by deleting each epoch of the renewal process with a fixed probability s_u, independently of the renewal process. The thinned process is a counting process that jumps at the time points $\tilde{T}_1 = \inf\{T_i \geq 0 : X_i > u\}$ and for $n \geq 1$, $\tilde{T}_{n+1} = \inf\{T_i \geq \tilde{T}_n : X_i > u\}$. This thinned process is again a renewal process, now generated by the mixture

$$\tilde{G}(x) = \sum_{j=1}^{\infty} s_u(1-s_u)^{j-1}\, G^{*j}(x).$$

(iv) Finally we return to the *(a,b,1)* class of Section 5.4. From the equations (5.6.38) and (5.4.24), some elementary algebra leads to

$$\tilde{Q}(z) = \tilde{p} + (1-\tilde{p})\frac{(1-\tilde{a}z)^{-\Delta}-1}{(1-\tilde{a})^{-\Delta}-1}, \tag{5.6.39}$$

where $\tilde{\Delta} = \Delta$ remains the same as before and where

$$\tilde{a} := \frac{as_u}{1-a(1-s_u)},$$

$$\tilde{b} := \frac{bs_u}{1-a(1-s_u)}$$

and

$$\tilde{p} = p + (1-p)\frac{(1-a(1-s_u)q)^{-\Delta}-1}{(1-a)^{-\Delta}-1} = \mathbb{P}(N_R = 0) = Q(1-s_u).$$

Hence the three subcases in Section 5.4 remain invariant under the given type of thinning, one just needs to adapt the parameters. Also, the binomial and the Engen cases remain invariant. Correspondingly, the distributions of claim numbers for the insurer and for the reinsurer then only differ by the applied parameters (and for this parameter change the quantity s_u plays a crucuial role). This is a further reason for the popularity of these claim count models in insurance and reinsurance modelling. A convenient general class of probability generating functions that remain invariant under thinning (including many zero inflated models) is discussed in Klugman et al. [492, Sec. 8.1], where also strong mixed Poisson characterizations of the class are given.

5.7 Notes and Bibliography

Classical models for claim numbers have been studied in various textbooks on actuarial science and statistics, for example Grandell [405, 406], Klugman et al. [491], Denuit et al. [276], and Mikosch [577]. In terms of the probability generating functions, the focus here was put on moments (i.e., $z = 1$), whereas one may also like to consider the probabilities ($z = 0$), for example Sundt and Vernic [722]. Linking discrete-time and continuous-time models is less commonly treated in actuarial circles even if it has a long tradition in collective risk theory. Ammeter [36] proposed a model where at equidistant time instants the value of the intensity is sampled anew from some given distribution. This model (often referred to as the Ammeter model) provides a simple form of a Cox process which is another alternative to embed discrete claim counts of mixed Poisson type within a continuous-time framework (with Cox processes directed by Lévy subordinators being a somewhat more general approach to that end). Björk and Grandell [138] considered an extension of [36] where the time instants of changing the intensity value are randomized as well, and Asmussen [56] studied a more general version in a Markovian framework, where subsequent intensities can possibly be dependent (see also Schmidli [673]). The concept of subordination is a classical topic in the theory and application of Lévy processes, for example Schoutens [683] or Kyprianou [521], but its explicit application to insurance claim processes was only taken up recently by Selch [695], even if infinitely divisible processes have been considered for many years. For an application of this joint subordination approach in mathematical finance, see Luciano and Schoutens [549].

An interesting argument for the use of mixed Poisson distributions is that if the data collected are thinned (e.g., due to a per claim deductible), then the mixed Poisson distribution is the only distribution that guarantees that also the resulting distribution of the original (non-thinned) number of claims is a valid probability distribution (see [492, p. 122]). A discussion on different interpretations of Pólya processes can also be found in Kozubowski and Podgorski [500]. The Pólya process also has been advocated in situations where there is some kind of claim contagion (cf. Panjer et al. [600]).

The order statistics property of the Poisson process actually holds more generally for the mixed Poisson process as well as the Yule process (but not for many further processes). It is very convenient for incorporating inflation and payment delays into the analysis (cf. [492, Ch. 9]) and also appears implicitly in density representations (see also Landriault et al. [527]). It can be useful to extend properties from the marginal to the conditional setting.

Shot-noise driven processes are classical examples of Cox processes used in the actuarial literature. For a recent reference in catastrophe modelling, see Schmidt [678] (see also Albrecher and Asmussen [14].

The $(a, b, 1)$ class has been introduced in an attempt to gather a variety of classical claim number distributions under the same umbrella. Willmot [784] reconsidered the equation and added a number of overlooked solutions. We point out that the recursion relation (5.4.22) can easily be generalized. For specific instances, see Willmot et al. [787], Schröter [684], and Sundt [719]. For a recent contribution and further references, see Hess et al. [438] and Gerhold et al. [388].

The section on statistical approaches is developed alongside the considered models and there is clearly a lot of potential for future research work.

There are conceptual links with approaches in the credit modelling context, for example Duffie et al. [311]. Maximum likelihood estimation for compound claim count distributions is quite tractable analytically with the help of generating functions (see Douglas [307] and Willmot and Sprott [788]). Badescu et al. [71] suggest modelling the claim number process together with its reporting delay as a marked Cox process, which allows for fluctuations in the exposure.

Simultaneous modelling of dependent portfolios becomes an increasingly important topic in practice. In addition to the recent contributions on multivariate Cox process modelling mentioned in this chapter, we also refer to Sundt and Vernic [722] for an overview of higher-dimensional approaches in terms of recursions. See also to Bäuerle and Grübel [91] and Genest and Neslehova [381] for early references to copula models. For common shock models, see, Wang [772], Lindskog and McNeil [545], and Meyers [576].

6

Total Claim Amount

After having modelled individual claim sizes and the number of claims, the next step is now to aggregate the individual claims and determine the resulting distribution. Even if the claim sizes are assumed to be independent (and independent of the number of claims), this is already a challenging yet classical actuarial task. In this chapter we will review the main techniques available for this purpose and then also discuss implications for the methods under reinsurance. Section 6.6 then gives some numerical illustrations and in Section 6.7 we will discuss some general aspects of the aggregation of dependent risks.

6.1 General Formulas for Aggregating Independent Risks

For most of this chapter we will assume that the two processes $\{N(t); t \geq 0\}$ and $\{X_i; i \in \mathbb{N}\}$ are independent. Recall from Chapter 2 that the *total* or *aggregate claim amount* at time t is defined by

$$S(t) = \begin{cases} \sum_{i=1}^{N(t)} X_i & \text{if } N(t) > 0, \\ 0 & \text{if } N(t) = 0. \end{cases}$$

Under the independence assumption, its c.d.f. then is

$$\mathbb{P}(S(t) \leq x) = \sum_{n=0}^{\infty} p_n(t) F_X^{*n}(x), \quad x \geq 0. \tag{6.1.1}$$

Whereas (6.1.1) fully specifies the total claim distribution, in this form it is not a workable expression to numerically evaluate $F_{S(t)}$ and we typically will have to look for ways to approximate $F_{S(t)}$. Since convolutions can efficiently be handled by Laplace transforms, a first step to simplify the expression is to look at the Laplace transform of $S(t)$:

$$\widehat{F}_{S(t)}(s) = \mathbb{E}(e^{-sS(t)}) = \sum_{n=0}^{\infty} p_n(t) \widehat{F}_X^n(s) = Q_t(\widehat{F}_X(s)), \tag{6.1.2}$$

Reinsurance: Actuarial and Statistical Aspects, First Edition.
Hansjörg Albrecher, Jan Beirlant and Jozef L. Teugels.

where $Q_t(z)$ is the generating function of the number of claims. From this expression one now obtains general formulas for the first few *moments* of $S(t)$.

- The most essential quantity related to the total claim amount is its expected value. As a function of time it offers a global picture of what happens to the portfolio over time. Taking the derivative with respect to s at $s = 0$ in (6.1.2) one easily obtains

$$\mathbb{E}(S(t)) = \mathbb{E}(X) \cdot \mathbb{E}(N(t)) \tag{6.1.3}$$

  for the mean total claim amount. This expression illustrates the complementary role of the average claim size and the average claim number. Clearly, this formula is pleasant and helpful when pricing according to the expected value principle, as dealing with the claim sizes and claim frequency can be separated.
- For the variance, Faa di Bruno's formula entails

$$\mathrm{Var}(S(t)) = \mathrm{Var}(X) \cdot \mathbb{E}(N(t)) + \mathbb{E}^2(X) \cdot \mathrm{Var}(N(t)). \tag{6.1.4}$$

  The role of the two ingredients comes out even better if we look at the dispersion. It easily follows from the above that

$$I_{S(t)} = I_X + \mathbb{E}(X) \cdot I_{N(t)} \cdot$$

  This illustrates that the variability is not only caused by the variability in claim sizes but also by that in the claim numbers. The above expression is useful in premium calculations that are based on the first two moments of $S(t)$ (cf. Chapter 7).
- Higher-order moments of S can be derived by taking the respective higher-order derivatives of $\widehat{F}_{S(t)}(s)$ at $s = 0$. In general, the nth moment of S can be expressed through combinations of the first n moments of X and of N.

Note that for the special case of a homogeneous Poisson process $\tilde{N}_\lambda$, one obtains

$$\mathbb{E}(S(t)) = \lambda t\, \mathbb{E}(X), \quad \mathrm{Var}(S(t)) = \lambda t\, \mathbb{E}(X^2)$$

and more generally $\kappa_n(S(t)) = \lambda t\, \mathbb{E}(X^n)$ for the nth cumulant of $S(t)$. The simplicity of these formulas is reason for the popularity of the compound Poisson model.[1]

Another general property of the total claim amount shows its direct dependence on the claim size distribution. Since a sum of non-negative random variables is always larger than the largest element in the sum, we note that $F_X^{*n}(x) \le F_X^n(x)$ and hence

$$\frac{1 - F_{S(t)}(x)}{1 - F_X(x)} \ge \frac{1 - Q_t(F_X(x))}{1 - F_X(x)}.$$

Letting $x \uparrow \infty$ we notice that

$$\liminf_{x \uparrow \infty} \frac{1 - F_{S(t)}(x)}{1 - F_X(x)} \ge \lim_{F_X(x) \uparrow 1} \frac{1 - Q_t(F_X(x))}{1 - F_X(x)} = \mathbb{E}(N(t)).$$

---

1 For $n = 3$ one observes, however, that the skewness of $S(t)$ in this model is always positive, which is a disadvantage in terms of model flexibility.

The above relation shows how the tail behavior of the total claim amount crucially depends on the heaviness of the individual claim size distribution and on the expected number of claims.

6.2 Classical Approximations for the Total Claim Size

A useful aggregate claim approximation should give reasonably accurate estimates and at the same time clarify the specific role of the two ingredients, the claim size and the claim number. We will now deal with several approximations for $F_{S(t)}$ that have been used in this context. In Section 6.6 we will implement these methods in R and compare their performance.

6.2.1 Approximations based on the First Few Moments

If the number of claims is large (which may be due to large t), then one can expect that for the sum of independent random variables, in case of finite variance $\mathrm{Var}(X)$, a central limit effect will be dominant. The *normal approximation*

$$F_{S(t)}(x) \approx \Phi\left\{\frac{x - \mathbb{E}(S(t))}{\sqrt{\mathrm{Var}(S(t))}}\right\} \tag{6.2.5}$$

may then be feasible. For this approximation to be useful, the number of terms in the sum should indeed be large, and it has to be the larger, the more skewed the distribution of the individual risks X_i is. Whereas this may work sufficiently well for large time horizons, within the typical one-year timeframe and appropriate choice of F_X, approximation (6.2.5) is usually too coarse in the regions of interest. If the skewness coefficient $\nu_{S(t)}$ is available, then a classical correction of (6.2.5) is the *normal-power approximation*

$$\mathbb{P}\left[\frac{S(t) - \mathbb{E}(S(t))}{\sqrt{\mathrm{Var}(S(t))}} \le z\right] \approx \Phi\left(\sqrt{\frac{9}{\nu_{S(t)}^2} + \frac{6z}{\nu_{S(t)}} + 1} - \frac{3}{\nu_{S(t)}}\right) \tag{6.2.6}$$

for $z \ge 1$.

Alternatively, if the first three moments of $S = S(t)$ are available, it has also been suggested that S can be approximated by a *shifted gamma distribution* with the same first three moments, that is,

$$S \approx \Gamma\left(\frac{4}{\nu_S^2}, \frac{2}{\nu_S\sqrt{\mathrm{Var}(S)}}\right) + \mathbb{E}(S) - \frac{2\sqrt{\mathrm{Var}(S)}}{\nu_S}.$$

Clearly, this can only be done if the skewness coefficient ν_S is positive.

Whenever there are higher moments available, one can use the rather elegant theory of *orthogonal polynomial expansions* to improve the approximation. For that purpose, starting from a reference density function $f_r(x)$ (defined on interval I) which one suspects to already be a reasonable approximation for the density $f_S(x)$ of the aggregate claim distribution, one adds correction terms based on moments of S. Concretely, if all moments for f_r exist, then the Gram–Schmidt orthogonalization identifies a set $(\pi_i)_{i\geq 0}$, where π_i is a polynomial of degree i, for which

$$\int_I \pi_i(x)\pi_j(x)f_r(x)\,dx = \begin{cases} 0 & \text{for } i \neq j, \\ 1 & \text{for } i = j. \end{cases} \tag{6.2.7}$$

If $\int_I e^{\alpha|x|}f_r(x)\,dx < \infty$ for some $\alpha > 0$, then this set of orthonormal polynomials is complete in the respective L^2 space, and hence if $\int_I (f_S(x)^2/f_r(x))\,dx < \infty$, one can express $f_S(x)/f_r(x)$ as $\sum_{i=0}^{\infty} A_i\pi_i(x)$, where

$$A_i = \int_I \pi_i(x)f_S(x)dx = \mathbb{E}(\pi_i(S)).$$

If K moments of S are available, one can correspondingly determine coefficients $A_0, \ldots, A_K$ and approximate f_S by

$$f_S(x) \approx f_r(x)\sum_{i=0}^{K} A_i\pi_i(x).$$

If the reference density f_r has m parameters, one can choose them in such a way that the first m moments corresponding to f_r coincide with those of $S(t)$. Let us consider two concrete examples:

- If we want to improve the normal approximation above, we can choose $f_r(x)$ to be the density of a normal $\mathcal{N}(\mathbb{E}(S(t)), \text{Var}(S(t)))$ random variable. Here $I = \mathbb{R}$ and the resulting polynomial family is

$$\pi_i(x) = \frac{1}{i!2^{i/2}} H_i\left(\frac{x - \mathbb{E}(S(t))}{\sqrt{2\,\text{Var}(S(t))}}\right), \quad i = 0, 1, \ldots,$$

  where $H_i(x) = \phi^{(i)}(x)/\phi(x)$ are the Hermite polynomials, with $\phi(x)$ denoting the density of the standard normal distribution. This is also known as the *Gram–Charlier approximation*. If we have (estimates for) the skewness $v_{S(t)}$ and the excess kurtosis $k_{S(t)} = \mathbb{E}(S(t)-\mathbb{E}(S(t)))^4/\text{Var}^2(S(t))-3$ of $S(t)$ available, the resulting approximation is

$$P(S(t) \leq x) \approx \Phi(z) - \frac{v_{S(t)}}{6}\Phi^{(3)}(z) + \frac{k_{S(t)}}{24}\Phi^{(4)}(z), \tag{6.2.8}$$

  where $z = (x - \mathbb{E}(S(t)))/\sqrt{\text{Var}(S(t))}$. That is, we get a refinement of (6.2.5) using the third and fourth moment. The three terms of (6.2.8) in fact coincide with the first three terms of the so-called *Edgeworth expansion*, which builds on saddlepoint approximations and also enjoys some popularity in actuarial circles.
- Choosing $f_r(x)$ to be a gamma(α, β) density, one can improve on the gamma approximation for $S(t)$. Here $I = (0, \infty)$ and the respective orthogonal polynomial family is

$$\pi_i(x) = (-1)^i \left[\frac{\Gamma(i+\alpha)}{\Gamma(i+1)\Gamma(\alpha)}\right]^{-1/2} L_i^{\alpha-1}(\beta x), \quad i = 0, 1, \ldots,$$

where $(L_i^{\alpha-1})_{i\geq 0}$ denote the generalized Laguerre polynomials. Furthermore $\alpha = \mathbb{E}^2(S(t))/\text{Var}(S(t))$ and $\beta = \mathbb{E}(S(t))/\text{Var}(S(t))$. By construction, again $A_1 = A_2 = 0$. The resulting approximation is referred to as *Bower's gamma approximation*. Observe that if we truncate the approximation after K terms, the tail of $f_{S(t)}$ is approximated by the dominant order $x^K f_{\Gamma(\alpha,\beta)}(x)$, which is itself a gamma-type decay. It will turn out that such a tail behavior is in fact quite appropriate under rather general conditions (cf. Section 6.2.2).

- For situations with heavy tails, it is tempting to choose a heavy-tailed reference density f_r and then improve the approximation of $S(t)$ with higher moments according to the recipe above. For a log-normal density f_r, the corresponding family of orthonormal polynomials has recently been worked out in Asmussen et al. [60]. Note, however, that the resulting family is not complete in the respective L^2 space, so that $f_{S(t)}(x)$ cannot be represented by the infinite series $f_r(x)\sum_{i=0}^{\infty} A_i\pi_i(x)$. This is of course related to the fact that the log-normal distribution is not determined by its moments. Nevertheless, an approximation based on a few terms can be useful (see [60] for an in-depth study of this and for further aspects of orthonormal approximations in general). Since Pareto distributions have only finitely many moments (and for many realistic parameter settings even only a few), the approach via orthonormal approximation lacks mathematical justification in that case.

Some further approximations with a similar flavor will be discussed in the Notes at the end of the chapter. In some of the mentioned approximations, it may happen that some probability mass is assigned to the negative half-line. However, since the focus usually is on large values of x, this problem may be considered negligible.

Approximations based on a few moments have the advantage that the simple expressions allow the influence of the individual moments (for both claim sizes and numbers) on the overall result to be assessed. This can be particularly useful if the estimates for those moments are not very reliable (therefore such, albeit rough, approximations can be a good complement to exact numerical values for the c.d.f. of $S(t)$, as determined in Sections 6.3 and 6.4, which do not give immediate insight to sensitivities). At the same time, moment-based approximations can perform quite poorly if one goes further into the tail (cf. Section 6.6 for illustrations), but these tail regions are of substantial importance in reinsurance applications. Moreover, if the number of claims that *de facto* determine the distribution of the entire portfolio is small (which is the case for large claims), a centralization of the claims is practically absent and approximations in the spirit of central limit theory will be inaccurate. In what follows we discuss as an alternative some asymptotic approximations which become increasingly accurate the further one goes into the tail. For that purpose, we have to distinguish the cases of light- and heavy-tailed claims.

6.2.2 Asymptotic Approximations for Light-tailed Claims

One of the important features of super-exponential distributions that distinguishes them from heavy-tailed ones is the existence of a family of *associated distributions* with arguments $s < 0$. These will enable a general procedure that leads to exponential estimates for the tail of $F_{S(t)}(x)$.

Recall from Chapter 3 that the Laplace transform $\widehat{F}(s)$ of a light-tailed claim has a strictly negative abscissa of convergence σ_F. Take any $s > \sigma_F$ and define

$$F_s(x) = \frac{1}{\widehat{F}(s)} \int_0^x e^{-st}\, dF(t).$$

Then it is easy to prove that F_s is a proper distribution on $\mathbb{R}_+$. The full family of such distributions $\{F_s; s > \sigma_F\}$ is called the class of distributions *associated* to F (also referred to as *exponential tilting* or *Esscher transform*). For each associated distribution one can compute its Laplace transform. It is easy to see that

$$\widehat{F}_s(u) = \frac{\widehat{F}(u+s)}{\widehat{F}(s)}\,. \tag{6.2.9}$$

This equation can be raised to the nth power where $n \in \mathbb{N}$, but $\widehat{F}_s^n$ is the Laplace transform of F_s^{*n}. By the uniqueness of the Laplace transform the relation

$$\widehat{F}^n(s) \int_0^\infty e^{-ux}\, dF_s^{*n}(x) = \widehat{F}^n(u+s) = \int_0^\infty e^{-ux}\, e^{-sx} dF^{*n}(x)$$

easily yields that for all $x \geq 0$

$$\widehat{F}^n(s)\, dF_s^{*n}(x) = e^{-sx}\, dF^{*n}(x).$$

Integration over the interval (y, ∞) gives the relation

$$1 - F^{*n}(y) = \widehat{F}^n(s) \int_y^\infty e^{sx}\, dF_s^{*n}(x). \tag{6.2.10}$$

The important feature of the above expression is that on the right-hand side there is a parameter s which is only restricted by the inequality $s > \sigma_F$. In practical applications a judicious choice of this parameter may rewrite intractable formulas into simpler ones. Up to now the sign of σ_F has been unimportant. In the super-exponential case, however, $\sigma_F < 0$ and one still can go a bit further in the above analysis. It is not hard to see that if $\sigma_F < \theta < 0$, then

$$1 - F^{*n}(x) = |\theta| \widehat{F}^n(\theta)\, e^{-|\theta| x} \int_0^\infty e^{-|\theta| v} \left\{ F_\theta^{*n}(x+v) - F_\theta^{*n}(x) \right\} dv, \tag{6.2.11}$$

which gives a rather explicit expression for the tail of the n-fold convolution of F in terms of a decreasing exponential.

As $1 - F_{S(t)}(x) = \sum_{n=0}^\infty p_n(t)(1 - F_X^{*n}(x))$, we get

$$1 - F_{S(t)}(x) = |\theta| e^{-|\theta| x} \int_0^\infty e^{-|\theta| v} \{M_\theta(x+v) - M_\theta(x)\}\, dv \tag{6.2.12}$$

for the so-called *weighted renewal function*

$$M_\theta(x) := \sum_{n=0}^{\infty} p_n(t)\, \widehat{F}_X^n(\theta) F_\theta^{*n}(x). \tag{6.2.13}$$

The goal is now to get an asymptotic approximation of (6.2.12) for $x \to \infty$. There is a generalization of *Blackwell's theorem* from renewal theory stating that for any weighted renewal function $M(x) = \sum_{n=1}^{\infty} a_n K^{*n}(x)$, under appropriate conditions on the function $a(x) := a_{[x]}$ as $x \uparrow \infty$, for any $y \in \mathbb{R}$ one has

$$M(x+y) - M(x) \sim \frac{y}{\mu_K} a\left(\frac{x}{\mu_K}\right) \quad \text{as } x \to \infty, \tag{6.2.14}$$

where μ_K is the (finite) mean of K. Among the many possible sufficient conditions on $a(x)$ we mention those given by Embrechts et al. [331]. There it is assumed that

$$\sum_{i=0}^{n} a_i \sim n^{\rho+1} \ell(n) \text{ as } n \to \infty,$$

where ℓ is slowly varying and in addition one of the following conditions has to hold (for weaker conditions and related results see Omey et al. [595]):

- $\rho > 0$
- $\rho = 0$ and for some $\delta > 0$, $x^{1+\delta}\{1 - K(x)\} \to 0$ when $x \uparrow \infty$
- $\rho < 0$ and $1 - K(x) = O(a(x))$.

In the applications below, the regular variation of $a(x)$ will be easily guaranteed, and the extra condition on $K(x) = F_\theta(x)$ is satisfied as well for $\theta > \sigma_{F_X}$, since then the tail of $1 - F_\theta(x)$ is still bounded above by a decreasing exponential. So the approximation (6.2.14) can then be applied and Lebesgue's theorem allows the limit to be brought inside the integral in (6.2.12), as the difference $M_\theta(x+v) - M_\theta(x)$ is bounded above by a regularly varying function which itself is integrable with respect to $\exp(-|\theta|x)$. That is, under the condition on $a(x)$ (which essentially translates into a condition on the decay of the claim number distribution) we obtain

$$1 - F_{S(t)}(x) \sim \frac{1}{|\theta|\mu_\theta} e^{-|\theta|x} a\left(\frac{x}{\mu_\theta}\right), \qquad x \to \infty, \tag{6.2.15}$$

where $\sigma_{F_X} < \theta < 0$ and $\mu_\theta = \int_0^\infty x\, dF_\theta(x) = \frac{|\widehat{F}_X'(\theta)|}{\widehat{F}_X(\theta)}$. The formula holds for an entire range of θ values, and the idea is to pick a value for which the desired asymptotic condition on $a(x)$ can be achieved.

Formula (6.2.15) is remarkable in several ways. First, if it is applicable, it shows that the exponentially bounded decay of the individual claim size distribution carries over to an essentially exponentially bounded decay of the total claim size distribution, and this explicit first-order term quantifies by how much the heaviness of the tail increases through the aggregation. Second, for regularly varying $a(x)$ one sees that the shape of the tail is essentially an exponentially decreasing term multiplied by a power term, which is the decay pattern of a gamma random variable. This gives an additional motivation to consider the Laguerre series expansion discussed in Section 6.2.1.

6.2.2.1 Examples

i) Consider first the *negative binomial distribution* with

$$p_n(t) = \binom{\alpha+n-1}{n}\left(\frac{b}{t+b}\right)^{\alpha}\left(\frac{t}{t+b}\right)^{n}.$$

One observes that the sequence $a_n = p_n(t)\,\hat{F}_X^n(\theta)$ indeed is of regularly varying type, if we choose $\theta = \theta^*$ in such a way that the geometric terms disappear, which is the case for

$$\hat{F}_X(\theta^*) = 1 + \frac{b}{t}. \tag{6.2.16}$$

For this value θ^* one then gets $a_n \sim n^{\alpha-1}\left(\frac{b}{t+b}\right)^{\alpha}$, while $\mu_{\theta^*} = \frac{t\hat{F}'(\theta^*)}{b+t}$. Correspondingly, (6.2.15) leads to

$$1 - F_{S(t)}(x) \sim \left(\frac{b}{|t\hat{F}_X'(\theta^*)|}\right)^{\alpha}\frac{1}{|\theta^*|\Gamma(\alpha)}e^{-|\theta^*|x}x^{\alpha-1}, \quad x\to\infty. \tag{6.2.17}$$

This estimate can be found in Embrechts et al. [332]. By elaborating on further refinements one even can give a sharpening of the above estimate of the form

$$1 - F_{S(t)}(x) = C_1 x^{\alpha-1}e^{-|\theta^*|x}\left\{1 + \frac{C_2}{x} + o\left(\frac{1}{x}\right)\right\}, \quad x\to\infty,$$

where C_1 is the constant resulting from the first approximation while C_2 is another constant depending on the variance of the associated distribution F_{θ^*}.

From (6.2.16) we see that the determination of θ^* can still be tricky, in practice one may need a rather precise information on the empirical Laplace transform corresponding to $\hat{F}_X(s)$. Note at the same time that the influence of the time horizon t on the estimate appears rather explicitly.

ii) The *logarithmic distribution* has appeared as a candidate in the $(a, b, 1)$ class. It can be treated similarly to the negative binomial distribution and is actually technically simpler. Recall from (5.4.26) that

$$p_n = (-\log(1-a))^{-1}\,a^n/n,\ n > 0$$

where $a < 1$. Here the essential power factor can again be removed by a choice of θ through $\hat{F}(\theta^*) = 1/a$. We then get

$$1 - F_S(x) \sim (-\log(1-a)|\theta^*|)^{-1}x^{-1}e^{-|\theta^*|x}, \quad x\to\infty.$$

iii) The *Paris–Kestemont distribution* appeared as an example of a claim counting process of the infinitely divisible type. More explicitly, the generating function for the number of claims is

$$Q_t(z) = \exp\{-\eta[1-(1+t(1-z))^{-\alpha}]\}.$$

From asymptotic analysis one can derive the following asymptotic expression for the probabilities

$$p_n(t) \sim \eta e^{-\eta}(1+t)^{-\alpha}\frac{1}{\Gamma(\alpha)}\left(\frac{t}{1+t}\right)^n n^{\alpha-1}, \quad n \to \infty.$$

As before we need $\theta = \theta^*$ to satisfy the condition $\widehat{F}(\theta^*) = 1 + t^{-1}$. A little algebra then yields the expression

$$1 - F_{S(t)}(x) \sim \frac{\eta e^{-\eta}}{|\theta^*|\Gamma(\alpha)}(t|\widehat{F}'(\theta^*)|)^{-\alpha}e^{-|\theta^*|x}x^{\alpha-1}, \quad x \to \infty.$$

iv) Take a *Sichel distribution* with probability mass function

$$p_n(t) = \frac{(\mu t)^n}{n!}(1+2\beta t)^{-\frac{1}{2}(\eta+n)}\frac{K_{\eta+n}\left(\frac{\mu}{\beta}\sqrt{1+2\beta t}\right)}{K_\eta\left(\frac{\mu}{\beta}\right)}, \quad n \to \infty.$$

In the asymptotic theory for Bessel functions we find an approximation for the modified Bessel function which leads to the asymptotic expression

$$p_n(t) \sim \left\{2(\mu t)^\eta K_\eta\left(\frac{\mu}{\beta}\right)\right\}^{-1}\left(\frac{2\beta t}{1+2\beta t}\right)^n n^{\eta-1}.$$

Following the same path as in the previous examples, we are forced to take θ in such a way that $\widehat{F}_X(\theta^*) = 1 + (2\beta t)^{-1}$. It then follows that

$$1 - F_{S(t)}(x) \sim \frac{1}{2|\theta^*|K_\eta\left(\frac{\mu}{\beta}\right)}\left(\frac{2\beta t^2\mu}{1+2\beta t}|\widehat{F}_X'(\theta^*)|\right)^{-\eta} e^{-|\theta^*|x}x^{\eta-1}, \quad x \to \infty.$$

For the special case when $\eta = -1/2$ we have

$$1 - F_{S(t)}(x) \sim \frac{\mu t}{|\theta^*|}\sqrt{\frac{|\widehat{F}_X'(\theta^*)|}{\pi(1+2\beta t)}}\, e^{-|\theta^*|x}x^{-\frac{3}{2}}, \quad x \to \infty,$$

a result due to Willmot [783].

While these expressions are simple and transparent, one should keep in mind that they only apply for large x values. Their appeal lies in the fact that one gets a first rough impression on the sensitivity of the aggregate claim size distribution with respect to the model parameters for regions far in the tail (whereas the approximations of Section 6.2.1 are more appropriate around the mean). This can help to quantitatively support the intuition, particularly in terms of robustness with respect to parameter uncertainty.

Note that the above approach does not apply for Poisson claim numbers, as the factorial term in the denominator of its probability function cannot be removed by some choice of θ, so that the condition on $a(x)$ is not fulfilled. In the Notes at the end of this chapter we mention other methods to derive exponential estimates for cases where the above method does not apply.

6.2.3 Asymptotic Approximations for Heavy-tailed Claims

Recall that for sub-exponential claims one has the asymptotic equivalence

$$1 - F_X^{*n}(x) \sim n(1 - F_X(x))$$

as $x \uparrow \infty$, which signifies that the largest claim asymptotically dominates an independent sum of n claims. One would hope that for a stochastic number $N(t)$ for the number of summands it is possible to simply shift the limiting operation through the summation to obtain the exceptionally simple result

$$1 - F_{S(t)}(x) \sim \mathbb{E}(N(t)) \cdot (1 - F_X(x)), \quad \text{as } x \to \infty. \tag{6.2.18}$$

This is indeed the case if the generating function $Q_t(z)$ is analytic in a neighborhood of the point $z = 1$. This simple condition of analyticity essentially means that $|p_n(t)| \leq r^{-n}$ for some value $r > 1$, that is, $p_n(t)$ has to decay geometrically fast with n. The asymptotic result (6.2.18) is very elegant and also rather robust with respect to variations in the model assumptions (even with respect to certain dependence among risks), which contributes to its enormous popularity in academic research (see the references at the end of the chapter). It also certainly adds to the intuition of dealing with large claims. However, for practical use the important question is whether for relevant finite magnitudes of x, the first-order asymptotic approximation (6.2.18) is already sufficiently accurate for practical purposes, and unfortunately this is typically not the case (see also Section 6.6).

One way to improve the approximation is to add higher-order terms in the asymptotic expansion. A classical result by Omey and Willekens [596, 597] establishes that if F_X has a density f_X with finite mean μ, $Q_t(z)$ is analytic at $z = 1$, and the refined asymptotic behavior $(1 - F_X^{*2}(x)) - 2(1 - F_X(x)) \sim 2\mu f(x)$ holds, then this implies

$$1 - F_{S(t)}(x) - \mathbb{E}(N(t))(1 - F_X(x)) \sim 2\mu f(x)\mathbb{E}\left(\binom{N(t)}{2}\right), \quad \text{as } x \to \infty. \tag{6.2.19}$$

More generally, under suitable conditions and if the respective quantities exist, higher-order correction terms are of the form

$$\sum_{j=0}^{k-2} \frac{(-1)^j \mathbb{E}\{N(t) \cdot (X_1 + \cdots + X_{N-1})^{j+1}\}}{(j+1)!} f^{(j)}(x).$$

Determining, however, the exact conditions under which such higher-order expansions are justified is a challenging research topic and has been studied intensively in the academic literature (e.g., see Grübel [410], Willekens and Teugels [780], Barbe and McCormick [81], and Albrecher et al. [23] for a recent account). In the latter reference it is also shown how additional higher-order terms can be replaced by adding the degree of freedom of shifting the argument x. For instance, the expression

$$1 - F_{S(t)}(x) \sim \mathbb{E}(N(t)) \cdot \left\{ 1 - F_X \left(x - \left(\frac{\mathbb{E}(N^2(t))}{\mathbb{E}(N(t))} - 1 \right) \mathbb{E}(X) \right) \right\}$$

achieves (roughly) the same accuracy as (6.2.19). For Poisson claim numbers this simplifies to the intuitive formula

$$1 - F_{S(t)}(x) \approx \mathbb{E}(N(t)) \cdot (1 - F_X(x - \mathbb{E}(S(t)))),$$

and this shift by the mean total claim has indeed been used sometimes as a rough approximation in practice, even before the mathematical justification in [23].

6.3 Panjer Recursion

When the claim sizes are discrete integers, then the expression (6.1.1) for the aggregate claim size can be calculated explicitly, determining the convolutions directly. For an aggregate claim of size m one needs to go up to $n = m$. However, this is extremely inefficient and leads to a computational complexity of order $O(m^3)$.

If, however, the claim number distribution $p_n = p_n(t) = P(N(t) = n)$ is of $(a, b, 1)$ type (5.4.22), that is, it satisfies the recurrence relation

$$p_n = \left(a + \frac{b}{n} \right) p_{n-1} \quad , \quad n \in \{2, 3, \ldots, \}$$

with a and b constants (which contains the Poisson and the negative binomial distribution, see Chapter 5), then the famous *Panjer recursion* can be applied: for $m = 1, 2, \ldots$ one has

$$\mathbb{P}(S = m) = \frac{1}{1 - a\,\mathbb{P}(X = 0)} \sum_{k=1}^{m} \left(a + \frac{bk}{m} \right) \mathbb{P}(X = k)\mathbb{P}(S = m - k), \qquad (6.3.20)$$

initiated by $\mathbb{P}(S = 0) = Q(\mathbb{P}(X = 0))$. Here we exceptionally allow $\mathbb{P}(X = 0)$ to be positive, for reasons that become clear below (in any case, for $\mathbb{P}(X = 0) = 0$ one gets $\mathbb{P}(S = 0) = \mathbb{P}(N = 0)$). The recursion (6.3.20) speeds up the computations considerably, leading to a computational complexity $O(m^2)$, and is popular in practice still today.

As discussed in Chapter 3, in many situations the use of a continuous claim size distribution F_X is preferred. In that case, one can discretize F_X according to

$$h_j = F_X(j\Delta) - F_X((j - 1)\Delta), \; j = 1, 2, \ldots, M$$

for some $\Delta > 0$, so that the resulting c.d.f. $H(x) = \sum_{j \le x} h_j$ (with $H(0) = 0$) provides a lower bound for $F_X(x)$. Likewise,

$$\tilde{h}_j = F_X((j + 1)\Delta) - F_X(j\Delta), \quad j = 0, 1, 2, \ldots, M - 1$$

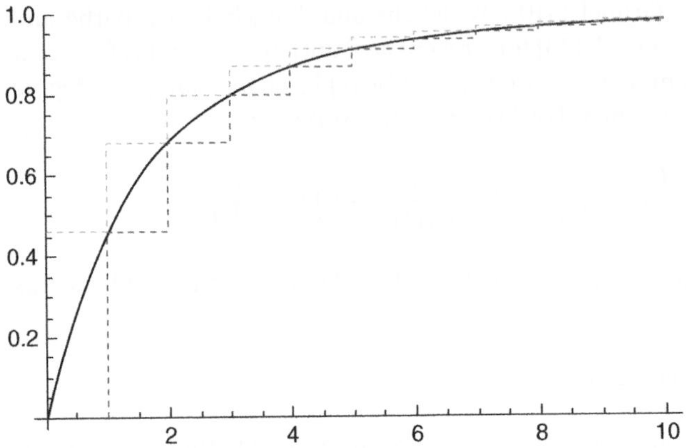

Figure 6.1 Discretizing the claim size distribution with $\Delta = 1$.

with c.d.f. $\tilde{H}(x) = \sum_{j \le x} \tilde{h}_j$,, which provides an upper bound for $F_X(x)$ (cf. Figure 6.1).

One can now use the Panjer recursion for the discrete bounds $H(x)$ and $\tilde{H}(x)$, leading to a lower and upper bound for the true value $F_S(x)$:

$$\mathbb{P}_L(S \le x) \le F_S(x) \le \mathbb{P}_U(S \le x).$$

By decreasing Δ one can then bring the upper and lower bound together with arbitrary accuracy. This approach is very popular in practice, and there is a tradeoff between sharpness of the bounds and increased computation time (for choosing smaller Δ). Employing the estimate

$$F_S(x) \approx (\mathbb{P}_L(S \le x) + \mathbb{P}_U(S \le x - 1))/2 \tag{6.3.21}$$

in fact can give a quite satisfactory estimate already for medium-range values of M. There is an enormous literature on problems with this numerical approach and suggestions how to resolve them (for a survey paper including many references on the history, refer to Sundt [720]). When F_X is heavy tailed, an efficient implementation can be quite tricky (e.g., see Hipp [443, 445] and also Gerhold et al. [388] for a recent treatment).

6.4 Fast Fourier Transform

As before, consider a discretized claim size random variable X with probability function $f_X(x_j)$, $j = 0, \dots, M-1$ for some fixed M. Since the goal is to obtain the total claim distribution $f_S(x_j) = P(S = x_j)$ $(j = 0, .., M-1)$, one could also start with the explicit form of the characteristic function of S (cf. (6.1.2))

$$\mathbb{E}\left(e^{izS}\right) = Q\left(\mathbb{E}\left(e^{izX}\right)\right),$$

where $i = \sqrt{-1}$ is the imaginary unit, and aim to obtain f_S directly by numerically inverting this characteristic function. This can in fact be done very efficiently using fast Fourier transform (FFT), the origin of which can be traced back to Gauss in 1805 before it was popularized and developed further by Cooley and Tukey [225].

Consider the discrete Fourier transform of the probability function $f_S(x_j)$ with $x_j = j\Delta$ for some $\Delta > 0$:

$$\tilde{f}_S(k\Delta) = \sum_{j=0}^{M-1} f_S(j\Delta)e^{\frac{2\pi i}{M}jk}, \quad k = 0, 1, \ldots, M-1,$$

and the respective inverse transform

$$f_S(j\Delta) = \frac{1}{M}\sum_{k=0}^{M-1} \tilde{f}_S(k\Delta)e^{-\frac{2\pi i}{M}kj}, \quad j = 0, 1, \ldots, M-1. \tag{6.4.22}$$

The quantities $\tilde{f}_S(i\Delta)$ are given by $Q(\tilde{f}_X(i\Delta))$ for all $i = 0, .., M-1$, where $\tilde{f}_X(k\Delta) = \sum_{j=0}^{M-1} f_X(j\Delta)e^{\frac{2\pi i}{M}jk}$, $k = 0, 1, \ldots, M-1$, that is, one needs to calculate (6.4.22). By using symmetries in the complex plane, these computations can now be speeded up. If one chooses M a power of 2, then the inverse transform of length M in (6.4.22) can be written as two Fourier transforms of length $M/2$, etc. The resulting computational complexity is $O(M \log M)$, which is much faster than the Panjer algorithm for large values of M (see Section 6.6 for an illustration).

This method is nowadays the fastest available tool to determine total claim size distributions numerically, but the implementation of the algorithm is not straightforward, and there can be numerical challenges with this method when M is not large (for the aliasing problem, see Embrechts et al. [327] and Grübel and Hermesmeier [411]). A nice discussion of the advantages and disadvantages of the method and comparisons to the Panjer algorithm can be found in Embrechts and Frei [325].

A further numerical method to obtain an estimate for the aggregate claim distribution is stochastic simulation (see Chapter 9).

6.5 Total Claim Amount under Reinsurance

Since we have discussed in Chapters 2 and 5 how the claim size and claim number distributions, respectively, change under the respective reinsurance contracts, one can apply the techniques discussed in Sections 6.1–6.4 to the respective new distributions. We discuss some aspects in this context in more detail here.

6.5.1 Proportional Reinsurance

For a QS contract, as already indicated in (2.1.1), it is straightforward to translate a result on the distribution of $S(t)$ into the corresponding one for $R(t)$:

$$\mathbb{P}(R(t) := aS(t) > x) = 1 - F_{S(t)}\left(\frac{x}{a}\right)$$

and correspondingly for the retained total claim amount $D(t)$.

For surplus reinsurance, in principle it is also conceptually simple to write down the distribution of individual reinsured and retained claims for a given sum insured (see (2.2.2)). The formulas for the aggregation to $R(t)$ and $D(t)$ are now, however, more involved, and as discussed in Section 2.2 the most transparent approach may be to treat the insured sum Q as a random variable (i.e., apply a collective view), cf. (2.2.3) and (2.2.4).

6.5.2 Excess-loss Reinsurance

For an L xs M reinsurance contract, we discussed already in Chapter 2 that the total claim size is given by

$$R(t) = \sum_{i=1}^{N(t)} \min\{(X_i - M)_+, L\} = \sum_{i=1}^{N(t)} \tilde{X}_i, \tag{6.5.23}$$

$$D(t) = \sum_{i=1}^{N(t)} \left(\min\{X_i, M\}1_{\{X_i \le M+L\}} + (X_i - L)1_{\{X_i > M+L\}}\right),$$

for the two parties (cf. (2.3.7)). In Section 2.3.1 it was also given that

$$\tilde{F}(x) = \mathbb{P}(\tilde{X} \le x) = \begin{cases} F_X(M+x) & \text{if } 0 \le x < L, \\ 1 & \text{if } x \ge L. \end{cases}$$

Note that an alternative approach for the reinsurer is to only count those claims that lead to a positive reinsurance claim. Define $\breve{R}_i$ by its tail

$$\mathbb{P}(\breve{R}_i > x) = \mathbb{P}(\tilde{X}_i > x | \tilde{X}_i > 0), \quad x > 0,$$

then one can write

$$R(t) = \sum_{i=1}^{N_R(t)} \breve{R}_i, \tag{6.5.24}$$

(cf. Section 5.6.1). Recall that in this case, the $(a, b, 1)$ class for $N(t)$ is particularly attractive, since $N_R(t)$ is then of the same kind, just with modified parameters $\tilde{a}$ and $\tilde{b}$.

Sticking to the formulation (6.5.23) and assuming i.i.d. claims independent of $N(t)$, one immediately gets

$$\mathbb{E}(R(t)) = \mathbb{E}(N(t)) \cdot \tilde{\mu}_1 \tag{6.5.25}$$

and

$$\text{Var}(R(t)) = \mathbb{E}(N(t)) \cdot \text{Var}(\tilde{X}) + \text{Var}(N(t)) \cdot \tilde{\mu}_1^2, \tag{6.5.26}$$

where $\tilde{\mu}_1$ is given by (2.3.8).

Assume now, for illustration, that there is also a second reinsurer (third partner) involved who pays the excess of each claim above $M + L$ (in the absence of such a second reinsurer, the sum of the first and third partners represents the situation for the first-line insurer in the L xs M treaty). Denote by $U(t) = \sum_{i=1}^{N(t)} U_i$ with $U_i = (X_i - (M + L))_+$ the aggregate claim size of that second reinsurer. Since $\mathbb{E}(D_iR_i) = M\,\mathbb{E}(R_i)$, $\mathbb{E}(D_iU_i) = M\,\mathbb{E}(U_i)$ and $\mathbb{E}(R_iU_i) = L\,\mathbb{E}(U_i)$, one immediately sees that the three covariances $\mathrm{Cov}(D_i, R_i)$, $\mathrm{Cov}(D_i, U_i)$ and $\mathrm{Cov}(R_i, U_i)$ are all positive. For the aggregate level, by iterated expectations,

$$\mathbb{E}(D(t)\cdot R(t)) = \sum_{k=0}^{\infty} \mathbb{P}(N(t) = k)\left(\sum_{i=1}^{k} \mathbb{E}(D_iR_i) + \sum_{i=1}^{k}\sum_{i\neq j=1}^{k} \mathbb{E}(D_iR_j)\right)$$

and hence

$$\mathrm{Cov}(D(t), R(t)) = \mathbb{E}(N(t))\cdot \mathrm{Cov}(D_i, R_i) + \mathrm{Var}(N(t))\cdot \mathbb{E}(D_i)\cdot \mathbb{E}(R_i)\,.$$

In this equation one can replace the roles of D, R by D, U and R, U without any problem. Using the expressions for the individual covariances one finds the expressions

$$\mathrm{Cov}(D(t), R(t)) = \mathbb{E}(D_i)\mathbb{E}(R_i)\,[\mathrm{Var}(N(t)) - \mathbb{E}(N(t))] + M\mathbb{E}(N(t))\mathbb{E}(R_i),$$
$$\mathrm{Cov}(D(t), U(t)) = \mathbb{E}(D_i)\,\mathbb{E}(U_i)\,[\mathrm{Var}(N) - \mathbb{E}(N(t))] + M\mathbb{E}(N(t))\,\mathbb{E}(D_i),$$
$$\mathrm{Cov}(R(t), U(t)) = \mathbb{E}(R_i)\,\mathbb{E}(U_i)\,[\mathrm{Var}(N(t)) - \mathbb{E}(N(t))] + L\mathbb{E}(N(t))\,\mathbb{E}(U_i).$$

Let now A and B be any two distinct letters from the set $\{D, R, U\}$. For the *variances,* we get

$$\mathrm{Var}\,(A(t) + B(t)) = \mathbb{E}(N(t))\cdot \mathrm{Var}\,(A_i + B_i) + \mathrm{Var}(N(t))\cdot (\mathbb{E}(A_i) + \mathbb{E}(B_i))^2$$

and hence by a simple calculation

$$\mathrm{Var}\,(A(t) + B(t)) - (\mathrm{Var}(A(t)) + \mathrm{Var}(B(t))) = 2\,\mathrm{Cov}(A(t), B(t)) > 0.$$

That is, there is an increase in variance if one combines different layers in the reinsurance chain.

The situation is different for the *standard deviations.* Since

$$\mathrm{Cov}(A(t), B(t)) < (\mathrm{Var}(A(t))\,\mathrm{Var}(B(t)))^{1/2}$$

one obtains

$$(\mathrm{Var}(A(t) + B(t)))^{1/2} < (\mathrm{Var}(A(t)))^{1/2} + (\mathrm{Var}(B(t)))^{1/2},$$

and therefore the standard deviation decreases by combining layers.

Let us look at the *coefficient of variation* of $A(t)$. From general principles we know that

$$\frac{\mathrm{Var}(A(t))}{\mathbb{E}^2(A(t))} = \frac{1}{(\mathbb{E}(N(t))\,\mathbb{E}(A_i))^2}\left(\mathbb{E}(N(t))\mathrm{Var}(A_i) + \mathrm{Var}(N(t))\mathbb{E}^2(A_i)\right)$$

so that

$$\frac{\mathrm{Var}(A(t))}{\mathbb{E}^2(A(t))} = \frac{\mathrm{Var}(N(t))}{\mathbb{E}^2(N(t))} + \frac{1}{\mathbb{E}(N(t))}\frac{\mathrm{Var}(A_i)}{\mathbb{E}^2(A_i)}$$

or

$$\mathrm{CoV}^2(A(t)) = \mathrm{CoV}^2(N(t)) + \frac{1}{\mathbb{E}(N(t))}\mathrm{CoV}^2(A_i)\,. \tag{6.5.27}$$

Correspondingly, the coefficient of variation of the aggregate claim size differs among the partners only by the one of the single claim size entering the above formula in the last element.

6.5.3 Stop-loss Reinsurance

In the case of an unbounded SL contract

$$R(t) = \{S(t) - C\}_+,$$

it is again very simple to translate the tail of $S(t)$ to the tail of $R(t)$:

$$\mathbb{P}(R(t) > x) = 1 - F_{S(t)}(x + C).$$

This formula neatly shows that good estimates for the tail $1 - F_{S(t)}(x)$ are immediately relevant for the reinsurer, especially if C is large. Also, for unbounded SL, the property of super-exponentiality or sub-exponentiality is passed on to the reinsurer.

Consider the *stop-loss premium* $\Pi_1(y,t) = \mathbb{E}(S(t) - y)_+$ (the expected reinsured amount under a SL contract with retention y) and the *generalized stop-loss premium*

$$\Pi_m(y,t) := \mathbb{E}(S(t) - y)_+^m = \int_y^\infty (x-y)^m dF_{S(t)}(x), \quad y \geq 0,\ m \in \mathbb{N}. \tag{6.5.28}$$

Note that

$$\Pi_0(y,t) = \mathbb{P}(S(t) > y) = 1 - F_{S(t)}(y)$$

reduces to the tail of the total claim amount. Instead of (6.5.28) we can also use an alternative which is derived from an integration by parts. So assume that $\mu_m = \mathbb{E}(X^m) < \infty$, then it easily follows that for $m > 0$

$$\Pi_m(y,t) = m\int_0^\infty v^{m-1}[1 - F_{S(t)}(y+v)]dv = m\int_0^\infty v^{m-1}[\Pi_0(y+v,t)]dv. \tag{6.5.29}$$

So if we know a way how to handle Π_0 (i.e., the tail of $S(t)$), then a simple additional integration gives all required insight into the generalized stop-loss premium. This also applies to the asymptotic behavior (see below). For $m = 1$, detailed estimates have been obtained by Teugels et al. [740].

In Section 6.2.2 we could – for a number of situations – derive an estimate for $x \to \infty$ of the form

$$1 - F_{S(t)}(x) \sim \frac{e^{-|\theta^*|x}}{|\theta^*|} K(x) \tag{6.5.30}$$

where $K \in \mathcal{L}$. Let us define a function K_1 by the relation

$$1 - F_{S(t)}(x) = \frac{e^{-|\theta|x}}{|\theta|} K_1(x).$$

When we introduce this expression into (6.5.29), a simple calculation tells us that

$$\Pi_m(y,t) = m\frac{e^{-|\theta|y}}{|\theta|^{m+1}} K_1(y) \int_0^\infty b^{m-1} e^{-b} \frac{K_1(y + b|\theta|^{-1})}{K_1(y)} db.$$

Applying classical uniformity properties of functions from the class $\mathcal{L}$ we see that under the conditions for $K \in \mathcal{L}$ we get for $y \to \infty$

$$\Pi_m(y,t) \sim m! \frac{e^{-|\theta|y}}{|\theta|^{m+1}} K(y).$$

Alternatively, and for all $m \geq 0$

$$\Pi_m(y,t) \sim \frac{m!}{|\theta|^m} \{1 - F_{S(t)}(y)\}.$$

This relationship can now be applied to all the cases discussed in Section 6.2.2.

Likewise, the sub-exponential estimates from Section 6.2.3 lead to corresponding estimates for the generalized SL premium. For instance, start from

$$1 - F_{S(t)}(x) \;=\; \mathbb{E}(N(t))(1 - F_X(x)) + R_1(t)f(x) - R_2(t)f'_X(x) + o(f'_X(x))$$

where

$$R_1(t) = 2\mu\mathbb{E}\binom{N(t)}{2} \text{ and } R_2(t) = \left\{3\mu^2\mathbb{E}\binom{N(t)}{3} + \mu_2\mathbb{E}\binom{N(t)}{2}\right\}.$$

Introduce this into the expression for Π_m to find that

$$\Pi_m(y,t) = m\mathbb{E}(N(t)) \int_0^\infty v^{m-1}[1 - F_X(v+y)]dv$$
$$+ mR_1(t) \int_0^\infty v^{m-1} f_X(v+y)dv - mR_2(t) \int_0^\infty v^{m-1} f'_X(v+y)dv.$$

Now

$$
\begin{aligned}
m\int_0^{\infty} v^{m-1}[1-F_X(v+y)]dv &= m\int_0^{\infty} v^{m-1}dv \int_{v+y}^{\infty} dF_X(u)\\
&= m\int_y^{\infty} dF_X(u)\int_0^{u-y} v^{m-1}dv\\
&= \int_y^{\infty}(u-y)^m dF_X(u)\\
&= \mathbb{E}(X-y)_+^m.
\end{aligned}
$$

Similar calculations lead to

$$\int_0^{\infty} v^{m-1}f(v+y)dv = \mathbb{E}(X-y)_+^{m-1},$$

and

$$\int_0^{\infty} v^{m-1}f'(v+y)dv = -(m-1)\mathbb{E}(X-y)_+^{m-2}.$$

We ultimately find that

$$
\begin{aligned}
\Pi_m(y,t) = \mathbb{E}(N(t))\mathbb{E}(X-y)_+^m + mR_1(t)\mathbb{E}(X-y)_+^{m-1}\\
+ m(m-1)R_2(t)\mathbb{E}(X-y)_+^{m-2} + o(f_X'(y)).
\end{aligned}
$$

6.6 Numerical Illustrations

In this section we illustrate the use of the Panjer recursion and FFT, and also look into the performance of the different approximations discussed earlier in this chapter. We also investigate the quantitative influence of the choice of counting process model on the distribution of the aggregate claim size for two of the case studies.

Case study: MTPL data, Company A. We first apply these techniques to the MTPL case study from Company A, approximating the compound distribution S_{365} for one year (i.e., here t is in units of days). For the claim size distribution we use the model from Figure 4.17, while the claim number distribution is Poisson with $\lambda = 0.15$, as discussed in Figure 5.9. Note that here the claims are heavy tailed with $\gamma = 0.506$, so that the variance and the third and fourth moment do not exist (and normal approximations cannot be applied). In order to assess the sensitivity of the choice of counting model, in Table 6.1 we contrast the expected values of N and S as well as resulting VaR levels (based on FFT), using the Poisson model and a fitted Cox model based on an inverse Gaussian subordinator, respectively. □

Case study: Dutch fire claim data. Table 6.2 gives the corresponding figures for the models fitted to the Dutch fire claim data in Chapter 5, again based on FFT calculations (for the claim size distribution we use the model from Figure 4.14). Figure 6.2 depicts the upper and lower bound for the c.d.f. of S_{365} based on Panjer recursions together with the average (6.3.21) for a lower and a higher discretization ($M = 2^{14}$ and $M = 2^{17}$).

Table 6.1 MTPL data for Company A: Key figures for models of X, N_{365} and S_{365} using Poisson and Cox models with inverse Gaussian subordinator

	$\mathbb{E}(\cdot)$	Var$(\cdot)$	VaR$_{0:95}$	VaR$_{0:99}$	VaR$_{0:995}$
X	388,438	/	1,065,451	2,405,538	3,416,123
N_{365} Poisson	55.27	55.27			
N_{365} Cox-IG	56.12	58.35			
S_{365} Poisson	21,467,676	/	31,100,000	41,420,000	48,760,000
S_{365} Cox-IG	21,798,923	/	31,570,000	41,960,000	49,340,000

Table 6.2 Dutch fire insurance claim data: Key figures for models of X, N_{365} and S_{365} using a Poisson model ($\lambda = 0.28$) and a Cox model with inverse Gaussian subordinator (cf. Figure 5.8).

	$\mathbb{E}(\cdot)$	Var$(\cdot)$	VaR$_{0:95}$	VaR$_{0:99}$	VaR$_{0:995}$
X	4,269,789	1.548163×10^{14}	12,962,991	45,000,000	60,499,575
N_{365} Poisson	102.53	102.53			
N_{365} Cox-IG	104.30	110.57			
S_{365} Poisson	437,795,514	1.774312×10^{16}	637,475,000	812,825,000	919,100,000
S_{365} Cox-IG	445,338,406	1.81631×10^{16}	647,650,000	824,300,000	931,175,000

One sees that the average (6.3.21) provides a very good estimate already for $M = 2^{14}$. Using the same discretizations for the respective FFT procedure, one obtains a plot that is practically identical to Figure 6.2 (but, just to have an impression in terms of computation time, for $M = 2^{17}$ the computation via the Panjer recursion took 19 seconds, whereas the FFT computation took 0.2 seconds on a usual PC, using an R code in each case). One observes that the resulting values for the higher discretization can safely be considered the exact values of $F_S(x)$ within the model assumptions. In order to assess the accuracy of the approximations from the earlier subsections, Figure 6.3 compares the $F_S(x)$ obtained by FFT with the normal approximation as well as first- and second-order approximations for heavy-tailed distributions. As mentioned in Section 6.2.3, one sees that both asymptotic approximations are rather coarse for practical use, even for values around the 99.5% quantile (see the zoom on the right-hand side). It is also quite clear from the plot that even if S_{365} is a sum of many i.i.d. random variables with finite variance, the normal approximation is quite inappropriate, once the distance to the mean is considerable. □

Note that on the basis of the splicing model conditional on the development period as given in Section 4.4.3, it is also possible to calculate the compound distribution for a particular accident year and development year. Here a thinning of the number of claims for accident year i towards a given development year d can then be based on Table 1.3.

Illustration: Light-tailed claims. Finally, we would like to give an illustration of the performance of the total claim approximations for exponentially bounded claim sizes. None of the case studies introduced in Chapter 1 seem to fall into this class. Rather than fitting an inappropriate model to the data, we therefore prefer to instead consider a

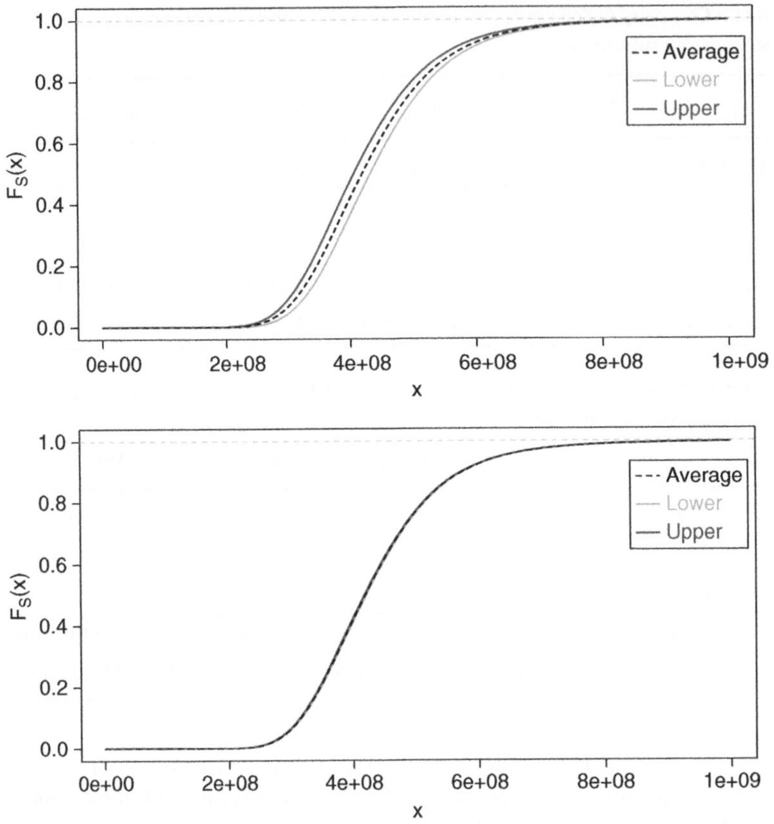

Figure 6.2 Dutch fire insurance claim data: Panjer bounds (and equally FFT bounds) for F_S based on interval splicing loss model and Poisson counting model: $M = 2^{14}$ and $\Delta = 250\,000$ (top); $M = 2^{17}$ and $\Delta = 25\,000$ (bottom).

simple example with a negative binomial claim number with parameters $t = 1, \alpha = 5$ and $b = 1$, and gamma-distributed individual claim sizes with $\alpha = 2$ and $\lambda = 2$. Figure 6.4 compares several approximations to the "exact" values of the c.d.f. $F_S(x)$ (approximated by FFT). One sees that using all first four moments, the Gram–Charlier approximation considerably improves on the normal approximation, both in the center and further in the tails. As for heavy tails, the light-tailed asymptotic approximation only becomes accurate for very large values of x, way beyond the 99% quantile. □

6.7 Aggregation for Dependent Risks

In the above sections, the focus was on the aggregation of i.i.d. claim sizes, which are also independent from the claim number process. This is the classical and very popular approach and a natural benchmark for all alternatives. For a discussion of the assumption of identically distributed claim sizes, see the Notes at the end of the chapter. The independence assumption is, however, a more challenging one. While in many

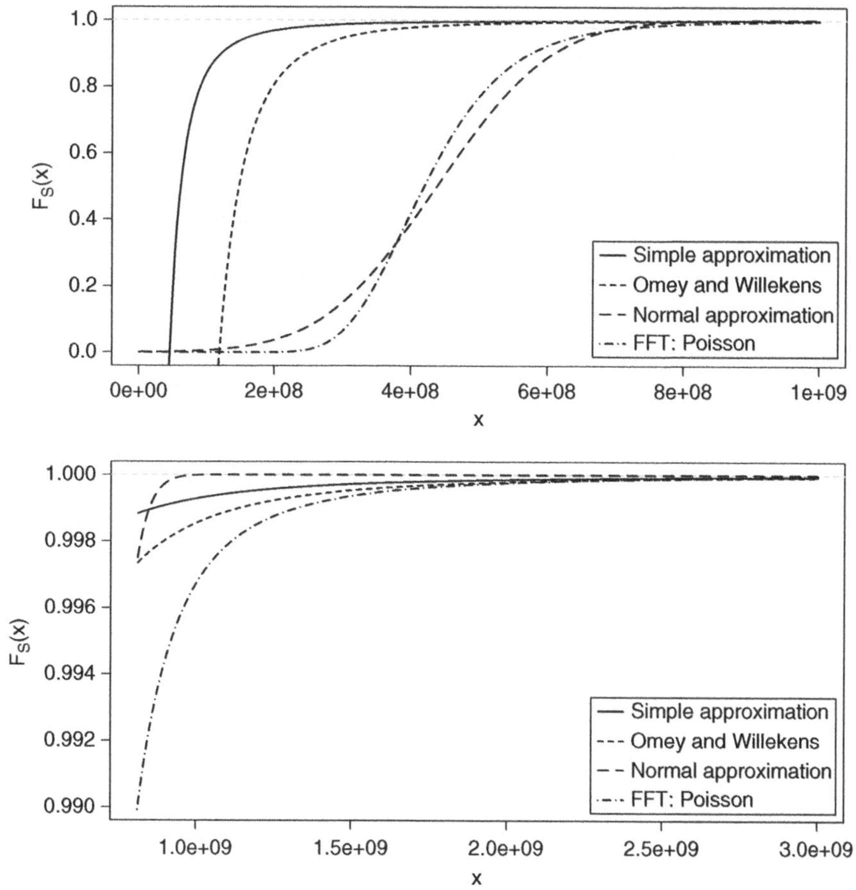

Figure 6.3 Dutch fire insurance claim data: application of (6.2.5), (6.2.18) and (6.2.19), and FFT as approximations of F_S based on interval splicing loss model and Poisson counting model (top); restricted graph with cumulative probabilities larger than 0.99 (bottom).

practical situations assuming independence will suffice the purpose, there will also often be situations in reinsurance practice where it is essential to consider dependence among the risks in the process of their aggregation. Dependence can enter in various ways:

- **Dependence between claim sizes within a portfolio (line of business)**
  It may happen that subsequent claims in a line of business (LoB) are dependent because their distribution is influenced by one or several shared external factor(s), but conditional on these factors the risks are independent (such factors could, for instance, be economic, weather or legal conditions). If such a dependence can be described explicitly, then the aggregation can be done for each choice of the factor under the independence assumption and the results then have to be mixed according to the distribution of the factor. If only some of the claims are affected by such a common factor, it may be appropriate to regroup (sum) them together to one large

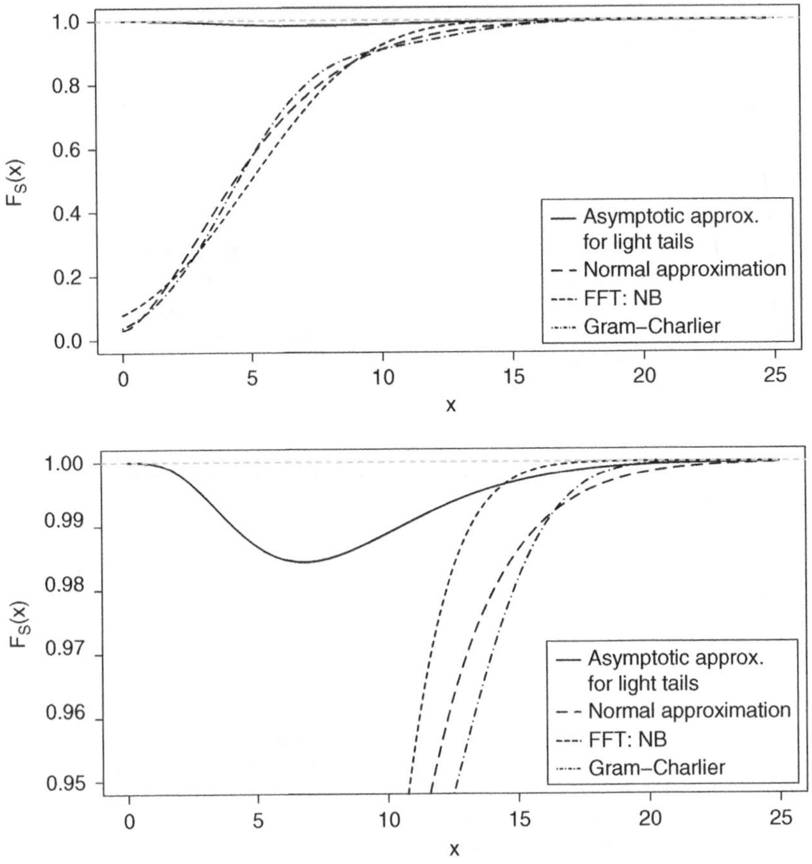

Figure 6.4 Compound negative binomial model with gamma-distributed claim sizes: normal approximation (6.2.5), Gram–Charlier approximation (6.2.8), asymptotic approximation (6.2.17), and FFT as approximations of F_S (top); restricted graph with cumulative probabilities larger than 0.95 (bottom).

claim, which can then again be considered independent of all the other ones (like in the per-event XL setting, where one does not consider individual claims, but the total claim due to each external event). In that case, the independent aggregation can again be applied for a correspondingly modified claim (and claim number) process.

Another possibility could be a causal dependence, for example that a large claim size influences the distribution of the next claim size (like an afterquake claim after an earthquake claim, or a storm loss after a previous storm loss when buildings are already destroyed). In the presence of many data points, it may be possible to formalize such dependence structures and then the aggregation procedure has to be adapted according to the identified model.

Finally, one may "statistically" see dependence in the claim data (like in an auto-correlation function), but not be able to attribute this to a concrete reason. In such a case, great care is needed, as many of the statistical fitting procedures for marginal distributions crucially rely on independence of the data points, so that then the consideration of dependence should already enter the model formulation procedure.

- **Dependence between claim numbers within an LoB**
  The algorithms discussed above in fact only use the distribution of the final value $N(t)$, so that dependence between claim numbers along the time interval is neither a problem nor a restriction (and most of the claim number processes discussed in Chapter 5 involve some sort of contagion or dependence). In fact, from the viewpoint of inter-occurrence times between claims, only the renewal model (with the homogeneous Poisson process being a special case) exhibits independence.
- **Dependence between claim sizes and frequency within an LoB**
  It is easy to imagine situations where the distribution of claims depends on the frequency of their occurrence, indicating a different underlying nature or mechanism to cause them. In Chapter 5 we discussed some recent approaches to use such information in the model calibration in connection with GLM techniques. On a causal level, one can again think that the length of the time between claims has an influence on the next claim size (like in the above earthquake or storm scenario). The implications for the aggregation of such structures will depend on the particular assumption on the dependence. Particular examples are Shi et al. [698] and Garrido et al. [371].
- **Dependence between claim sizes and claim numbers of different LoBs**
  A scenario that will occur quite frequently in reinsurance practice is the occurrence of events that simultaneously trigger claims in several LoBs (leading to dependence between the respective claim number processes), and then the respective claim sizes will most likely also not be independent of each other. In Chapters 4 and 5 we discussed the fitting of multivariate models to claim data. It is indeed preferable that in the presence of dependence of claim sizes the fitting is directly done on the multivariate level. If this is not feasible, however, then one popular approach is to combine the fitted marginals through some copula (which may be chosen according to some structural understanding of the nature of the dependence like common risk drivers or based on previous empirical dependence patterns).

  The final goal is often to obtain the distribution of the aggregate sum of all claims of those dependent LoBs. If in fact all claim occurrence times in these LoBs coincide (i.e., $N(t)$ represents the number of claims for all those LoBs), one can then first determine (or approximate) the distribution of the sum of the dependent components, and for this resulting (again scalar) claim distribution apply the techniques discussed in the above sections, particularly Panjer recursion and FFT.
- **Dependence in the company level aggregation**
  For solvency considerations, insurance and reinsurance companies finally have to aggregate the portfolios of different LoBs and of different countries to obtain one overall aggregate claim distribution. This is often done by determining the aggregate claim distribution $S_1(t), \dots, S_n(t)$ of each of the n risk units using the above techniques under the independence assumption. In a second step the resulting (marginal) sums $S_1(t), \dots, S_n(t)$ are combined with a copula, which is then used to determine the distribution of $S_1(t) + \cdots + S_n(t)$. This procedure may in fact be iterated several times within the company hierarchy, leading to hierarchical copula

structures. Clearly the choice of copula function will be crucial for the outcome of this procedure.[2]

While independence is a well-defined and unique concept, there is a (literally) infinitely large class of dependence models available and the choice is inevitably also a matter of purpose and taste. It is beyond the scope of this book to give a detailed exposition of concrete dependence models proposed and used in (re)insurance practice, but we point to comments and references in the Notes below.

6.8 Notes and Bibliography

The collective risk model with a random claim number $N(t)$ already explicitly postulates the identical distribution of all claim sizes. For an originally heterogeneous portfolio one can think of F_X as a mixture distribution with respective weights (see the classical textbooks on risk theory mentioned in Chapter 1 for details). Correspondingly, the assumption of identically distributed claims (within the same LoB) is not that restrictive. When claim data are spread over longer time horizons, one needs, however, to be particularly careful to adjust them so as to make them comparable (e.g., corrections for inflation, market share, or – as in the case studies of aggregate storm and flood claims – normalization by total building value).

In individual risk models (where the aggregate claim is determined as the sum of losses of each policy and hence the number of summands is fixed), it is a decisive computational advantage if the chosen individual loss distribution is closed under convolution (like for the gamma and inverse Gaussian case). In the collective model we have seen in this chapter that this property is not essential (in most cases the distribution will anyway be discretized for the computational purposes of aggregation).

For surveys on classical aggregate claim approximations, see Teugels [739], Chaubey et al. [199] and Papush et al. [604]. An early reference to the method of orthogonal polynomials in actuarial science is Gerber [383].

Approximations in the spirit of *Berry–Esseen theory* but using compound Poisson distributions have, for example, been given in Panjer et al. [602] and Hipp [443]. Also, Stein's method can be relevant for this purpose, see, for example, Barbour and Chen [82]. Further, the *Wilson–Hilferty approximation* and the *Haldane approximation* have been advocated by Pentikäinen in [611], where more references can be found. Refer also to Willmot [785], where even time-dependent claims are used. For an alternative gamma series expansion, see [30].

We indicate briefly here some other methods of deriving exponential estimates for the tail of $F_{S(t)}(x)$:

- A traditional approach to get exponential estimates relies on the *saddle-point method* that probabilistically can be viewed as an approximation by a normal distribution whose mean and variance are chosen in some optimal way. This method is popular

2 Note that the (in principle preferable) multivariate modelling of all these LoBs is, for practical reasons, often not possible. In the standard model of Solvency II, this step is in fact even coarser, using prescribed correlation coefficients between the business units.

and its use is widespread. It also applies to the simple Poisson case (for which the procedure of Section 6.2.2 does not apply). An excellent source of information on this method is Jensen [464].

It should be mentioned that the asymptotic expressions obtained by the saddle-point method can be slightly different from those obtained by the method described in Section 6.2.2. For example, for the Pólya model the saddle point method offers an analogous but different estimate for the total claim distribution (e.g., see Embrechts et al. [328]). For extensions of the saddle-point approximation to other risk models, see Jensen [463].

- In Den Isefer et al. [272], *projection pursuit methods* are advocated to calculate the distribution of the aggregate claim amount. For approximations using *Lévy processes*, see Furrer et al. [363]. One may in fact interpret the use of a Lévy process as the opposite approach: whereas the classical actuarial approach is to model individual claims and aggregate them, in the Lévy setup one typically starts with the aggregate distribution and strives to ponder the implications on the jump behavior on smaller time intervals.
- An approximation of the distribution of the total claim amount by an *inverse Gaussian distribution* has been advocated in Chaubey et al. [199]. We briefly show how one may obtain results for specific distributions when the method of Section 6.2.2 fails. For a slightly different approach, see Gendron et al. [380]. Recall that the inverse Gaussian distribution has the Laplace transform

$$\widehat{F}_X(s) = \exp\left\{-\beta\mu\left(\sqrt{1+\frac{2s}{\beta}}-1\right)\right\}.$$

Hence $\sigma_{F_X} = -\beta/2$ and $\widehat{F}_X(\sigma_{F_X}) = e^{\beta\mu}$ (and due to the finiteness of this value there will not always exist an appropriate θ value to apply the method of Section 6.2.2). However, with the explicit form of the n-fold convolution F_X, by reversing summation and integration, one finds with $\eta := (\beta\mu^2)/2$

$$1 - F_{S(t)}(x) = \sqrt{\frac{\eta}{\pi}} \int_x^\infty y^{-3/2} e^{-\beta y/2} \left\{ \sum_{n=1}^{\infty} n p_n(t) e^{\beta\mu n} e^{-n^2 \frac{\eta}{y}} \right\} dy.$$

Write $g(y)$ for the summation inside the integral. When x and henceforth y is large enough, one can expand the function $g(y)$ around the point y^{-1}. We find $g(y) = \sum_{k=0}^{\infty} c_k(t)\, y^{-k}$ where

$$c_k(t) = \frac{(-\eta)^k}{k!} \sum_{n=1}^{\infty} n^{2k+1} p_n(t) e^{\beta\mu n}.$$

The latter quantities can of course be written in terms of the generating function $Q_t(z)$. For example, $c_0(t) = e^{\beta\mu} Q_t'(e^{\beta\mu})$. After a couple of easy calculations one finally obtains for $x \uparrow \infty$

$$1 - F_{S(t)}(x) = \mu\sqrt{\frac{2}{\beta\pi}} x^{-3/2} e^{-\frac{\beta x}{2}} \left\{ c_0(t) - \frac{1}{x}\left(c_1(t) + \frac{3c_0(t)}{\beta}\right) + \dots \right\}.$$

The above expansion is in contrast with an approximation suggested by Seal [692], where a moment fit is applied.

- If the claim number process is a *mixed Poisson process*, then Willmot developed another procedure that works very well if the mixing distribution has a certain asymptotic behavior (see [786]).

Early references to higher-order asymptotic approximations for heavy tails are Embrechts et al. [326, 337] and Taylor [731], see also Willmot [785]. As illustrated in Section 6.6, the first-order asymptotic approximation (6.2.18) is rather inaccurate for not extremely large values of x, but it also turns out to be remarkably robust with respect to dependence (e.g., see Asmussen et al. [66], Albrecher et al. [15]), which may be considered both an advantage and a disadvantage. Segers and Robert [651] show that if the claim number distribution is more heavy tailed than the claim size distribution, then in certain situations one can replace the claim size (rather than the claim number) by its mean in the first-order approximation.

The mathematical principle underlying the Panjer recursion can be traced back all the way to Euler [341] and has also reappeared early in computer science (see Exercise 7.4 in Knuth [495]). For its first application in actuarial science, see Williams [781] and Panjer [600, 601]. For generalizations, see, for example, Willmot et al. [787] and Dhaene et al. [287]. For a detailed survey of recursive methods we refer to Klugman et al. [491] and Sundt and Vernic [722], who also discuss recursive methods in higher dimensions. For a recent contribution see Rudolph [658]. Numerical inversion of characteristic functions have a long history in risk theory (e.g., see Bohman [144] and Seal [693]). Early references for the FFT method in aggregate claim approximations are Bertram [131] and Bühlmann [167].

Chains of XL treaties are covered, for example, in Albrecher and Teugels [29].

We would like to emphasize that the calculation of the VaR values in Tables 6.1 and 6.2 here for simplicity was done for the fitted distribution, as if this fitted distribution is the true one. However, if the parameter uncertainty due to the estimation procedure were taken into account, the resulting VaR values would be different. In fact, in most applications this effect is overlooked or ignored. However, one should be aware that this can lead to systematic underestimation of VaR values and other risk measures. See Pitera and Schmidt [621] for a recent analysis and illustration.

For an embedding of risk models with causal dependence between claim sizes and their inter-occurrence times into a semi-Markovian framework, see, for example, Albrecher and Boxma [16]. Albrecher and Teugels [28] propose a dependence structure between claim sizes and inter-occurrence times that still preserves a certain random walk structure and is hence quite tractable. For an illustration on how conditional independence of risks given a common random parameter can lead to Archimedean dependence structures, see Albrecher et al. [18] and Constantinescu et al. [224]. A possible source of dependence in data is also a hidden common trend (like neglected inflation etc.); for a study of the consequences of failing to detect such trends on quantiles of compound Poisson sums with heavy-tailed summands, see Grandits et al. [407].

Classical references for copula techniques are Joe [467] and Nelsen [589] and for an early insurance application Frees and Valdez [359]. A rich source for the application of copulas in risk management is McNeil et al. [568]. For a general approach to a hierarchical pair-copula construction, see, for example, Aas et al. [1]. Since particularly in the reinsurance context often the number of available data points is not sufficient to identify an appropriate copula for given marginal distributions, it is natural to look for best-case and (particularly) worst-case bounds for quantiles (the VaR) of the sum of dependent risks, see, for example, Embrechts et al. [334–336] and Bernard et al. [124]. For the practical computation of such bounds, Puccetti and Rüschendorf [635] developed the so-called *rearrangement algorithm*. Since these bounds will in general be quite wide, there has been quite a lot of research activity in how to narrow these bounds under additional information such as higher moments of the sum, see, for example, Bernard et al. [121, 126, 128] and Puccetti et al. [636]. Bignozzi et al. [134] investigated the asymptotic behavior of the VaR of the sum of risks relative to the sum of the individual VaRs under dependence uncertainty. For an assessment of the robustness of VaR and ES with respect to dependence uncertainty, see Embrechts et al. [338] and also Cai et al. [177] for more general risk measures. The tradeoff between robustness and consistent risk ranking is investigated in Pesenti et al. [613]. Filipović [349] compares a bottom-up approach for risk aggregation to the multi-level aggregation structure inherent in some standard models of regulators.

In some situations with dependence one can avoid formulating and calibrating the dependence model directly. As an example, for storm risk modelling in Austria, Prettenthaler et al. [633] used a building-value-weighted wind index for each region for which aggregate claim data were available to relate the losses to wind speed during the storm. Using the suitably normalized claim data, and implementing a region-specific correction factor (taking into account the topography and type of buildings of each region) as well as a storm-specific correction factor (considering the different duration, torsion etc. of each storm), such a relationship could be established. One could then use previous wind field data (that go back in time much longer than the storms for which claim data are available) to create additional loss scenarios and then use resampling techniques to determine global and local loss quantiles. For flood loss modelling, one can alternatively sometimes use concrete risk zonings (produced by experts on-site) that assign return periods of floods to each local region throughout a country (such zonings exist with remarkable resolution). Together with a map of buildings one can assign losses to each return period, and the dependence modelling then reduces to model the joint occurrence of return periods across these regions, see Prettenthaler et al. [631] for a case study in Austria.

Finally, note that for most of the models exhibiting some sort of dependence between the individual risks, the only way to approximate the aggregate claim distribution or, more particularly, its quantiles, is stochastic simulation, which is discussed in Chapter 9.

7

Reinsurance Pricing

An important question when setting up a reinsurance contract with

$$R(t) = S(t) - D(t) \tag{7.0.1}$$

is to determine the appropriate premium $P_R(t)$ that the cedent has to pay to the reinsurer for entering the contract. If $P(t)$ is the premium that the cedent received from the policyholders for risk $S(t)$, then $P_D(t) = P(t) - P_R(t)$ is the part of $P(t)$ that compensates the cedent for keeping $D(t)$, that is, we have to identify the premium sharing that goes along with the risk-sharing mechanism (7.0.1). Since reinsurance is after all a form of insurance, one can expect similar principles to be applied for reinsurance pricing as there are for first-line insurance. There are, however, also substantial differences to pricing for first-line insurance, for instance with respect to

- the available data situation underlying a decision about the premium
- the degree of asymmetry of information between the two parties in the contract on the underlying risk and the corresponding data
- the way in which the long-term relationship between the parties influences the pricing (profit participation, adaptations of risk-sharing rules in case of severe claim experience, etc.)
- the demand and supply pattern for finding reinsurance coverage (particularly also the availability of alternatives to reinsurance, cf. Section 10.2)
- the risk of moral hazard for certain reinsurance forms.

Also, there has to be an agreement of how to share administrative costs for the acquisition of insurance policies and the settlement of claims, which is usually done by the cedent. In proportional reinsurance, this is taken into account in the form of a *reinsurance commission*, which will reduce the reinsurance premium (cf. Section 7.3). Due to the nature of the reinsurance market (with less available loss experience, fewer companies, limited diversification possibilities etc.), premiums will in general be adapted much faster to the loss experience than in the primary insurance market, and correspondingly market cycles (e.g., caused by the occurrence or the lack of catastrophic events) play a prominent role (e.g., see Meier and Outreville [569, 570]).

Reinsurance: Actuarial and Statistical Aspects, First Edition.
Hansjörg Albrecher, Jan Beirlant and Jozef L. Teugels.

As for premiums in primary insurance markets, an actuarial reinsurance premium will typically consist of the *expected loss* of the underlying risk plus a *safety loading*.[1] In Section 7.1 we review the classical principles to determine such a safety loading, and Section 7.2.2 discusses how to include the cost of capital and solvency requirements in these considerations. We will then deal with pricing issues for proportional reinsurance in Section 7.3, and study challenges of non-proportional treaties in Section 7.4.

The final reinsurance premium also has to contain a margin for additional costs like administrative costs, participation in acquisition expenses, runoff expenses, taxes, asset management fees, and finally *profit*.

We mention that there are further factors particular to reinsurance that will influence the size of the loading in the premium. In Chapter 1, a list of possible motivations to take reinsurance was given, and the relative importance of these in a particular situation will clearly have a considerable impact on the premium level that can be agreed on. In addition to that, there is a possible remaining *basis risk* for the cedent, that even if the reinsurance treaty is designed on an indemnity principle, there can be clauses to fix a participation amount for future corrections of the original claim size, and so the actual costs may differ from the respective reinsured amount. This risk may also be priced in. Another issue is *counterparty risk*: the necessary solvency capital for a cedent with respect to counterparty risk (e.g., as required in Solvency II) is influenced by the rating of the reinsurance company, and if this rating is not favorable, the cedent may ask for a premium reduction or a deposit (e.g., see Sherris [697]). Also, the concrete risk appetite of the top management on both the cedent's and the reinsurer's side will have an impact on the willingness to enter a treaty for a certain premium. Finally, under competitive conditions the market demand will determine which premiums can be charged, and the premium calculation procedures based on the stochastic nature of the risks will then mainly serve as lower limit prices at which offering the product becomes viable for the reinsurer. In the sequel we will not explicitly deal with these additional factors (but see Antal [43] for more details).

In this chapter we mainly focus on the (traditional) actuarial approach to pricing, dealing with the liabilities from the underwriting of risks. For the final implementation, this approach will have to be combined with financial pricing techniques, also taking into account the capital investments and its management, potential further regulatory constraints as well as economic valuation principles. Among the key drivers to determine prices will then be to maximize the resulting market value of the company, for instance using capital asset pricing models and their adaptations to insurance or option pricing techniques explicitly taking into account the possible default of the company (e.g., see Bauer et al. [89]). Some further comments on this will be given in Section 10.3.

---

1 Not employing such a safety loading will lead to bankruptcy in the long run, and this is also the classical argument to justify it. In recent years, however, due to changes in the regulatory framework, the focus has gradually moved towards considerations involving the cost of capital, which one may consider as a change of paradigm (see Section 7.2.2).

7.1 Classical Principles of Premium Calculation

Before selling insurance and buying reinsurance, the insurer has to adopt some kind of assessment of the riskiness of the overall position. In particular, if a rule can be formulated that assigns to the claim $S(t)$ a premium $P(t) = \psi(S(t))$ for some functional ψ under which the position becomes acceptable (in terms of both profitability and safety), then this defines a *premium calculation principle*. After reinsurance, the same analysis has to be done for the remaining position $D(t)$. The reinsurer will have to do a similar analysis for the risk $R(t)$, albeit with a possibly different function ψ reflecting a different risk attitude, different access to markets and hence diversification possibilities etc. In order to discuss this in general, consider the risk Y with c.d.f. F_Y, where Y could be any of the above quantities or also a single risk that is considered. If we denote the premium by $P(Y)$, then this functional might satisfy a variety of properties that make the premium principle acceptable. Examples include:

- positive loading: $\mathbb{E}(Y) \leq P(Y)$
- no-ripoff: $P(Y) \leq Q_Y(1) = \inf\{x : F_Y(x) = 1\}$
- no unjustified risk loading: $P(Y = b) = 1$ implies that $P(Y) = b$
- positive homogeneity: $P(aY) = aP(Y)$ for any real constant $a > 0$
- translation invariance: $P(Y + b) = P(Y) + b$ for any real constant b.

Other properties are phrased in terms of two risks Y_1 and Y_2 with c.d.f. F_{Y_1} and F_{Y_2}, respectively:

- monotonicity: $P(Y_1) \leq P(Y_2)$ if $F_{Y_1}(x) \geq F_{Y_2}(x)$ for all x
- subadditivity: $P(Y_1 + Y_2) \leq P(Y_1) + P(Y_2)$
- additivity: $P(Y_1 + Y_2) = P(Y_1) + P(Y_2)$ if Y_1 and Y_2 are independent
- comonotonic additivity: $P(Y_1 + Y_2) = P(Y_1) + P(Y_2)$ if Y_1 and Y_2 are comonotonic (i.e., they can both be expressed as an increasing function of one single random variable)
- convexity: $P(p\,Y_1 + (1-p)\,Y_2) \leq p\,P(Y_1) + (1-p)\,P(Y_2)$ for any $0 < p < 1$.

Not all of these properties are necessarily desirable in all situations. As an example, while positive homogeneity may look natural in terms of switching between currencies, it will often not make sense when one thinks of a as a huge number reflecting a scaled-up risk.[2]

We now list some popular examples of premium principles.

- The *expected value principle* is given by

$$P(Y) = (1 + \theta)\,\mathbb{E}(Y), \tag{7.1.2}$$

  where $\theta > 0$ is a fixed constant. So here the safety loading is proportional to the expected claim size. This principle is very popular, both due to its transparency and simplicity, and especially in reinsurance portfolios the reliability of information on

2 Note that in the mathematical finance literature, such axiomatic properties were reconsidered intensively in the context of risk measures, where positive homogeneity, translation invariance, monotonicity, and subadditivity together became popular under the term *coherence* (see Artzner et al. [49]). While subadditivity for coherent risk measures is considered crucial, the unconditional appropriateness of subadditivity in the insurance context is not so clear (see Dhaene et al. [285]).

the risk beyond the first moment is often limited. For $\theta = 0$ one talks about the *pure premium*.[3]

- If information on the variance of Y is available, then another common principle is to choose the safety loading proportional to that variance:

$$P(Y) = \mathbb{E}(Y) + \alpha_V \operatorname{Var}(Y), \tag{7.1.3}$$

which is called the *variance principle*, where α_V is a positive constant (for a guideline on how to choose this constant see Section 7.5).

- If the variance is replaced by the standard deviation, we get the *standard deviation principle*

$$P(Y) = \mathbb{E}(Y) + \alpha_S \sqrt{\operatorname{Var}(Y)} \tag{7.1.4}$$

where again α_S is a positive quantity. The switch from the variance to the standard deviation may be considered natural in view of underlying (currency) units: if the expectation is in €, the variance will be in €^{2}, whereas the standard deviation is also in units of € (on the other hand, one may also assume the constant α_V to be in units of 1/€, see Section 7.5 for a concrete example). Eventually, measuring variability by variance or by standard deviation gives just different weight to "additional" risk and in that sense is a matter of taste and choice. In any case, the resulting properties of the two principles differ substantially. Both are in fact used extensively, depending on the context.

- The above principles use only the first two moments of Y. If information on the entire Laplace transform of Y is available, then the *exponential principle*

$$P(Y) = \frac{1}{a} \log \mathbb{E}(e^{aY}) \tag{7.1.5}$$

(for some fixed risk aversion coefficient $a > 0$) has many appealing properties. One notices, however, that such a principle can only be applied to risks Y with an exponentially bounded tail, as the Laplace transform has to be finite at the positive value a.

- Classical utility theory suggests using the *zero utility principle*, under which the premium is determined as the solution of the equation

$$u(w) = \mathbb{E}(u(w + P(Y) - Y)), \tag{7.1.6}$$

where $u(x)$ is the utility (for the entity offering insurance) of having capital x, whereas w is the (deterministic) current surplus and $P(Y)$ is the premium for which this entity is indifferent about entering the contract in the sense of expected utility. The utility function is usually non-decreasing, indicating that larger risks should be counteracted

3 Depending on the context, the terms *net premium* and *risk premium* are also often used to denote the expected value of the claim size, but the terminology is sometimes ambiguous and we will use the term *pure premium* throughout.

by larger premiums; it is also typically concave, indicating *risk aversion*. If Var (Y) is small, one can approximate

$$P(Y) \approx \mathbb{E}(Y) + \frac{|u''(w)|}{2\,u'(w)}\,\mathrm{Var}\,(Y),$$

that is, the utility principle has some similarity to the variance principle (cf. [383]).

For a linear utility function u, (7.1.6) gives back the pure premium, and for $u(x) = -e^{-ax}$ the value of w becomes irrelevant and one finds the exponential principle again. Other frequent choices for u are power functions and logarithmic functions.

If u is a refracted linear function with $u(x) = (1+\theta)x$ for $x < 0$ and $u(x) = x$ for $x \geq 0$ for some parameter $\theta > 0$, then for $w = 0$ the criterion (7.1.6) gives the *expectile principle*

$$P(Y) = \mathbb{E}(Y) + \theta\,\mathbb{E}((Y-P)_+),$$

that is, the loading is proportional to the expected loss. Another way to write this is $(1+\theta)\mathbb{E}((Y-P)_+) = \mathbb{E}((P-Y)_+)$, that is, the premium is determined in such a way that when entering the contract, the "expected gain" is a multiple $(1+\theta)$ of the "expected loss" (e.g., see Gerber et al. [386]).

In the presence of other risks in the portfolio, it is actually more natural to assume that the current surplus is not a deterministic quantity, but itself a random variable W. In this case (7.1.6) translates into

$$\mathbb{E}(u(W)) = \mathbb{E}(u(W + P(Y) - Y))$$

and the resulting premium will depend on the joint distribution of Y and W (i.e., on the diversification potential of adding Y to the portfolio). For instance, exponential utility $u(x) = -e^{-ax}$ then gives

$$P(Y) = \frac{1}{a}\log\frac{\mathbb{E}(e^{a(Y-W)})}{\mathbb{E}(e^{-aW})},$$

which for small risk aversion a leads to the approximation

$$P(Y) \approx \mathbb{E}(Y) + \frac{a}{2}\mathrm{Var}\,(Y) - a\,\mathrm{Cov}(Y, W),$$

(e.g., see Gerber and Pafumi [387]).

While the idea of this approach is theoretically quite attractive (since under the weak assumptions stated in the classical von Neumann–Morgenstern theory a preference among two random future cash-flows can always be expressed as a larger respective expected utility), the determination of the appropriate utility function describing the risk attitude for all magnitudes will in practice typically not be feasible. For a recent approach for insurance pricing involving utility functions beyond a concave shape see Bernard et al. [122].

- The *mean value principle* is defined by

$$P(Y) = v^{\leftarrow}(\mathbb{E}(v(Y))) \tag{7.1.7}$$

where v an increasing and convex *valuation function* with inverse $v^{\leftarrow}$. For $v(x) = e^{rx}$ we get again the exponential principle, while for $v(x) = x$ we find the pure premium.

- Another approach is to distort the distribution of Y by transforming some of its probability mass to the right, in that way making the risk more dangerous. Then the premium can be obtained by taking the expected value of the modified random variable. If this distortion is done by an exponential function, one obtains the *Esscher premium principle*

$$P(Y) = \frac{\mathbb{E}\left(Y\, e^{hY}\right)}{\mathbb{E}\left(e^{hY}\right)}$$

  for some positive parameter h. It is clear that the Esscher principle can only be invoked for claims which have an exponentially bounded tail.
- A more general toolkit arises from distorting the distribution tail of Y. Let g be a non-negative, non-decreasing, and concave function such that $g(0) = 0$, $g(1) = 1$. Then the *risk-adjusted premium principle* is defined by

$$P(Y) = \int_0^{\infty} g(1 - F_Y(x))\, dx\,.$$

  The choice $g(t) = t^{\beta}$ is referred to as *proportional hazard transform* (e.g., see Wang [769]). Note that this principle can also be applied for heavy-tailed risks.

Distorting the loss random variable provides a natural link to *arbitrage-free pricing* when integrating financial aspects in the pricing mechanism (cf. Section 10.3).

Needless to say there are many other premium principles, each with some merits and defects (see the further references at the end of the chapter).

7.2 Solvency Considerations

The premium principles discussed in Section 7.1 are guidelines that can be applied to the individual risks, but also to the aggregate risk in the portfolio. However, the question of premium calculation is also intimately connected with solvency considerations, and on the aggregate level there has to be an appropriate interplay between received premiums and additionally available capital to ensure a solvent business.[4] Then, in a next step, the aggregate premium can be sub-divided on the individual policies. The premium principle used for this second sub-dividing step does not necessarily have to coincide with the one applied for determining the aggregate premium (cf. Section 7.5 below).

We will now discuss two solvency criteria for determining the aggregate premium, the first one mainly for intuitive and historical reasons, the second one being the currently relevant one for many regulatory systems in practice.

4 After appropriate incorporation of this interplay, one may finally interpret (and justify) a premium principle for the aggregate risk along such lines, for example the constants in the principles above being determined by the available capital and the target capital for solvency purposes (see also Section 7.5).

7.2.1 The Ruin Probability

A traditional approach to look at the safety of an insurance portfolio is to consider its surplus process $C(t)$ as a function of time, and observe whether this process ever becomes negative (an event that is referred to as *ruin*).[5,6] If at time $t = 0$, one starts with capital w, then the *(infinite-time) ruin probability* is defined as

$$\psi(w) = \mathbb{P}(\inf_{t\geq 0} C(t) < 0 \mid C(0) = w),$$

and the *finite-time ruin probability* up to time T is correspondingly

$$\psi(w, T) = \mathbb{P}(\inf_{0\leq t\leq T} C(t) < 0 \mid C(0) = w).$$

Concretely, the ruin probability quantifies the risk of "running out of money" in the portfolio within the considered time period. In order to determine this quantity, one needs to specify the properties of the stochastic process $C(t)$. The simpler the chosen model for $C(t)$, the more explicit expressions one can expect for $\psi(w)$ and $\psi(w, T)$. If one assumes stationarity (or some sort of regenerative property) for $C(t)$ as well as a certain degree of independence between increments of this process, then one can often apply recursive techniques, leading to integral (or integro-differential) equations for $\psi(w)$ and $\psi(w, T)$, which in some cases can be solved explicitly. Such explicit formulas are then helpful to determine model parameters (in particular the invoked premium) in such a way that a target value ϵ for the ruin probability can be achieved.

The classical model for $C(t)$ in this context is the *Cramér–Lundberg model*

$$C(t) = w + c\,t - \sum_{i=1}^{N(t)} X_i, \tag{7.2.8}$$

where the premiums are assumed to arrive continuously over time according to a constant premium intensity c and the aggregate claim size up to time t follows a compound Poisson process with rate λ (see Figure 7.1 for a sample path). Infinite-time ruin can only be avoided if the drift of the process is positive, that is,

$$c = (1+\theta)\lambda\,\mathbb{E}(X) \tag{7.2.9}$$

for some (relative) safety loading $\theta > 0$. This model is clearly very simplistic, but it enables an explicit expression for the ruin probability of the form

$$\psi(w) = \left(1 - \frac{\lambda\mathbb{E}(X)}{c}\right) \sum_{n=0}^{\infty} \left(\frac{\lambda\mathbb{E}(X)}{c}\right)^n (1 - F_I^{*n}(w)) \tag{7.2.10}$$

5 Clearly, *ruin* in this context does not necessarily mean bankruptcy, as there may be ways to still bail out the situation, but it is a natural and helpful decision criterion to assess the safety of the portfolio over time.
6 The ruin theory approach discussed in this section can in principle be applied to both a portfolio of an insurer (seeking for reinsurance to modify the resulting safety $\psi(w, T)$) as well as to the portfolio of a reinsurer addressing the question which reinsurance premiums can be offered so that the resulting safety for the reinsurer is acceptable.

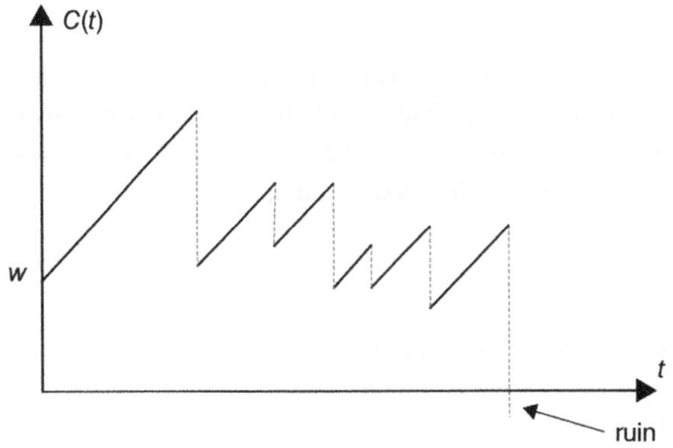

Figure 7.1 Sample path of a Cramér–Lundberg process $C(t)$.

for each $w \geq 0$, where F_I^{*n} is the n-fold convolution of the *integrated tail* (or *equilibrium*) c.d.f.

$$F_I(x) = \frac{1}{\mathbb{E}(X)} \int_0^x (1 - F_X(y))dy. \tag{7.2.11}$$

For particular models for the claim size distribution, this expression can simplify further significantly, for example if X is exponentially distributed with parameter v, then for each $w \geq 0$

$$\psi(w) = \frac{\lambda}{vc} e^{-(v-\lambda/c)w},$$

which is a simple and fully explicit formula in terms of the model parameters. In particular, for a fixed bound $\psi(w) \leq \epsilon$, one can – for given initial capital – determine the necessary (relative) safety loading θ in the premiums in view of (7.2.9).[7] Alternatively, for a fixed premium c one can determine the necessary initial capital w to ensure the bound ϵ.

If the so-called *adjustment equation*

$$\lambda + rc = \lambda\, \mathbb{E}(e^{rX}) \tag{7.2.12}$$

has a positive solution $r = \gamma > 0$ (which requires an exponentially bounded tail of the claim size distribution[8,9]), then one can show that for all $w \geq 0$

$$\psi(w) \leq e^{-\gamma w}. \tag{7.2.13}$$

7 Note that one can interpret the premium principle implied by this approach as an expected value principle (cf. Section 7.1).

8 This is particularly the case if all policies have an upper coverage limit per claim.

9 The quantity γ in this context should not be confused with the extreme value index in other chapters (this notation is very common for both objects).

This is the famous *Lundberg inequality* and γ is referred to as the *adjustment coefficient* (or *Lundberg coefficient*).[10] If the Lundberg coefficient exists, by combining (7.2.12) and (7.1.5) one can write

$$P(t) = c\,t = \frac{1}{\gamma} \log \mathbb{E}(e^{\gamma S(t)})$$

for the aggregate claim $S(t) = \sum_{i=1}^{N(t)} X_i$ up to time t, that is, the premium principle in this model can be interpreted also as an exponential principle with risk aversion coefficient γ.

If one now wants to ensure an upper bound $\psi(w) \le e^{-\gamma w} = \epsilon$, then one can translate the given security level ϵ and the initial capital w into the corresponding necessary adjustment coefficient $\gamma_\epsilon = |\log \epsilon|/w$, which leads to the premium requirement

$$P(t) = \frac{1}{\gamma_\epsilon} \log \mathbb{E}(e^{\gamma_\epsilon S(t)}).$$

Consequently, the use of the exponential premium principle and the concrete choice of its risk aversion coefficient have a motivation in the framework of infinite-time ruin probabilities (cf. Dhaene et al. [284]).

The above considerations rely on the assumptions of the Cramér–Lundberg model, which – for the sake of simplicity and mathematical elegance – ignores many aspects that are relevant in insurance practice, including the claim settlement procedure and loss reserving (claims are not paid immediately), inflation and interest, investments in the financial market, varying portfolio size, dependence between risks, later capital injections, dividend payments etc. Many variants of the Cramér–Lundberg model which include such features have been developed in the academic literature over recent decades, leading to nice and challenging mathematical problems. However, the resulting formulas (or algorithms) typically become much more complicated, even if some driving principles can be extended to models with surprising generality (see Asmussen and Albrecher [57] for a detailed survey).

If the model incorporates so many factors that analytical or even numerical tractability is lost, one can still simulate sample paths of the resulting process $C(t)$ and approximate $\psi(w)$ or $\psi(w, T)$ numerically (see also Chapter 9). However, apart from the complexity of the resulting calculations, there are other reasons that make the direct application of the ruin theory approach for solvency considerations less appealing. One of them is that it may be very difficult to formulate the dynamics of the stochastic process $C(t)$ sufficiently explicitly so that the result can reflect the company's situation (and the resulting model uncertainty may be considerable). This also includes the discrete nature of many model ingredients (the data situation may be sufficient to model a one-year aggregate view, but not the dynamics within the year). Furthermore, for infinite-time

---

10 In fact, under weak conditions the (Cramér–Lundberg) approximation

$$\psi(w) \sim C\,e^{-\gamma w}, \quad w \to \infty \tag{7.2.14}$$

holds for some constant $C < 1$, such that the inequality captures the qualitative behavior of $\psi(w)$ in terms of w and is often reasonably sharp (e.g., see [57]).

ruin considerations, there may not be a good reason to believe that business will continue in the long term in the same way as it does in the short term. Yet, the ruin probability can still provide insight into the portfolio situation: to value a certain strategy says something about how robust such a strategy is in the long run if it were continued that way (and this is a relevant question for sustainable risk management, even if one adapts the strategies later). In addition, the ruin probability quantifies the diversification possibilities of risks over time. Whereas the accounting standards in place in many countries nowadays do not allow for time diversification (e.g., equalization reserves), they can be an indispensable tool in the management of risks with heavy tails or risks where the diversification possibilities in space are limited, a situation that can be quite typical for reinsurance companies. In that sense, ruin probability calculations (or simulations) can be a valuable complement in the assessment of reinsurance solutions and their consequences.

We will come back to ruin theory considerations in Chapter 8 when the choice of reinsurance is discussed, particularly because it played a considerable role historically and is still implicitly behind the intuition of certain choices of reinsurance treaties nowadays.

7.2.2 One-year Time Horizon and Cost of Capital

The current regulatory approach employed in most countries is simpler and considers a time horizon of one year only. Let $Z(t)$ denote the available capital (assets minus liabilities) at time t. In broad terms, the criterion at time 0 is to hold sufficient capital so that the risky position $Z(1)$ becomes acceptable according to some risk measure ρ. Concretely, if $L(1) = Z(0) - Z(1)$ denotes the loss during the following year (leaving aside discounting), then the capital requirement is $\rho(L(1))$. For Solvency II, which has been implemented in the European Union since 1 January 2016, this risk measure determining the required capital is the *Value-at-Risk*

$$\rho(L(1)) = \text{VaR}_\alpha(L(1)) = Q_{L(1)}(\alpha)$$

for $\alpha = 0.995$. That is, under this requirement the probability that the loss of next year can not be covered by the available capital is bounded by 0.005. The Swiss Solvency Test, in contrast, uses the *conditional tail expectation*

$$\rho(L(1)) = \text{CTE}_\alpha(L(1)) := \mathbb{E}[L(1)|L(1) > \text{VaR}_\alpha(L(1))] \tag{7.2.15}$$

for $\alpha = 0.99$ (see also Section 4.6). In this case, even when one of the scenarios beyond the 99% quantile happens, the expected loss then still can be covered by the available capital.[11]

There are many sources of risk that influence the distribution of $L(1)$ in practice, so that its determination is complex. This includes also the challenge of the valuation of

11 In the technical document [344] of the SST the risk measure (7.2.15) is referred to as the *expected shortfall*, which usually is defined as the average $\text{ES}_\alpha(L(1)) := \int_\alpha^1 \text{VaR}_\eta(L(1))\,d\eta/(1-\alpha)$ (also often called Tail-VaR). While for some discrete random variables $L(1)$ the two concepts can lead to different numbers, they coincide for continuous random variables $L(1)$, which is the case of our interest (for details see [615]).

the assets and of the liabilities, which is supposed to be done in a market-consistent way. Among the risk sources are market risk (volatility of equity prices, interest rates, exchange rates, etc.), counterparty risk (the risk of default or rating changes of counterparties), and insurance risk. Typically these risks are investigated separately and then combined using some assumptions on their dependence.

For the purpose of pricing, we now focus on the *insurance risk* in a non-life portfolio and for simplicity assume that there is only new insurance business, which is also settled within the year (the arguments can then be adapted to include the loss development of business from previous years). Let us hence assume that the loss $L(1)$ here is determined by the aggregate insurance loss $S(1)$ minus the collected premium $P(1)$ received for the respective policies. This will lead to a regulatory solvency capital requirement $\mathrm{SCR}(1) = \rho(L(1))$, which the insurance company needs to hold. The capital SCR(1) will typically be provided by the shareholders of the company or external investors. For this risky investment they will demand a certain return rate r_{CoC}, leading to costs $r_{CoC} \cdot \mathrm{SCR}(1)$.[12]

One can now interpret $P(1)$ as being composed of the expected claim size $\mathbb{E}(S(1)) = \mathrm{BE}(1)$ (referred to as *best estimate*) and the safety loading RM(1) (referred to as *risk margin*). Following the regulatory view, one can see the necessary safety loading as the amount to finance the needed capital, that is,

$$\mathrm{RM}(1) = r_{CoC} \cdot \mathrm{SCR}(1),$$

since then the regulatory requirement for pursuing the insurance business is fulfilled (see also Section 8.3). If the investors are the shareholders of the company and any premium in excess of claim payments at the end of the year is their profit, then r_{CoC} can also be interpreted as the *expected* return rate on their investment SCR(1).

Summarizing, with such a premium $P(1)$ both the expected scenario and "adverse" claim situations up to the safety level of the regulatory capital requirement are covered. The need and size of the safety loading can accordingly be interpreted this way (rather than considerations of long-term safety, as in classical ruin theory).[13]

Entering a reinsurance treaty, the insurer's aggregate claim switches from $S(1)$ to $D(1)$, and correspondingly he will face a reduced expected claim size $\mathrm{BE}_D(1)$ and correspondingly smaller $\mathrm{SCR}_D(1)$, with a respective capital cost reduction. If this cost reduction exceeds the reinsurance premium which one has to invest to achieve this reduction, the resulting contract is worthwhile entering for the insurer.

Let us now look more closely at the pricing of different reinsurance forms.

---

12 Both Solvency II and the Swiss Solvency Test suggest a rate of $r_{CoC} = 6\%$ to be used in the calculations, particularly for loss reserving (e.g., see [344, 456]), whereas rates in the market are typically higher. See Albrecher et al. [19] for an approach to justify the size of r_{CoC} from an economic equilibrium perspective and in view of the limited liability of investors.

13 In fact, since (in the VaR case) every year the solvency capital is set to the level that avoids ruin with probability α, one can consider consecutive years as independent realizations of a random variable that each year leads to ruin with probability $1 - \alpha$. Correspondingly, the resulting infinite-time ruin probability is 1, that is, the company will become bankrupt with certainty, and (as a geometric random variable) the expected lifetime of the company is $1/(1 - \alpha)$. However, the actual capital held by the insurance company will typically be larger than this minimal requirement SCR(1) (i.e., the so-called *solvency ratio* will be larger than 1). Nevertheless, discussing the matter in terms of the minimally prescribed SCR(1) provides a basis for reasoning, and adaptations can then be made as seen fit.

7.3 Pricing Proportional Reinsurance

For proportional reinsurance treaties, the premium calculation is a priori quite simple, as it is natural to share the premium $P(t)$ according to the same proportion as the risk $S(t)$, and this is indeed the driving guiding principle. For a QS treaty $R(t) = a \cdot S(t)$ one correspondingly has

$$P_R(t) = a \cdot P(t),$$

and for surplus treaties one can determine the premium share in line with the respective proportionality factor that applies for each policy (or rather class of policies). However, a number of additional items have to be considered in practice. The most important is the *reinsurance commission* that will be subtracted from the reinsurance premium. It compensates the cedent for the fact that acquisition costs of policies, costs for the estimation, and settlement of the claims as well as other administrative costs are carried by the cedent, and the reinsurer will participate in those to some extent. The concrete amount of participation is often made dependent on the actual loss experience:

- *sliding scale commissions:* after setting a provisional commission (in terms of a percentage) and a reference loss ratio[14], for each percentage point that the actual loss ratio deviates from that reference point, the commission percentage is adapted (not necessarily 1:1) inversely with the loss ratio, but within upper and lower limits
- *profit sharing provisions:* if the reinsurer's participation in a given year is very successful (i.e., there are few losses), the reinsurer passes back some of the premium, again along predefined terms
- *loss corridors:* to further protect the reinsurer, an agreement may be that the reinsurer only covers $a\%$ of the reinsurer's loss ratio, and then again on from $b\%$ $(b > a)$, whereas the cedent retains the part in between.

When these loss-dependent features are defined in a piecewise linear fashion, then the calculation of their impact is still fairly straightforward, but depend on the distribution of the loss ratio. So to settle the amount of the commission, this distribution needs to be estimated from information on historical data. Here some care is needed. If the treaty is of a "losses occurring" type, then the earned premium and the accident year losses are relevant. On the other hand, a treaty can be of the "risks attaching" type, in which case losses on policies written during the treaty period are covered, and the written premium and the losses of all those policies should be considered. In a next step, catastrophe losses (which caused many claims) and shock losses (which caused single very large claims) are then most often removed from the set of collected data. Then the remaining historical losses have to be developed to ultimate values, both for the number of claims (IBNR) and the sizes of claims (IBNR and IBNER) (see also Section 7.4.2.3). In addition, the historical premiums have to be adjusted to the future level, including the rate changes that are expected during the treaty period. Finally, the losses also have to be trended to the future period. From all these obtained

14 The *loss ratio* is defined as the incurred claims (together with expenses associated to their investigation and settling) divided by the earned premium and is a popular performance measure in practice.

data points, the expected loss ratio is then estimated by the arithmetic average of the respective historical loss ratios adjusted to the future level. At this point, a catastrophe loading then has to be added to the expected non-catastrophe loss ratio, for which various methods are used, including simulation models (e.g., see Clark [215, 216] for details).

Finally, in certain lines of business the reinsurer may actually have a lot, sometimes more, experience in the nature of the claims, and may want to adapt or correct the premium amount collected by the cedent to the size that he deems more appropriate.

7.4 Pricing Non-proportional Reinsurance

From an actuarial point of view, the pricing of non-proportional reinsurance is considerably more involved than that for proportional treaties. As mentioned before, the more reliable information one has about the claim size distribution, the more flexibility one gains in terms of using a premium principle like the ones mentioned in Section 7.1. In typical non-proportional reinsurance treaties, one often has difficulties in determining more than the first one or two moments, so that an expected value principle or variance principle is quite common. In addition, a number of clauses (including reinstatement constructions etc.) often even make the determination of the first two moments a non-straightforward assignment. We will therefore discuss some guiding principles for determining the pure reinsurance premium. Two main approaches can be distinguished. The first is mainly based on the reinsurer's own experience and assessment of the underlying risks, normalized to the volume of the exposure (reflected by the premium that the cedent collected from policyholders for this portfolio (*exposure method*)). The second relies on the previous loss experience of that particular reinsured portfolio (*experience method*).

We focus here on XL treaties (for SL the same principles apply). For more information on large claim reinsurance forms, refer to Section 10.1.

7.4.1 Exposure Rating

7.4.1.1 The Exposure Curve

Consider an individual claim X, which is subdivided into $X = D + R$. Recall from (2.3.9) that under an ∞ xs M contract we have

$$\mathbb{E}(D) = \int_0^M (1 - F_X(z))\, dz, \qquad \mathbb{E}(R) = \int_M^\infty (1 - F_X(z))\, dz. \tag{7.4.16}$$

This shows that if one is interested in the pure reinsurance premium in such a treaty, the function

$$r_X(M) := \frac{1}{\mathbb{E}(X)} \int_0^M (1 - F_X(z))\, dz$$

is particularly useful, as it gives for each retention $M > 0$ the fraction $\mathbb{E}(D)/\mathbb{E}(X)$ of the risk X that stays with the cedent. $r_X(M)$ is called the *exposure curve*. Mathematically, one immediately recognizes $r_X(M)$ as the *equilibrium distribution function* of X, which leads to nice properties. For instance, for the Laplace transform one has

$$\int_0^\infty e^{-sx}\,dr_X(x) = \frac{1-\hat{F}_X(s)}{s\,\mathbb{E}(X)},$$

where $\hat{F}_X(s)$ is the Laplace transform of X. In particular, there is a one-to-one correspondence between F_X and r_X. From $r_X'(x) > 0$, $r_X''(x) < 0$ one sees that the exposure curve is an increasing and concave function (and, as the c.d.f. of a positive random variable, it starts in 0 and tends to 1 for $M \to \infty$).

Even if in light of this one-to-one correspondence one does not gain directly from modelling the exposure curve (instead of the distribution of X), it will turn out to be a quite useful description below, and actuaries have developed a remarkable intuition about the shape of this curve for certain lines of business. Note that for the aggregate claim size $S(t) = \sum_{i=1}^{N(t)} X_i$ in a portfolio with i.i.d. risks X_i, one can use the same exposure curve as for X:

$$r_S(M) := \frac{\mathbb{E}(D(t))}{\mathbb{E}(S(t))} = \frac{\mathbb{E}(N(t))\mathbb{E}(D)}{\mathbb{E}(N(t))\mathbb{E}(X)} = r_X(M),$$

as the influence of $N(t)$ cancels out. For the direct modelling of exposure curves, one often uses discrete shapes. A continuous one-parameter family of such curves that still enjoys some popularity was given by Bernegger [129]:

$$G_{b,g}(x) = \begin{cases} x & g = 1 \text{ or } b = 0 \\ \frac{\log(1+(g-1)x)}{\log g} & b = 1, g > 1 \\ \frac{1-b^x}{1-b} & bg = 1, g > 1 \\ \frac{\log \frac{(g-1)b+(1-gb)b^x}{1-b}}{\log(gb)} & b > 0, b \neq 1, bg \neq 1, g > 1 \end{cases}$$

with $b(c) = e^{3.1-0.15(1+c)c}$ and $g(c) = e^{(0.78+0.12c)c}$ for a parameter c (which is often chosen to be one of the values $c \in \{1.5, 2, 3, 4\}$, but other values are used as well). Note that here x is in % of the size of the risk, measured, for example, by the sum insured or the PML (see Section 7.4.1.2).

The exposure curve plays a crucial role in determining the pure reinsurance premium.

7.4.1.2 Pure Premiums

The idea in exposure rating for the reinsurer is to consider the insurer's pure premium P_j for each different risk group j and then use own experience (that the reinsurer may have gained through previous treaties in that market), experience from related portfolios or market statistics to determine the premium. That "experience" is reflected in the choice of the exposure curve. For an ∞ xs M contract on risk group j one then gets

$$P_{R,j} = P_j(1 - r_j(M))$$

and for an L xs M contract correspondingly

$$P_{R,j} = P_j(r_j(M + L) - r_j(M)).$$

That is, one has intuitive and simple adaptations of the insurer's premiums available. Such an approach is only feasible if the chosen exposure curve r_j applies to all risks of that risk group, an assumption that may only be fulfilled if the sizes Q_i of the risks in the policies underlying the claims X_i are sufficiently similar. If this is not the case, there are two common adaptations:

a) If the loss degree $V_i = X_i/Q_i$ can be assumed to be i.i.d. across the policies (which seems a reasonable assumption in *property, marine hull* and *personal accident* insurance[15]), then one can estimate (or use from past sources) the exposure curve r_V for the loss degree directly to get for an ∞ xs M treaty

$$P_{R,i} = P_i(Q_i)\left(1 - r_V\left(\frac{M}{Q_i}\right)\right).$$

A particular advantage in this case is the fact that r_V is inflation-invariant and currency-invariant (as inflation will affect M and Q_i in the same way).

b) In *liability* insurance the sum insured will often be chosen arbitrarily and is typically much smaller than the maximum claim size, so that the above assumption is not reasonable. One may then, however, assume that the original risk X is identically distributed within the same risk class, and that the claim for each policy is just truncated at different values Q_i, that is, $X_i = \min(X, Q_i)$ for some generic non-truncated X with exposure curve r_X. One then obtains

$$r_{Q_i}(M) = \begin{cases} \frac{\mathbb{E}(\min(X,M))}{\mathbb{E}(\min(X,Q_i))} = \frac{r_X(M)}{r_X(Q_i)}, & 0 < M \leq Q_i, \\ 1, & M > Q_i, \end{cases}$$

which is referred to as the *increased limits factors curve* (*ILF curve*). But here r_X is not inflation- and currency-invariant, which is undesirable. An alternative reasoning going back to Riebesell [648] proposes that when doubling any insured sum, the pure premium should be multiplied by $1 + z$ for some fixed $0 < z < 1$, that is, $\mathbb{E}(\min(X, 2Q)) = (1+z)\mathbb{E}(\min(X, Q))$ (in contrast to property insurance above where the insured sum is a measure of the size of the risk, here the pure premium increases less than the sum insured). This rule then implies

$$r_{Q_i}(M) = \left(\frac{M}{Q_i}\right)^{\log_2(1+z)},$$

which is again inflation- and currency-invariant (see Mack and Fackler [556] for a characterization of distributions of X for which this logarithmic scaling rule of

15 In property insurance the risk size is typically measured by the PML or (for small risks) by the sum insured. In personal accident and marine hull the sum insured is used.

$\mathbb{E}(\min(X, Q))$ can indeed be fully justified). The resulting reinsurance premium then again is

$$P_{R,i} = P_i(Q_i)\left(1 - r_{Q_i}(M)\right).$$

For further extensions and discussions see Riegel [649] and Fackler [343].

As for proportional treaties, in addition the reinsurer may want to correct the used values of P_i if he does not consider them appropriate.

7.4.1.3 Safety Loadings

To determine the safety loading, it is often the variance that is of interest (if available). For this purpose, for an L xs M contract a natural possibility is to use the representation

$$R(t) = \sum_{j=1}^{N_R(t)} \breve{R}_j$$

introduced in (6.5.24), where $N_R(t)$ is the number of claims that the reinsurer faces, and $\breve{R}_j$ is the reinsured amount of the jth claim that concerns the reinsurer. Let $\tilde{\mu}_{R,k} := \mathbb{E}(\breve{R}_j^k)$ (note that this differs from $\tilde{\mu}_k$ introduced in Section 2.3.1, as $\breve{R}_j$ does not have an atom at 0). If one assumes i.i.d. claim sizes, then

$$\begin{aligned}
\mathrm{Var}(R(t)) &= \mathbb{E}(N_R)\tilde{\mu}_{R,2} + (\mathrm{Var}\,(N_R) - \mathbb{E}(N_R))\tilde{\mu}_{R,1}^2 \\
&= \mathbb{E}(N_R)\tilde{\mu}_{R,1}\left(\frac{\tilde{\mu}_{R,2}}{\tilde{\mu}_{R,1}} + \left(\frac{\mathrm{Var}\,(N_R)}{\mathbb{E}(N_R)} - 1\right)\tilde{\mu}_{R,1}\right) \\
&= P_R^{pure}\left(\frac{2\int_M^{M+L}(r_X(M+L) - r_X(x))dx}{r_X(M+L) - r_X(M)} + \left(\frac{\mathrm{Var}\,(N_R)}{\mathbb{E}(N_R)} - 1\right)\tilde{\mu}_{R,1}\right)
\end{aligned}$$

where P_R^{pure} is the pure premium from the previous section. Hence, the variance can be expressed in terms of the exposure curve in the region of the layer. Note that (particularly for higher layers) $\mathrm{Var}\,(N_R)/\mathbb{E}(N_R)$ will often be close to 1, so the second term in the above sum can be quite small.

In Section 7.5 we will deal with the determination of safety loadings from an aggregate point of view.

7.4.2 Experience Rating

In contrast to the exposure method, in *experience rating* one bases the calculations on the loss experience of the concrete portfolio. This can work if there is sufficient credible claim experience available. In this context the techniques introduced in Chapter 4 can be very useful. As already discussed there, the insurer will typically not pass on the entire series of previous claims, but only those that are immediately relevant for the layer under consideration (so that the reinsurer faces the statistical challenge of data analysis under truncation). For instance, only past claims larger than half of the deductible over the last 5–10 years would be communicated. In order to make those past data points comparable, they have to be suitably adjusted. Each considered factor may influence the size of the claims, the number of the claims or both. Besides changes in legislation, technical changes or changing insurance market conditions (which are sometimes not easy to incorporate), the following three factors need to be considered and can typically be quantified to a satisfactory degree.

7.4.2.1 Inflation

Data points spreading over several years need to be inflation-corrected. Depending on the line of business, there may also be other adjustment indices that are more suitable (like the building cost index). This makes the data points comparable, but there is also another important aspect of inflation in XL treaties: a claim that previously did not touch the layer under consideration may nowadays be above M, that is, relevant for the treaty (this is the main reason for the reporting threshold for past claim sizes to be considerably below M and not at M). To see this, consider the expectation $\mathbb{E}(R_i^{M,L}) = \int_M^{M+L}(1-F_X(z))dz$ in an L xs M contract. After inflation with factor $\delta > 1$ this changes to

$$\int_M^{M+L} (1 - F_{\delta X}(z))dz = \delta \int_{M/\delta}^{(M+L)/\delta} (1 - F_X(z))dz = \delta\, \mathbb{E}(R_i^{M/\delta,L/\delta}),$$

that is, the latter corresponds to the (scaled) expected reinsured claim size in the layer $[M/\delta, (M+L)/\delta]$. In typical cases this resulting expectation is larger than the inflation-corrected original expected value. For instance, if X is strict Pareto with parameters $\alpha > 1$ and x_0, a simple calculation gives $\delta\, \mathbb{E}(R_i^{M/\delta,L/\delta}) = \delta \cdot \delta^{\alpha-1}\, \mathbb{E}(R_i^{M,L}) > \delta \cdot \mathbb{E}(R_i^{M,L})$.

The inflation may affect the claim sizes and the claim number in different ways, which complicates the analysis. Inflation can be a major issue in MTPL lines (for a detailed discussion see Fackler [342]).

7.4.2.2 Portfolio Size Changes

In a next step, all data points have to be adjusted to the portfolio volume of the present year. After correcting for tariff adjustments, the premium amount is often considered a suitable measure for the volume, particularly in fire and motor liability portfolios (although inflation may affect premiums and claim sizes differently, which also has to be taken into account). For fire, the sum insured (or PML) is also used (see also Section 7.4.1). In other lines of business, other measures may be considered more natural (like passenger kilometers flown in aviation liability). Whether the claim sizes are assumed to increase proportionally with volume or with respect to some other functional relationship, and whether this affects both the sizes and the number of claims, strongly depends on the line of business and the type of cover (e.g., for per-risk XL, volume will typically affect the number of claims and not the size, whereas for cumulative XL it will rather influence the size of the (aggregate) claims per event).

7.4.2.3 Loss Development

In some lines of business (and particularly so in liability), it can take many years until a claim payment is finally settled. In such a case, one has – for all claims that are not fully developed yet – only development patterns and current estimates of the final loss burden available. The reinsurer then needs to use loss reserving techniques on an individual claim basis in order to make the data points of different years comparable.

This needs to be done both for the number of claims (IBNR) and the sizes of the claims (IBNR and IBNER) (see Chapters 1 and 4, where this is discussed in detail and illustrated with data). One may finally also have to discount the data according to the typical payment patterns, as later payments enable to invest the premiums until the payments are due.

If all these adaptations are done and one is left with comparable data points, then the simplest procedure is to build the empirical c.d.f. and use it for the pricing (this is called *burning cost rating*). The resulting expected claim size (i.e., the arithmetic mean of the data points, correspondingly referred to as *burning cost*) is often considered quite useful, but clearly the empirical distribution will typically not be sufficient to model the entire risk (using the empirical c.d.f. implies that the largest possible claim size has already occurred!). Also, one often has rather few data points in the layer. We refer to Chapter 4 where we have discussed and illustrated respective EVT techniques extensively. Also, a Bayesian approach with an a priori guess for model parameters (such as the extremal index) is sometimes implemented. Finally, *credibility techniques* can be quite useful, where a weighted average is taken between the own claim experience and the one of related portfolios. The weights are chosen according to how "credible" the respective claim information is, and the development of respective algorithms is a classical actuarial technique (see the references at the end of the chapter).

If there is no relevant claim experience available at all, but one has an estimate of how frequently a loss occurs (say once in n years), then a crude way to state a pure premium is to divide the layer size L by n. This leads to a *payback tariff* (the contract is said to have an n-years *payback*, see also Section 2.3.3).

7.4.3 Aggregate Pure Premium

If one finally has distributions for the individual reinsured claim sizes R_i and the claim number $N(t)$ available, then the aggregate claim size distribution for the reinsurer can be determined just as for the first-line insurer. In the presence of an aggregate (typically annual) deductible AAD and an aggregate limit AAL we have

$$R(t) := \min\left\{\left(\sum_{i=1}^{N(t)} \min\{(X_i - M)_+, L\} - AAD\right)_+, AAL\right\}$$

(cf. (2.3.14)). For instance, if all risks are i.i.d. and $N(t)$ is Poisson or negative binomial, then the distribution of $S_R = \sum_{i=1}^{N(t)} \min\{(X_i - M)^+, L\}$ can be determined by Panjer recursion (cf. Section 6.3). The aggregate pure premium for the XL treaty is then obtained through

$$\mathbb{E}(R(t)) = \int_{AAD}^{AAD+AAL} (1 - F_{S_R}(x))\,dx. \tag{7.4.17}$$

As discussed before, in XL practice often the premium is not fixed in advance, but dependent on the loss experience during the contract. In that case, the (pure) premium rule is then the function P which satisfies

$$\mathbb{E}(P(X_1, X_2, \ldots)) = \mathbb{E}(R(t)). \tag{7.4.18}$$

Examples are contracts with the following:

- *Slides:* For a slide with a *fixed loading* L_s, the agreement is of the form

$$P(R(t)) = \min\{P_{\min} + (R(t) + L_s - P_{\min})_+, P_{\max}\},$$

that is, the reinsurance premium actually is the aggregate claim experience $R(t)$ of the reinsurer plus the loading L_s, but capped at a minimal and maximal premium values $P_{\min}$ and $P_{\max}$, respectively. One way to implement this is that for given $P_{\min}$, loading L_s and distribution of $R(t)$, the value of $P_{\max}$ is determined such that (7.4.18) holds.
Alternatively, a slide with *proportional loading* b is a contract of the form

$$P(R(t)) = \min\{P_{\min} + (b \cdot R(t) - P_{\min})_+, P_{\max}\}.$$

- *Reinstatements:* This variant of XL contracts has already been discussed in Section 2.3.2. The sequence $(c_n)_{1 \le n \le k}$ is fixed in advance and called a *premium plan*, that is, further liabilities may have a different price than the first one. The values c_n may be fixed *(pro-rata capita)* or depend on the time when the reinstatements are paid *(pro-rata temporis)*. The latter variant is intuitive, since close to the expiry of a contract it will be less likely that the next liability will be used up, but this is nowadays not so common anymore. For the reinstatement of the nth such liability ($n \le k$) one then has

$$P_j = c_n \frac{P_0}{L} \min\{\text{reinstatement of liability } n,\ L\}.$$

For pricing under different premium schemes, see Mata [562] and Hess and Schmidt [439].

7.5 The Aggregate Risk Margin

The risk margin that the reinsurer needs to put on top of the pure premium and the expenses will of course also depend on the other risks in the reinsurer's portfolio (in view of diversification possibilities etc.). We will take here the viewpoint of Section 7.2.2 and assume that the reinsurer has a required cost-of-capital rate r_{COC} (i.e., a return target on risk-adjusted capital), which in view of his overall situation leads to the aggregate risk margin $\mathrm{RM} := \mathrm{RM}(1)$. This amount now has to be sub-divided onto the risk margins of all m treaties that the reinsurer has in the portfolio:

$$\mathrm{RM} = \sum_{j=1}^{m} \mathrm{RM}_j,$$

where RM_j denotes the risk margin of treaty j with risk R_j (so $R(1) = \sum_{j=1}^{m} R_j$). Theoretically, the classical rule

$$\mathrm{RM}_j = \frac{\mathrm{RM}}{\mathrm{Var}\,(R(1))} \mathrm{Cov}(R_j, R(1))$$

appears natural. However, knowledge about the covariance between treaty j and the aggregate risk $R(1)$ will most often be out of reach, so one has to resort to simpler rules.[16] A typical compromise for such a simpler rule is to consider the fluctuations of each treaty stand-alone, for example in terms of the *variance principle*

$$\mathrm{RM}_j = \frac{\mathrm{RM}}{\sum_{k=1}^{m} \mathrm{Var}\,(R_k)} \mathrm{Var}\,(R_j) := \alpha_V \cdot \mathrm{Var}\,(R_j)$$

(which is exactly the above covariance principle for independent reinsurance treaties) or in terms of the *standard deviation principle*

$$\mathrm{RM}_j = \frac{\mathrm{RM}}{\sum_{k=1}^{m} \sigma(R_k)} \sigma(R_j) := \alpha_S \cdot \sigma(R_j).$$

A further possibility is the *square root rate on-line (ROL) principle:* if L_j denotes the aggregate upper limit of the reinsurer in treaty j, then for the ROL $r_j := \mathbb{E}(R_j)/L_j$ (cf. Section 2.3.3), one divides according to

$$\mathrm{RM}_j = \frac{\mathrm{RM}}{\sum_k L_k \sqrt{r_k}} L_j\sqrt{r_j} := \alpha_R \cdot L_j\sqrt{r_j}.$$

Recall that the philosophy behind the ROL is that it approximates the probability of a total loss L_j of the layer. If one assumes that either no claim or the total loss L_j occurs for treaty j, then one would get $\sigma(R_j) = L_j\sqrt{r_j(1-r_j)}$, and (for small r_j) the principle then resembles the standard deviation principle above.[17]

A challenge in the implementation of the above approach in practice is that at the time of the pricing of treaty j, one often does not yet know which other treaties will be added to the portfolio, so that one has to estimate the constant α_V, α_S or α_R, respectively, for the future portfolio on the basis of the current composition and planned modifications. The resulting constant can then be used to price each of the treaties according to the respective premium principle (so this provides a concrete guideline for the choice of the constants α_V and α_S in Section 7.1).

One issue that remains is the following inconsistency: if the layer L_j of treaty j were subdivided into two layers, then (due to the positive dependence of the loss in these two layers) the variance of the loss of the entire layer would be larger than the sum of the variances of the two sublayers (and similarly for standard deviation and ROL). That is, there is (at least theoretically) an incentive for the cedent to chop the layers into smaller and smaller pieces whenever such a premium principle is applied. One suggestion to

16 For an exception for Cat-XL see Bernegger [129].

17 Indeed, it is observed in practice that the square root ROL principle approximates the standard deviation principle remarkably well, particularly if the layer is not too large and the distribution is heavy tailed – in which case claim payments are more likely to be close to L_j.

mitigate this non-additivity is the so-called *infinitesimal ROL principle*: starting again from the aggregate pure premium (7.4.17)

$$\mathbb{E}(R_j) = \int_{M_j}^{M_j+L_j} (1 - F_{S_R^{(j)}}(x))\, dx,$$

where $S_R^{(j)}$ is the aggregate reinsured claim size in treaty j in the absence of aggregate deductible and limit, the idea is to apply the ROL principle separately for each small part $(z, z + \Delta z) \in [M_j, M_j + L_j]$, that is,

$$r_j(z) = \frac{1}{\Delta z} \int_{z}^{z+\Delta z} (1 - F_{S_R^{(j)}}(x))\, dx,$$

and in the limit $\Delta z \to 0$ one gets $r_j(z) = 1 - F_{S_R^{(j)}}(z)$. The infinitesimal contribution to the premium according to the ROL principle then is $\alpha_R^* \sqrt{1 - F_{S_R^{(j)}}(z)}\, dz$, so that the overall contribution amounts to

$$\mathrm{RM}_j = \alpha_R^* \int_{M_j}^{M_j+L_j} \sqrt{1 - F_{S_R^{(j)}}(z)}\, dz,$$

and the constant α_R^* is analogously RM divided by the sum of all these m integrals. By construction this method to allocate the risk margins does not suffer from the non-additivity problem.

The idea of deriving the individual risk margins (loadings) from its marginal capital requirements can be traced back to Kreps [507]. The capital allocation for the individual layers is often done using the so-called *capacity appetite limit* (CAL) curve, which assigns the risk weights to different tranches of capital. This CAL curve can be seen as the reinsurer's implicit utility function and is provided to the pricing team by overall capital considerations. For more refined pricing techniques in view of capital considerations see [689].

7.6 Leading and Following Reinsurers

In many realizations of reinsurance contracts there are in fact several reinsurers involved in a treaty. A *leading reinsurer* negotiates the premium, but finally only takes a certain proportional share of both the premium and the risk, and other reinsurers take the remaining proportions. If we again assume that there are m treaties and the reinsurer's share in treaty j is $a_j \leq 1$, then the aggregate risk for the reinsurer is

$$R = \sum_{j=1}^{m} a_j R_j.$$

As a consequence, the risk margin then is determined by the premium calculation of the leading reinsurer.[18]

If an overall premium P_j for a reinsurance treaty R_j is already negotiated, a (following) reinsurer has to determine on the basis of his internal premium principle, if and to what extent a participation in that treaty is feasible. For fixed costs k involved in entering the contract, one gets the condition

$$P(a_j \cdot R_j) + k \leq a_j P_j$$

for the internal premium rule P, on the basis of which one can determine which share a_j is appropriate (or optimal). When the reinsurer internally employs a standard deviation principle, this leads to a linear inequality for a_j (such that the optimal share is the maximum available share), whereas for a variance principle the resulting inequality is quadratic in a_j.

In terms of leading and following reinsurers it is quite common that all reinsurers equally participate in all layers (*"across the board"*). If one of the reinsurers does not want to take the equal share in the upper layers (e.g., due to internal limits) or lower layers (because of high administrative costs due to the large number of claims there), then this can be passed on to one of the other reinsurance partners or a reinsurance broker looks for an external additional reinsurer to take part in only those layers.

7.7 Notes and Bibliography

For general classical accounts on premium calculation in risk theory we refer to Gerber [383], Goovaerts et al. [400], and Kaas et al. [476]. For a more applied view see Mack [555] and Parodi [606]. An early discussion of convexity in the context of premium calculation is Deprez and Gerber [279]. An ordering using Lorenz curves was discussed in Denuit et al. [278], see also Heilmann [434]. An extension of the expected value principle to a quasi-mean value principle with its properties is given in Hürlimann [454].

Reich [642] has shown that the standard deviation principle enjoys some fundamental properties. Benktander [111] suggests a linear combination of expected value, variance, standard deviation principles.

The distortion principle can be found in Denneberg [273] and turns out to be quite general, in the sense that many pricing rules (and risk measures) can be expressed that way for an appropriate distortion function (see Wang [770, 771], and Pflug and Römisch [615] for a general overview in the context of risk measures). Duality concepts play a major rule in this context (see Yaari [799] for a classical influential contribution). Furman and Zitikis [362] discuss the general and intuitive concept of actuarial weighted pricing functionals, which contain many pricing rules (also when related risks are considered) and propose implications for capital allocation.

18 If, for instance, the present reinsurer under consideration is the leading reinsurer for treaty j with share a_j and uses a variance principle with some constant α_V, then the risk margin is $\mathrm{RM}_j = \alpha_V\, a_j^2 \mathrm{Var}(R_i)$, so that the total risk margin for that treaty amounts to $\mathrm{RM}_j / a_j = \alpha_V\, a_j \mathrm{Var}(R_j)$. That is, the degree of participation of the leading reinsurer here impacts the risk margin.

Principles for calculating premiums are of course choices to measure risk, and the research area of risk measures, their axiomatic foundations, and entailed capital allocations has seen an enormous activity over the last two decades, boosted by the paper of Artzner et al. [49] on coherent risk measures. We do not give an overview of the correspondingly vast academic literature, which is often primarily targeted towards financial applications. An early axiomatic approach for insurance pricing can be found in Wang et al. [773], Venter [755], and Young [801]. An interpretation of the ruin probability concept in the context of risk measures can be found in Dhaene et al. [284], Cheridito et al. [203], and Trufin et al. [746]. As mentioned earlier, for (re)insurance purposes the VaR plays a central role in regulation, and also the CTE is a much discussed measure (and implemented in the Swiss insurance regulation). A survey of what can happen to VaR calculations when the claim size distribution has heavy-tailed or time-dependent behavior is given in Bams et al. [77]. Danielsson [246] covers the role of the VaR in connection with extreme returns, see also Neftci [588] and Luciano et al. [548]. The performance of extreme value theory in VaR calculations is compared to other techniques by Gençay et al. [379]. For risk measures with a pricing undertone, see Van der Hoek et al. [751]. For more details on ES as a risk measure, see Acerbi et al. [8] and Fischer [352]. Necessary and sufficient conditions for coherence are covered by Wirch and Hardy [793]. A rather general risk measure for the super-exponential case has been developed in Goovaerts et al. [401] and is based on a generalization of the classical Markov inequality from probability theory. Siu and Yang [702] introduced a set of subjective risk measures based on a Bayesian approach. For other alternatives, see, for example, Balbás et al. [75]. Powers [625] suggests the use of the third and fourth moments to measure risk.

Despite the toolkit of techniques discussed throughout this book, the data situation on the losses of certain reinsurance portfolios often does not provide sufficient reliable distributional information beyond a few moments (if at all), and this is one of the main reasons why simple rules based on one or two moments are still abundantly used in pricing. We point out, however, that one has to be very careful when estimating empirical moments from the data points directly. As amply illustrated in Chapter 4, it may easily happen in reinsurance applications that the underlying moments that one wants to estimate do in fact not exist. For instance, when one tries to estimate the sample dispersion or (particularly) the sample coefficient of variation directly from a set of i.i.d. data points with a distribution for which the first or second moment do in fact not exist, the erratic behavior of their estimates may somewhat cancel out in the estimator of the ratio, and so one may not "see" the problem immediately and proceed with the estimate, even if the true value is not finite. This problem is illustrated in some detail in Albrecher et al. [25, 27], where asymptotic properties of the corresponding estimators in such situations are also worked out.

Brazauskas [162] deals with the effects of data uncertainty on the estimation of the expected reinsured amount under various reinsurance treaties from the viewpoint of robust statistics.

Historically, Beard [92] claimed that the most troublesome premium calculation likely to arise in practice would be the determination of an XL reinsurance premium based only on the largest claims experience. A first attempt on the use of extreme value techniques for pricing of XL treaties was made by Jung [473]. For an early practical approach including IBNR techniques, see Lippe et al. [546]. For a treatment of large claims within credibility, see BÃ¼hlmann and Jewell [171] and Kremer [504]. For information and references on credibility techniques, refer to Dannenburg et al. [247], Kaas [475], and Bühlmann and Gisler [169]. Alternative estimation procedures can be found in Schnieper [682]. The effect of contract terms on the pricing of a reinsurance contract is discussed in Stanard et al. [708].

A standard reference on stochastic ordering is Shaked et al. [696], see also Kaas et al. [475] and in particular Denuit et al. [274]. For a general account on the concept of comonotonicity, refer to Dhaene et al. [281, 282].

A rich source for practical issues in pricing of proportional reinsurance is Clark [216]. Antal [43] discusses pricing from the reinsurance perspective in detail. For refinements of exposure rating techniques for fire portfolios, see Riegel [650]. Mata [563] gives more information on burning cost methods. Verlaak et al. [761] develop a regression technique to form a benchmark market price out of individual MTPL XL prices. Desmedt and Walhin [280] suggest a method of combining exposure rating and experience rating for pricing rarely used layers through using exposure techniques on the experience rates of working layers.

The final aggregate *combined ratio* (defined as the ratio of the sum of incurred losses and operating expenses divided by earned premium) of reinsurance companies naturally varies widely across reinsurance companies and years, but a realistic average magnitude reported in practice is 80–90% (e.g., see [354]). Loading factors θ (cf. (7.1.2)) of individual reinsurance treaties in practice again can vary considerably, the range from $\theta = 0.1$ in lower layers to $\theta = 0.6$ in more extreme layers (and also outer contracts when combining reinsurance forms) is, however, typical in competitive reinsurance markets (e.g., see Verlaak and Beirlant [760]).

Pricing is also conceptually quite different between property and casualty lines: whereas the former is dominated by prefunding losses, the idea of postfunding losses underlies the pricing of casualty losses, see [375] for details on this and many other practical matters.

In life reinsurance, the standard model for pricing Cat XL layers goes back to Strickler [712], for a recent refinement see Ekheden and Höossja [319].

Note that the stop-loss premium (defined as the expected claim size of a reinsurer in an unlimited SL cover, cf. (6.5.28)) corresponds to $\mathbb{E}(R_i)$ from (7.4.16), but applied to the c.d.f. of the aggregate claim size. Even if such unlimited covers are not often applied in reinsurance practice, the stop-loss premium has been studied intensively in terms of its theoretical properties, and these results can be used as benchmarks. For instance, bounds on stop-loss premiums have been studied by Bühlmann et al. [168], and Runnenburg et al. [659] under heavy- and light-tailed assumptions (see also Kremer [503]). For bounds where the claim distribution is unknown but in the proximity of the empirical distribution of past claims, see Xu et al. [798]. For numerical and algorithmic aspects, see Kaas [474], and recursive methods are discussed in Dhaene et al. [287]. Also, it is of particular interest to look into robustness of stop-loss premiums with respect to dependence of the individual claims. First studies in this direction include Dhaene and Goovaerts [283], Albers [12], and Denuit et al. [275]. The interplay between maximal stop-loss premiums and comonotonicity is dealt with by Dhaene et al. [286]. Various approximations are compared in Reijnen et al. [643].

Pure premiums for drop-down XL covers as introduced in Chapter 2 can be found in Kremer [506]. For a general analysis of various risk measures for reinsurance layers, see Ladoucette and Teugels [522].

8

Choice of Reinsurance

An insurer has to make a choice among all feasible reinsurance treaties. For example, he might want to maximize his expected profit after reinsurance. Or he could minimize the probability of ruin after reinsurance. Or he may want to modify the risk profile in such a way that the needed solvency capital after reinsurance becomes affordable.

It is impossible to completely formalize the decision process on the choice of a reinsurance form and its concrete specification. Many factors will influence such a decision, which involves experience in the market (and with the contract partner) as well as availability of requested contract forms for a reasonable premium. Eventually there may even be an element of personal taste involved. On the other hand, when the objective function (to be maximized or minimized) and possible constraints can be defined together with a premium rule that assigns a reinsurance premium to every available contract form, the identification of the optimal reinsurance treaty becomes a purely mathematical problem, and sometimes leads to quite tractable and at times even simple solutions. Over the last few decades there has been an enormous amount of academic activity on this topic, and this could easily be the subject of an entire book on its own. For a direct implementation of such results in practice, however, the involved assumptions in such theoretical results will usually be a too coarse description of the real situation. Also, reinsurance will often be only one tool in a more general framework of optimal capital and risk transfer between market participants, where other than actuarial principles can play a prominent role (see the Notes at the end of the chapter). However, the mathematical results described below can help to foster the intuition and comprehension of the consequences of certain choices of contracts, and hence may serve as guiding principles and possible justifications of treaties.

The goal of this chapter is to present some classical lines of reasoning for rationalizing the choice of reinsurance forms, link them to some more recent contributions and provide pointers to the specialized academic literature.

From a cedent's perspective, the choice of a reinsurance form will intrinsically depend on the aggregate portfolio risk $S(t)$, on the premium $P(t)$ that he gets for bearing $S(t)$, on the reinsurance premium and on the costs involved in the transaction of the potential reinsurance contract. Let us put together the relevant quantities.

Reinsurance: Actuarial and Statistical Aspects, First Edition.
Hansjörg Albrecher, Jan Beirlant and Jozef L. Teugels.

- Recall that the total claim amount $S(t)$ is subdivided into the retained amount $D(t)$ and the reinsured quantity $R(t)$:

  $$S(t) = D(t) + R(t).$$

  For simplicity of notation we will drop the reference to t in the notation of this chapter and write

  $$S = D + R.$$

  Implicitly, one may think of $t = 1$, as most (non-life) reinsurance treaties are signed on a yearly basis (note that correspondingly D and R here refer to the aggregate retained and ceded amount, and not to the respective parts of single claims as in other parts of this book). Considering R as a function of S, we assume throughout $0 \leq R(S) \leq S$ for any reinsurance form of interest (see the Notes for a discussion on this). From a moral hazard perspective, it is also reasonable to only look for forms $R(S)$ that do not increase faster than S itself (i.e., $0 \leq R'(x) \leq 1$, where $R(x)$ is differentiable).
- The total premium P collected by the first insurer will be subdivided into

  $$P = P_R + P_D,$$

  where P_R refers to the premium required by the reinsurer while P_D is the premium retained by the first insurer to cope with D. In Chapter 7 we discussed in detail possible guidelines on how to determine P_R for a given risk R. In this chapter we will typically assume the premium rule P_R as given, and also the premium P that the first insurer received from the policyholders for covering S (note that the principles behind the calculation of P and P_R may differ substantially). That is, the viewpoint here is that the first-line business is already underwritten before a reinsurance solution is considered (while in practice certain first-line policies may only be accepted together with a connected reinsurance arrangement). We also assume here that P is already the "actuarial premium", that is, the part of P that concerns administrative expenses like acquisition costs of policies etc. has already been subtracted; likewise P_R is here already net of commissions that the reinsurer pays to participate in these administration costs (cf. Chapter 7). There are, however, also *transaction costs* involved in the transfer of R from the first-line insurer to the reinsurer that arise through the instalment of the contract (acquisition and administration of the reinsurance treaty, etc.). These costs are possibly also shared between the first insurer and the reinsurer, and for simplicity we tacitly assume here that such transaction costs are already included in the specification of P_R (i.e., in the respective safety loading, even if that part of P_R will not arrive at the reinsurer). Since such transaction costs are basically lost in the reinsurance process, the (joint) benefit from reinsurance has to exceed the total involved transaction costs for a treaty to make sense.

Whereas it is typically the first-line insurer (possibly through a broker) who approaches the reinsurer for a particular treaty, seeking protection R, it will mainly be the reinsurer who decides about the amount P_R for which he is willing to offer this protection. In view of the limited number of reinsurers in the market, the issue of market competition driving the pricing rules is less prominent than in the primary insurance business (although it is still present of course). Accordingly, when deciding about optimal reinsurance forms it therefore seems reasonable to assume that for each possible shape R, a rule (or premium principle) P_R is available and the first insurer then determines which form of D will be the most suitable. This is the viewpoint pursued for most parts of this chapter. It should be noted, however, that the portfolio composition, and hence diversification possibilities, of the reinsurer will finally also play a role (cf. Section 7.5) so that the fixing of a rule P_R for all forms of R is a considerable simplification of reality. Yet, as discussed in Chapter 7, often the reinsurer will only assume a part of that risk by himself and search for participating (following) reinsurers, so that a desirable treaty for the cedent may be organized by looking for optimal proportional participation of reinsurance partners.

Finally, note that here we do not consider interest rates. This makes the exposition more transparent and respective adaptations will not significantly influence the results (in particular when the time horizon is only one year).

8.1 Decision Criteria

There is a natural compromise between the complexity of the considered decision criteria and the mathematical tractability of a possible solution of the resulting optimization problem. We will start with the criteria that have typically been considered in the academic literature so far and that will form the basis for most results discussed in the rest of this chapter. We will then discuss how to possibly complement or modify these criteria to bring them closer to the decision processes that are employed in current actual reinsurance practice.

A reasonable criterion for the first insurer is to choose the reinsurance form R (with premium P_R) which maximizes his expected income after reinsurance, that is, $\mathbb{E}(P_D - D)$. Of course, this should not be done without considering at the same time the safety of the resulting strategy. The latter can, for instance, be realized by introducing a respective penalty term in the objective function or in a side constraint for the choice of R. Examples include the following:

(i) The *security level condition* asks to ensure $\mathbb{P}(D - P_D \geq w) \leq \epsilon$ for a predetermined small constant ϵ. The quantity w will then typically be related to the capital position (and the costs to hold the resulting regulatory solvency capital) of the insurer (see also Section 7.2.2).

(ii) The *variance condition* requires that $\mathrm{Var}(D) \leq d$ for some fixed threshold d. Here d replaces the role of the security level; this criterion is attractive because of its simplicity (and often there may be a reasonable estimate for $\mathrm{Var}(D)$ available, whereas the entire tail is hard to estimate, particularly if there are only a few data points available). If a normal approximation for D is justified, then the variance condition is closely related to the security level condition above. Note that in many situations a certain duality holds that maximizing the expected value under a variance constraint is equivalent to minimizing the variance under a constraint on the expected value.

(iii) In the spirit of utility theory, instead of maximizing $\mathbb{E}(P_D - D)$ one can consider maximizing the expected utility $\mathbb{E}(u(w + P_D - D)))$ of the insurer, where w is the present capital position. Here one unites the profitability and safety aspect in the analysis, as (un)favorable scenarios are weighed differently when determining D. This approach is classical and quite elegant. However, while theoretically (under very mild conditions) it is always possible to express the risk preferences by comparing expected utilities, it may be difficult to actually determine the utility function u which reflects the risk attitude of the insurer. The usual assumption is that the insurer is risk-averse, that is, that u is increasing and concave. A particularly popular assumption then is exponential utility $u(x) = -e^{-\alpha x}$ for a *risk aversion coefficient* $\alpha > 0$, leading to transparent calculations and dropping the influence of the current surplus w (see also the exponential premium principle in Section 7.1).

(iv) Whereas all the above criteria are based on a one-year time horizon, it may also be interesting to consider the long-term solvency of a reinsurance strategy R. This can be done by considering a *ruin condition* $\psi_D(w) \leq \epsilon$ for some fixed small ϵ, where $\psi_D(w)$ is the ruin probability of the insurer after entering the reinsurance contract (cf. Section 7.2.1). Alternatively, one may also consider the finite-time ruin probability $\psi_D(w, T)$. Whereas this is typically not the considered criterion in practical applications, it can still be quite useful to assess the riskiness of the resulting portfolio beyond the one-year time horizon. We will discuss this criterion further in Section 8.2.5.

In fact, many of the above safety criteria lead to comparable expressions for the optimal quantities R. Of course, if no satisfactory candidate for R can be found according to these criteria, either the offered premium P_R or the security level of the insurer may be too large, and one will have to look for compromises.

When looking at decision criteria from the perspective of practice, we may return to Section 1.2 where possible motivations for taking reinsurance were listed, and the relative importance of those criteria in a particular situation will help to suitably combine them and shape a final objective. We mention a few respective amendments of the criteria given above:

(v) A variant of (i) is to maximize the expected profit relative to the required solvency capital needed for running the portfolio, called the *RORAC criterion*. This criterion is considered very relevant in practical implementations nowadays, and we will deal with it in more detail in Section 8.3.

(vi) As a variant of (iv), in addition to absolute ruin (i.e., terminal death of the firm) shareholders of the company may be concerned about regulatory ruin, which is the event that the actual capital goes below the minimum solvency capital requirement, at which point the managers lose control over the company. Further variations of that may be that one is interested in the event that the capital falls below a higher threshold (like 1.5 times the minimum solvency capital requirement), which can lead to financial distress, changes in rating etc.). Also, one may look for strategies so that for a sequence of such triggers one specifies allowed small probabilities to be underrun. Finally, in combination with (v) one may be interested in maximizing RORAC subject to some of these regulatory trigger constraints. The corresponding optimization problems will of course quickly become very complex.

(vii) Rather than fixing the reinsurance premium P_R for each reinsurance form R a priori, one should also notice that the reinsurer himself will want to optimize his portfolio according to certain (and possibly similar) criteria. This may lead to a multi-objective optimization problem. An analytical solution is then beyond what one can hope for, but numerical implementations can be feasible. For instance, a simplified optimization procedure may consider various shapes R and their interplay with the (already existing) remaining portfolio of the reinsurer, who then will optimize his resulting portfolio and assign premiums to each R according to a procedure like the one described in Section 7.5. If a utility approach is used for this step, the utility function of the reinsurer may for instance be approximated by the CAL curve.

As mentioned before, the actual choice of a reinsurance form will finally often be triggered by intuition and experience as well as simplicity and transparency (rather than concrete calculations according to some of the above criteria). In particular, one may look for concrete protection against many claims or against large claims and choose a corresponding form among the (few) choices offered. Also, the actual criteria driving a reinsurance decision will often not be as "formal" or simple as the above list suggests (additional factors will include consequences of a reinsurance treaty on taxation, dividends etc.). These consequences can then serve as an additional guideline in the choice, and also help to pin down the parameters within the agreed reinsurance form (such as layer sizes, proportionality factors etc.).

In fact, a certain part of the academic theory on the topic was and is motivated also by the reverse direction: are there objectives and constraints under which one can identify a practically implemented reinsurance form as optimal? If yes, are the respective identified criteria indeed in line with what the insurer (or reinsurer) intends to use as guidelines? This approach may then also help to identify inconsistencies in implemented strategies.

8.2 Classical Optimality Results

A classical point of departure is the concept of Pareto-optimality that within this setup was proposed by Borch [148].

8.2.1 Pareto-optimal Risk Sharing

Reinsurance is a particular form of risk sharing, and one may view a reinsurance solution as a redistribution of random collective wealth between the partners. Assume in general m involved companies (if $m > 2$, then this is the case of several reinsurers involved in the contract) with a total wealth of $W = W_1 + \cdots + W_m$, where W_i denotes the wealth of company i (comprising current capital and future random cash-flows, i.e. a random variable). The risk-sharing mechanism will lead to the redistribution $W = Z_1 + \cdots + Z_m$, where Z_i is the new random position of company i (the total wealth stays the same). A risk sharing is then called *Pareto-optimal* if there cannot be an improvement for one party without worsening the situation of another. Such an equilibrium is a quite natural concept if there are no transaction costs for shifting risk and if the situation is symmetric (i.e., no priority for one party in the decision-finding process).[1] If we measure the value

---

1 In an efficient and complete risk market, such an optimum can be achieved in a decentralized way, see, for example Deelstra and Plantin [267]. One should keep in mind, however, that the reinsurance market cannot be described as being complete.

of each position in terms of expected utility (where u_i is the utility function of company i), then each solution $\tilde{Z}_1, \ldots, \tilde{Z}_m$ which maximizes

$$\sum_{i=1}^{m} k_i \mathbb{E}(u_i(Z_i)) \tag{8.2.1}$$

is Pareto-optimal (where $k_i > 0$ are the weight factors of each company in the process). *Borch's theorem* then states that $\tilde{Z}_1, \ldots, \tilde{Z}_m$ is Pareto-optimal if and only if

$$k_i u_i'(\tilde{Z}_i) \equiv \Lambda \tag{8.2.2}$$

is the same random variable for all $i = 1, \ldots, m$ (a simple proof for this is based on a perturbation argument, e.g. Gerber and Pafumi [387]).

Example. Assume that W is light-tailed and all companies have an exponential utility function

$$u_i(x) = -e^{-\alpha_i x}/\alpha_i, \; x \in \mathbb{R} \tag{8.2.3}$$

(dividing by α_i will not alter any decision, but simplifies the exposition). Then (8.2.2) translates into

$$\tilde{Z}_i = -\frac{\log \Lambda}{\alpha_i} + \frac{\log k_i}{\alpha_i}$$

and summing over all these terms yields

$$W = -\frac{1}{\alpha} \sum_{i=1}^{m} \log \Lambda + \sum_{i=1}^{m} \frac{\log k_i}{\alpha_i},$$

where $\alpha = (\sum_{i=1}^{m} 1/\alpha_i)^{-1}$ is the harmonic mean of the risk aversion coefficients. Combining the last two expressions one obtains for each $i = 1, \ldots, m$

$$\tilde{Z}_i = \frac{\alpha}{\alpha_i} W + b_i, \tag{8.2.4}$$

where $b_i = (\log k_i)/\alpha_i - \alpha/\alpha_i \cdot \sum_{j=1}^{m} (\log k_j)/\alpha_j$ are deterministic payments between the partners (for $b_i < 0$, company i pays this amount to the collective, for $b_i > 0$ it receives this amount) with $\sum_{i=1}^{m} b_i = 0$. That is, together with that *side payment* b_i, company i will take a fraction of the total wealth W, and this proportion is determined by the risk aversion of company i relative to those of the other companies, and is smaller for larger risk aversion. Remarkably, this proportion does not depend on the weights k_i. If one further chooses equal weights $k_1 = \ldots = k_m = 1$, then $b_i = 0$ for all $i = 1, \ldots, m$, that is, there are no side payments at all.

One interpretation of this result is now as follows: If a consortium of m companies has received the premium P for covering the total loss S, one can ask how to share the total wealth $W = P - S$. By the nature of exponential utilities, ignoring other (deterministic) capital positions of the individual companies does not influence the risk preferences.

The structure of the Pareto-optimal risk sharing (8.2.4) then exactly corresponds to a QS treaty, in which company i accepts the proportion α/α_i of both the received premium P and the risk S, plus deterministic side payments in case of $b_i \neq 0$. These side payments are hence corrections to the proportional premium $(\alpha/\alpha_i)P$ enforced by the different weight of companies in the identification of a Pareto-optimal solution.

As a further consequence, since $\sum_{i=1}^{m} b_i = 0$ and the proportions of each company do not depend on weights, the cheapest premium P across all Pareto-optimal solutions which policyholders can be offered for an aggregate risk S from a consortium of m companies with utility functions (8.2.3) is

$$P = \sum_{i=1}^{m} \frac{1}{\alpha_i} \log \mathbb{E}(e^{\alpha_i \frac{\alpha}{\alpha_i} S}) = \frac{1}{\alpha} \log \mathbb{E}(e^{\alpha S}),$$

that is, an exponential premium with risk aversion α. □

An implicit assumption in the above example was that any redistribution of W is accepted by all partners unconditionally (whereas in reinsurance applications there is often the asymmetry that the cedent has already accepted S and needs to find partners willing to participate in bearing S). Also, for other than exponential utility functions, the initial capital position will influence the result. If in the above example one considers power utility functions, the optimal risk sharing still turns out to be of QS type, whereas in that case not only the side payments but also the proportions depend on the weights k_i (cf. [387]). Nevertheless, the above result (here for $m = 2$) identifies a framework under which a QS treaty is an optimal solution, and this can serve as an intuitive background for such a treaty.

8.2.2 Stochastic Ordering

Since optimizing the shape of a reinsurance form is equivalent to identifying an extremal element in some class of c.d.f. (typically for the retained claim size), it is helpful to formalize the comparison of c.d.f. (respectively their underlying random variables) according to the needs of the situation. Among the many stochastic ordering concepts and results available (e.g., see Shaked et al. [696] and Denuit et al. [274] for excellent surveys), we restrict ourselves here to some basic notions that will be relevant in the later sections.

To start with, a random variable X is said to be smaller than a random variable Y in *stochastic dominance* ($X \prec_{\mathrm{st}} Y$), if $\mathrm{VaR}_\alpha(X) \leq \mathrm{VaR}_\alpha(Y)$ for all levels $\alpha \in [0,1]$. Hence $X \prec_{\mathrm{st}} Y$ is equivalent to $F_X(y) \geq F_Y(x)$ for all $y \in \mathbb{R}$, so this order compares the size of the random variables. Alternatively, a random variable X is said to be smaller than a random variable Y in *stop-loss order* ($X \prec_{\mathrm{sl}} Y$), if

$$\mathbb{E}(X - y)_+ \leq \mathbb{E}(Y - y)_+ \quad \text{for all } y \in \mathbb{R}, \tag{8.2.5}$$

that is, the (pure) stop-loss premium of risk X is smaller than the one for Y for all retentions y. This ordering concept is particularly useful for our purposes, and not only compares the size, but also the variability of the random variables. One can show that $X \prec_{\mathrm{sl}} Y$ is equivalent to

$$\mathbb{E}(v(X)) \leq \mathbb{E}(v(Y)) \tag{8.2.6}$$

for all non-decreasing convex functions v such that the expectations exist. Moreover, $X \prec_{\text{sl}} Y$ is equivalent to $\text{CTE}_\alpha(X) \leq \text{CTE}_\alpha(Y)$ for all levels $\alpha \in [0,1]$ (e.g. see [274, Prop. 3.4.8]). If (8.2.6) holds for *all* convex functions (such that the expectations exist), then X is said to be smaller than Y in *convex order* ($X \prec_{\text{cx}} Y$), and one can show that

$$X \prec_{\text{cx}} Y \;\Leftrightarrow\; X \prec_{\text{sl}} Y \text{ and } \mathbb{E}(X) = \mathbb{E}(Y).$$

Choosing the convex function $v(x) = x^2$ (and using the property of equal means), it immediately follows that

$$X \prec_{\text{cx}} Y \Rightarrow \text{Var}\,(X) \leq \text{Var}\,(Y), \tag{8.2.7}$$

so the convex order compares variability for random variables with equal mean.

A c.d.f. F_Y is said to be *more dangerous* than F_X, if $\mathbb{E}(X) \leq \mathbb{E}(Y)$ and there exists a constant c such that $F_X(y) \leq F_Y(y)$ for all $y < c$ and $F_X(y) \geq F_Y(y)$ for all $y \geq c$. In this case, one writes $X \prec_{\text{da}} Y$. This ordering concept serves as a sufficient criterion for stop-loss order, which will turn out to be very useful below:

$$X \prec_{\text{da}} Y \Rightarrow X \prec_{\text{sl}} Y; \tag{8.2.8}$$

(this result it is also known as Ohlin's Lemma or the Karlin–Novikov cut criterion). Correspondingly, $X \prec_{\text{da}} Y$ and $\mathbb{E}(X) = \mathbb{E}(Y)$ imply $X \prec_{\text{cx}} Y$.

By choosing different functions $v(x)$, we will see in the next sections that the search for an optimal shape of the retained risk $D = S - R$ often translates into looking for D that is minimal in terms of stop-loss order, which provides a unifying concept for several of the safety criteria applied below.

8.2.3 Minimizing Retained Variance

Assume in the following that $\text{Var}\,(S) < \infty$. Before we embark in more specific results, a simple argument shows that, for given $\mathbb{E}(R)$, any R that minimizes $\text{Var}\,(D)$ must depend on the aggregate risk S (and not in a more complicated form on the individual claims X_i). If this were not the case, then one could define

$$\underline{R} := \mathbb{E}(R|S), \quad \underline{D} := \mathbb{E}(D|S).$$

But clearly $\mathbb{E}(\underline{R}) = \mathbb{E}(R)$ and $\mathbb{E}(\underline{D}) = \mathbb{E}(D)$, whereas

$$\text{Var}\,(\underline{R}) = \text{Var}\,(\mathbb{E}(R|S)) \leq \text{Var}\,(\mathbb{E}(R|S)) + \mathbb{E}(\text{Var}\,(R|S)) = \text{Var}\,(R).$$

8.2.3.1 Optimality of a SL Contract

For a fixed available premium amount P_R and an expected value principle $P_R = (1 + \theta)\mathbb{E}(R)$ for determining the reinsurance premium (i.e., $\mathbb{E}(R)$ is fixed as well), a classical result of Borch, Kahn, and Pesonen (e.g., see [614]) states that if the insurer wants to minimize $\text{Var}\,(D)$ after reinsurance, a SL contract is the best possible choice.

A direct way to see this goes as follows. Note first that we only have to look for a reinsurance form R that depends on S (see discussion above). Denote with $R_a := (S-a)_+$ and

$$D_a := \min(S, a) \tag{8.2.9}$$

the reinsured and retained amount, respectively, of a SL treaty with retention a. Then we can choose a such that

$$\mathbb{E}(R_a) = \mathbb{E}(R) = P_R/(1+\theta), \tag{8.2.10}$$

the given value (this is always possible since $\mathbb{E}(R_a)$ is a decreasing function of a, $\mathbb{E}(R_0) = \mathbb{E}(S) \geq \mathbb{E}(R)$ and $\mathbb{E}(R_\infty) = 0 \leq \mathbb{E}(R)$). Condition (8.2.10) automatically also entails $\mathbb{E}(D_a) = \mathbb{E}(D)$. It remains to be shown that $\mathrm{Var}\,(D_a) \leq \mathrm{Var}\,(D)$ for any other form D. But since the first moment of D and D_a coincide, this is equivalent to

$$\mathbb{E}(D_a - a)^2 \leq \mathbb{E}(D - a)^2. \tag{8.2.11}$$

The inequality $|D_a - a| \leq |D - a|$ even holds for each realization of S: it is obvious for $S \geq a$, and for $S < a$ we have $D_a = S$ and so $D - a \leq S - a = D_a - a < 0$, establishing (8.2.11).

Another way to prove the optimality of a SL contract, but using the considerations of Section 8.2.2, is to observe that under a fixed P_R which is calculated according to an expected value principle (hence also $\mathbb{E}(D)$ is fixed), for D_a in (8.2.9) (with a necessarily determined by (8.2.10)) one has $D_a \prec_{\mathrm{da}} D$ for any alternative D. Indeed, for $x < a$ and any D it holds that $F_{D_a}(x) \leq F_D(x)$, whereas for $x \geq a$ we have $F_{D_a}(x) = 1 \geq F_D(x)$. Hence D_a is the smallest feasible retained D in stop-loss order: $D_a \prec_{\mathrm{sl}} D$, cf. (8.2.8). Since the premium is fixed, so is $\mathbb{E}(D_a) = \mathbb{E}(D)$, which entails $D_a \prec_{\mathrm{cx}} D$, and by (8.2.7) D_a minimizes the variance of the retained risk under the given restrictions.

8.2.3.2 Optimality of a QS Contract

If, instead, the reinsurance premium is calculated according to a variance principle, and the safety loading (i.e., $\mathrm{Var}\,(R)$) is fixed, then a QS contract minimizes $\mathrm{Var}\,(D)$ (e.g., see [614]). To see this, note first that for any candidate R we have $\mathrm{Var}\,(R) < \mathrm{Var}\,(S)$ (otherwise a QS treaty with $a = 1$ is optimal anyway, since that would lead to $\mathrm{Var}\,(D) = 0$). Choose $a < 1$ with $\mathrm{Var}\,(a\,S) = \mathrm{Var}\,(R)$, that is, the proportionality factor for which the QS treaty satisfies the $\mathrm{Var}\,(R)$ condition on the safety loading. Using the Cauchy–Schwarz inequality, one then realizes that for any reinsurance form D

$$\begin{aligned}\mathrm{Var}\,(D) &= \mathrm{Var}\,(S - R) = \mathrm{Var}\,(S) - 2\,\mathrm{Cov}(S, R) + \mathrm{Var}\,(R) \\ &\geq \mathrm{Var}\,(S) - 2\sqrt{\mathrm{Var}\,(S)\,\mathrm{Var}\,(R)} + \mathrm{Var}\,(R) \\ &= (1 - a)^2\,\mathrm{Var}\,(S),\end{aligned} \tag{8.2.12}$$

so that, for the same safety loading, the QS contract $R = aS$ yields the smallest retained variance.

8.2.3.3 Optimality of a Change-loss Contract

In Section 8.2.3.2, $\mathrm{Var}(R)$ was fixed, which due to the employed variance principle prespecified the safety loading $P_R - \mathbb{E}(R)$, but not the premium amount P_R itself (in contrast to Section 8.2.3.1, where P_R was fixed by specifying $\mathbb{E}(R)$ under the expected value principle). If we want to find the optimal choice among all reinsurance forms for fixed P_R under a variance (or related) principle, one can proceed in a different way.

For any $b > 0$ we have

$$\begin{aligned}\mathrm{Cov}(S,R) &= \mathrm{Cov}(S-b,R)\\ &= \mathrm{Cov}((S-b)_+,R) - \mathbb{E}((b-S)_+\cdot R) + \mathbb{E}((b-S)_+)\mathbb{E}(R)\\ &\le \mathrm{Cov}((S-b)_+,R) + \mathbb{E}((b-S)_+)\mathbb{E}(R),\end{aligned}$$

with equality if $R(S) = 0$ whenever $0 \le S \le b$. Using this bound in (8.2.12), and applying the Cauchy–Schwartz inequality for the resulting covariance, we get

$$\mathrm{Var}(D) \ge \mathrm{Var}(S) - 2\sqrt{\mathrm{Var}\,(S-b)_+\,\mathrm{Var}\,(R)} - 2\mathbb{E}((b-S)_+)\mathbb{E}(R) + \mathrm{Var}\,(R). \quad (8.2.13)$$

If we now postulate

$$\mathbb{E}(R) = f(P_R, \sqrt{\mathrm{Var}(R)}) \quad (8.2.14)$$

for a sufficiently regular function f, then R only appears through its variance in the above inequality and with

$$t = \sqrt{\mathrm{Var}(R)}/\sqrt{\mathrm{Var}\,(S-b)_+} \quad (8.2.15)$$

it can be reexpressed as

$$\begin{aligned}\mathrm{Var}(D) \ge\; &\mathrm{Var}(S) + (t^2 - 2t)\,\mathrm{Var}\,(S-b)_+\\ &- 2\mathbb{E}(b-S)_+\cdot f\left(P_R, t\sqrt{\mathrm{Var}\,(S-b)_+}\right).\end{aligned} \quad (8.2.16)$$

At the same time, we see from the Cauchy–Schwartz inequality that equality is achieved in (8.2.16) when

$$R(S) = a(S-b)_+ \quad (8.2.17)$$

for some positive constant a, which from the requirement $R(S) \le S$ is also bounded by 1. Hence we have identified the optimal reinsurance form and it only remains to determine the constants $0 < a \le 1$ and $b \ge 0$, which we get by minimizing the right-hand side of (8.2.16) w.r.t. t. Under weak conditions on f (which are fulfilled for our cases of interest here), one can show that the right-hand side is convex in t attaining a minimal value, which then by (8.2.15) has to be equal to a.

We hence obtained that whenever P_R is fixed and the premium principle can be described by (8.2.14), then a change-loss contract (8.2.17) minimizes the retained variance, where the optimal constants a and b are determined by the two equations

$$a\mathbb{E}((S-b)_+) = f\left(P_R, a\sqrt{\mathrm{Var}(S-b)_+}\right)$$

and

$$(a-1)\sqrt{\mathrm{Var}\,(S-b)_+} = \mathbb{E}(b-S)_+ \left.\frac{\partial f(P_R,t)}{\partial t}\right|_{t=a\sqrt{\mathrm{Var}\,(S-b)_+}}. \tag{8.2.18}$$

Let us consider three examples:

- For the *expected value principle* we have $f(P_R, t) = P_R/(1+\theta)$ and we obtain $R(S) = (S-b)_+$ with $\mathbb{E}(S-b)_+ = P_R/(1+\theta)$, which brings us back to Section 8.2.3.1 (cf. (8.2.10)).
- For the *standard deviation principle* we have $f(P_R, t) = P_R - \alpha_S\, t$, and (8.2.18) simplifies to $(1-a)\sqrt{\mathrm{Var}\,(S-b)_+} = \alpha_S\mathbb{E}(b-S)_+$.
- Finally, for the *variance principle* the function is $f(P_R, t) = P_R - \alpha_V\, t^2$, and (8.2.18) simplifies to $1-a = -2a\alpha_V\mathbb{E}(b-S)_+$.

The rigorous proof of the above result can be found in Kaluszka [478], where also an adaptation for identifying the optimal XL treaty under minimizing the aggregate retained variance is given (which is again of change-loss type) and further premium principles satisfying (8.2.14) are discussed (for the standard deviation principle (see also Gajek and Zagrodny [364])).

If, in addition to the above problem formulation, we ask for a prespecified target value $\mathbb{E}(D) = m$ (and hence also $\mathbb{E}(R) = \mathbb{E}(S) - m$ is fixed), then in (8.2.13) the expression for $\mathbb{E}(R)$ can be replaced by that fixed quantity directly, and when now implementing the substitution (8.2.15), the last term in (8.2.16) does not depend on t. Correspondingly, the right-hand side is minimized either by $t = 1$ or the largest possible t value that still is in line with the constraint $\mathbb{E}(D) = m$ (if that value is smaller than 1). Correspondingly, the optimal reinsurance form minimizing Var (D) then is

$$R(S) = \frac{\mathbb{E}(S) - m}{\mathbb{E}(S-b)_+}(S-b)_+,$$

and hence again of change-loss type, where the constant b depends on the distribution of S and is determined by a suitably adapted equation (see Kaluszka [480] for details).

There are many variants of these types of results (see the Notes at the end of the chapter for more information).

8.2.4 Maximizing Expected Utility

In a number of cases the utility framework of Section 8.2.1 is considered for one party marginally, and in particular one may ask for the reinsurance treaty that maximizes the expected utility of the cedent, that is,

$$\max_R \mathbb{E}[u(w - P_R - (S-R))], \tag{8.2.19}$$

where u denotes the utility function of the cedent, and w his current wealth (here considered a deterministic number which includes already the received premiums from the policyholders). The concrete optimal solution will now depend on the imposed conditions (e.g., the premium rule P_R and the class of admissible reinsurance forms R under consideration).

For instance, if the cedent has a risk-averse (i.e., concave and increasing) utility function $u(x)$, and if the reinsurance premium P_R is fixed, then the function $v(x) = -u(w - P_R - x)$ is increasing and convex. As a consequence, in (8.2.19) we in fact look for $D = S - R$ that is minimal in terms of stop-loss order. If the reinsurance premium is calculated according to the expected value principle, the argument of Section 8.2.3.1 in terms of the $\prec_{\text{da}}$-order applies in the same way, again identifying the SL contract as optimal. The optimality of a SL treaty in the context of risk-averse utility functions was already established by Arrow [46] (see also Borch [150]). If in addition there is an upper limit on the reinsurance coverage, then a SL contract with that upper limit is optimal (cf. Cummins and Mahul [241]). For an adaptation of Arrow's result when the reinsurer imposes an upper limit of its expected loss, see Zhou and Wu [808].

As another example, consider a cedent facing total risk S and receiving an offer P_S as reinsurance premium for the entire risk S. On that basis the cedent now decides about the extent (fraction a) to which he wants to enter this treaty (assuming that the reinsurance premium scales according to the proportionality factor a of the resulting QS treaty). Then (8.2.19) turns into

$$\max_a \mathbb{E}(u(w - a \cdot P_S - (1-a)S)).$$

For exponential utility $u(x) = -e^{-\alpha x}$ and normally distributed risk $S \sim \mathcal{N}(\mu, \sigma^2)$, this leads to a remarkably simple expression for the optimal retained fraction:

$$1 - a^* = \frac{P_S - \mu}{\alpha \sigma^2}.$$

It is proportional to the safety loading of the reinsurance premium offer, and inversely proportional both to the risk aversion coefficient and the variance of the risk S. Even if for QS treaties the pricing often works slightly differently (cf. Section 7.3), this formula is of particular interest, as it has a striking resemblance to the Merton ratio in optimal portfolio theory (although there S is log-normally distributed and u is a power function) (cf. Gerber and Pafumi [387]).

Deprez and Gerber [279] consider (8.2.19) for convex, Gâteaux-differentiable premium rules $P_R = H(R)$, and show for any risk-averse utility function u by a simple perturbation argument that the general solution of this maximization problem then is the reinsurance form R^* that satisfies

$$H'(R^*) = \frac{u'(w - H(R^*) + R^* - S)]}{\mathbb{E}[u(w - H(R^*) + R^* - S)} \tag{8.2.20}$$

(the argument in fact also applies when w is replaced by a random initial position W that is independent of S). Note that the random variable $H'(R)$ is a gradient for which

$$\frac{d}{dt} H(R + tV)|_{t=0} = \mathbb{E}(H'(R)V)$$

for any random variable V and hence measures the sensitivity of the premium for small changes of the underlying risk.

Particular principles of premium calculation for which the above result applies are the pure premium principle $H(R) = \mathbb{E}(R)$ with $H'(R) = 1$, the variance principle (7.1.3) with $H'(R) = 1 + 2\alpha_V(R - \mathbb{E}(R))$, the standard deviation principle (7.1.4) with Gâteaux derivative $H'(R) = 1+2\alpha_S(R-\mathbb{E}(R))/\sqrt{\mathrm{Var}\,(R)}$ and the exponential principle (7.1.5) with $H'(R) = e^{aR}/\mathbb{E}(e^{aR})$. For example, if the utility function of the insurer is exponential with risk aversion b and the premium principle of the reinsurer is (7.1.5) with risk aversion a, then (8.2.20) turns into

$$\frac{e^{bR^*}}{\mathbb{E}(e^{bR^*})} = \frac{e^{a(S-R^*)}}{\mathbb{E}(e^{a(S-R^*)})},$$

which specifies R^* up to an additive constant. Choosing the latter in such a way that $R^*(0) = 0$, we then get

$$R^*(S) = \frac{a}{a+b}S,$$

that is, a QS contract where the proportion is determined by the risk aversion of the insurer and reinsurer. Note that this exactly corresponds to the risk-sharing agreement of the example in Section 8.2.1, as both the insurer and reinsurer here essentially make decisions based on exponential utility.

Finally note that if the reinsurer offers a pure premium principle $H(R) = \mathbb{E}(R)$, then by (8.2.20) $R^* - S$ must be a constant, for example leading to $R^*(S) = S$. That is, with risk averse utility, the insurer should exploit the "cheap" premium offer of the reinsurer to the largest possible extent.

8.2.5 Minimizing the Ruin Probability

8.2.5.1 One-year Time Horizon View

Let us first stick to the discrete setting of the previous sections and look for the reinsurance form that minimizes the ruin probability at that future time point (say, one year from now), that is,

$$\min_R \mathbb{P}(w - P_R - (S-R) \leq 0) = \max_R \mathbb{E}(1_{\{w-P_R-(S-R)\geq 0\}}). \tag{8.2.21}$$

Clearly, if this is the criterion and we can afford full protection, then a SL contract will lead to a ruin probability equal to 0. If for a premium P_R we can purchase the cover $R = (S - (w - P_R))_+$, then this is optimal. However, there are two issues to note here. First, in many situations this will not be feasible, that is, one may only be able to afford a SL contract $R = (S - b)_+$ with $w - P_R < b$, and in that case the cover does not help at all to improve the survival probability. And secondly, if such a cover costs already more than the earned premium P on the original policies and if one's goal is to minimize ruin, then it is preferable to stay out of those policies altogether, as one then stays positive (with the capital $w - P$ available before writing policies) with probability 1 (one may indeed question in general the suitability of the criterion of minimizing ruin from this perspective).

Let us still take the viewpoint that the first-line business is already written (with the collected premium being already contained in w) and the goal is to minimize the remaining ruin probability. Assume also that the reinsurer offers protection according to an expected value principle, that the reinsurance premium amount P_R is fixed and the above full protection can not be afforded (note that Arrow's result on the optimality of an SL treaty does not apply here, since the indicator function on the right-hand side of (8.2.21), interpreted as a utility, is not concave). Gajek and Zagrodny [366] showed that then, if S is a continuous random variable, the best possible contract is of the form

$$R^*(S) = \begin{cases} 0, & S \le w - P_R \\ S - (w - P_R), & w - P_R \le S \le c^* \\ 0, & S \ge c^*, \end{cases} \tag{8.2.22}$$

where c^* is the largest value that can be afforded for this contract with the amount P_R, that is, a truncated SL contract minimizes the ruin probability. The result is somewhat intuitive, as one tries to get full protection for an as large claim size as possible (for empirical evidence and a discussion of the popularity of such treaties in catastrophe reinsurance, see Froot [360]). The proof is based on an application of the Neyman–Pearson lemma (cf. [366] for details. See also Bernard and Tian [127]).

8.2.5.2 Infinite-time Ruin Probability

Let us continue the viewpoint that first-line business is already written and the insurer wants to identify for a given premium rule of the reinsurer the reinsurance strategy that minimizes the probability of getting ruined. However, now consider the long-term view of the probability of never getting ruined (the infinite-time ruin probability). The underlying risk model can be in discrete time (aggregate claims and premiums are checked at the end of each time period, say a year) or in continuous time when the monitoring is done continuously. The idea is to fix a reinsurance strategy in the beginning that is then kept through time. As argued in Chapter 7, the underlying philosophy is not to indeed follow that same strategy forever, but to get a feeling of the effects of such a strategy on the safety in the long run. Note again that the problem is only non-trivial, when one assumes that reinsurance premiums are higher than first-line premiums (otherwise the entire portfolio could be passed on to the reinsurer, leading to zero ruin probability).

The Lundberg bound (7.2.13) was shown to hold for a continuous-time model with adjustment coefficient defined in (7.2.12). For a risk model in discrete time, (7.2.13) also holds, where the adjustment coefficient is then the positive solution γ of

$$\int_0^\infty e^{-\gamma(P-S)}\, dF_S(x) = 1,$$

with P the premium income and S the aggregate claim per time unit (e.g., see [57]).

The bound (7.2.13) turns out to be a quite good approximation for the true value of the ruin probability $\psi(w)$ in many cases and due to its simplicity it is often used as a (conservative) approximation $\psi(w) \approx e^{-\gamma w}$. Then, for both continuous- and discrete-time models the problem of minimizing the ruin probability is translated into maximizing the adjustment coefficient.

In a number of situations, maximizing the adjustment coefficient can be translated back to minimizing convex order. For instance, consider the Cramér–Lundberg process and its adjustment equation (7.2.12). Then one sees immediately that by the convexity of the function $v(x) = e^{rx}$, the retained risk that maximizes γ in (7.2.12) is the one that is minimal with respect to convex order (under the assumption that the reinsurance premium is calculated according to an expected value principle with fixed loading, which leaves the left-hand side of (7.2.12) invariant for all considered forms D). By the arguments of Section 8.2.3.1 the latter brings us to the result that an XL treaty maximizes γ (as the optimality of the SL treaty translates into XL, when only individual treaties are considered) (e.g., see Gerber [383]). Centeno [189] gives an algorithm to calculate the optimal retention for this XL treaty. The result by Gerber is extended by Hesselager [440] in the sense that, if the insurer can freely choose among global and individual reinsurance contracts with the same pure premium and reinsurance loading, then the adjustment coefficient will be maximized by a SL treaty. Waters [775] investigates the dependence of the adjustment coefficient on the retention. He does this for proportional, SL and XL reinsurance contracts and under a variety of different premium schemes (see also Hald and Schmidli [418]).

In [188], Centeno considers change-loss contracts on individual claims, that is,

$$R(t) = a \sum_{i=1}^{N(t)} (X_i - M)_+,$$

and for a compound Poisson model studies the combination of parameters a and M that maximizes the adjustment coefficient for the first-line insurer, also when a target value on the expected income is fixed.

An intimate connection between maximization of the adjustment coefficient and maximizing expected utility for an exponential utility function is exploited in Guerra and Centeno [413, 414], who establish optimal reinsurance treaties when the premium principle of the reinsurer is a convex functional. It turns out that for an expected value principle, the optimal form is of SL type, whereas for variance and standard deviation principles the optimal $R(S)$ is a non-linear function of S, which is not among the reinsurance forms typically employed in practice. For strategies restricted to the individual claim level, see [194].

For heavy-tailed claims, the adjustment coefficient does not exist, but with an unlimited XL cover the maximum retained claim size is upper-bounded by the retention, so that then γ exists again for the first-line insurer and the above approach to maximize γ still makes sense. One should keep in mind, however, that maximizing γ is only an approximate solution for minimizing the ruin probability $\psi(w)$. Under certain model assumptions one can go in fact much further. In the following, we give an illustration.

Consider the Cramér–Lundberg model and compare any two c.d.f. F_1 and F_2 for the claim size through convex ordering. From the Pollaczeck–Khintchine formula (7.2.10), which gives an exact expression for $\psi(w)$ for any claim size distribution, one then sees that $F_1 \prec_{cx} F_2$ implies $\psi_1(w) \leq \psi_2(w)$, since convex order is preserved under convolution and compounding (e.g., see [57, Prop. IV.8.2] for details). As for previous instances, this result can then be used to show in a simple way that if the reinsurance premium is calculated with an expected value principle, an unlimited XL treaty minimizes the ruin probability among all reinsurance forms applied to individual claims. Indeed, compare any reinsurance form candidate $D = X - R$ with an XL treaty, the retention M of which is chosen in such a way that $\mathbb{E}(D(X)) = \mathbb{E}(\min(X, M))$ (i.e., the two treaties have the same premium). Denote by F_D the c.d.f. of $D(X)$ and by F_{XL} the c.d.f. of $\min(X, M)$. Indeed, by the same arguments as in Section 8.2.3.1, one gets $F_{\text{XL}} \prec_{\text{da}} F_D$ and by the identical mean subsequently $F_{\text{XL}} \prec_{\text{cx}} F_D$, so that

$$\psi_{\text{XL}}(w) \le \psi_D(w)$$

for all capital values w.

With the same technique one can also show that if the reinsurer poses the additional constraint that $R(X) \le L$, then an L xs M treaty is optimal. Indeed, consider again any alternative $D = X - R$ for a premium P_R and choose M such that $\mathbb{E}(D(X)) = \mathbb{E}(\min(X, M) + (X - M + L)_+)$, so that the two reinsurance premiums coincide. Let F_D and F_{XL} again denote the c.d.f. of the retained claim. Then $F_{\text{XL}} \prec_{\text{da}} F_D$, because for $x < M$, $F_{\text{XL}}(x) = F_X(x) \le F_D(x)$ as above, whereas for $x \ge M$ one has $F_{\text{XL}}(x) = F_X(x+L)$. However, the limit $X - D \le L$ leads to $D \ge X - L$, that is, $F_D(x) \le F_X(x + L)$ (even for all x). Consequently, $F_{\text{XL}} \prec_{\text{sl}} F_D$ and since the retained amounts have the same mean, further $F_{\text{XL}} \prec_{\text{cx}} F_D$. However, the latter again implies that for all $w \ge 0$

$$\psi_{\text{XL}}(w) \le \psi_D(w)$$

that is, an XL treaty with layer L minimizes the ruin probability $\psi(w)$. This result holds for any claim size distribution (which is relevant to note, as due to the finite layer L a heavy-tailed claim stays here also heavy tailed after reinsurance, so the adjustment coefficient does not exist at all).

Section 8.7 will discuss dynamic reinsurance strategies for continuous-time risk models.

8.2.6 Combining Reinsurance Treaties over Subportfolios

The first insurer often needs to deal with many (sub)portfolios simultaneously. It is then natural to look for an optimal combination of reinsurance forms to achieve an overall objective. In this section we will consider two classical approaches (one on a combination of proportional treaties and one on the non-proportional case) that originally go back to de Finetti [255].

Assume that the insurer has to deal with I subportfolios with total claim amounts $\{S_i, 1 \le i \le I\}$ that will be divided into deductible and reinsured amounts by $S_i = R_i + D_i, 1 \le i \le I$. For simplicity we assume that the portfolios are independent and that the type of reinsurance is the same for all of them. If the premium received for covering S_i is P_i and the reinsurance premium is P_{R_i}, then the total income amounts to

$$U := \sum_{i=1}^{I} \left(P_i - D_i - P_{R_i}\right).$$

The following results are obtained now under the objective to maximize $\mathbb{E}(U)$, given a constraint on $\text{Var}\,(U)$.

8.2.6.1 Proportional Reinsurance

If we look for the best combination of QS treaties across the subportfolios, then $R_i = a_i S_i$. For the premium principle of the reinsurer consider an expected value principle

$$P_{R_i} = (1 + z_i)\,\mathbb{E}(R_i)\,, 1 \le i \le I \tag{8.2.23}$$

for (possibly different) positive constants $z_i, 1 \le i \le I$. Maximizing $\mathbb{E}(U)$ under the condition $\text{Var}\,(U) = c$ for a fixed constant c can now be done by the method of Lagrange multipliers. Let λ be such a multiplier. We need to maximize the expression

$$\mathbb{E}\left(\sum_{i=1}^{I} \{P_i - (1-a_i)S_i - (1+z_i)\,\mathbb{E}(a_i\,S_i)\}\right)$$
$$+ \lambda\left(c - \text{Var}\sum_{i=1}^{I} \{P_i - (1-a_i)S_i - (1+z_i)\,\mathbb{E}(a_i\,S_i)\}\right)$$

by choosing the proportions $a_i, 1 \le i \le I$ properly. The expression becomes

$$\sum_{i=1}^{I} \left(P_i - (1+z_i a_i)\mathbb{E}(S_i)\right) + \lambda\left(c - \sum_{i=1}^{I}(1-a_i)^2\,\text{Var}\,(S_i)\right).$$

Equating the partial derivative with respect to a_i with zero leads to

$$z_i\mathbb{E}(S_i) - 2\lambda(1-a_i)\,\text{Var}\,(S_i) = 0\,,\ 1 \le i \le I,$$

so that

$$1 - a_i = \min\left(\frac{z_i\,\mathbb{E}(S_i)}{2\lambda\text{Var}\,(S_i)}, 1\right),\ 1 \le i \le I, \tag{8.2.24}$$

where the constant λ is determined by the side condition on the variance. Concretely, if $a_i > 0$ for all $i = 1, \ldots, I$, then

$$\lambda^2 = \frac{1}{4c}\sum_{i=1}^{I}\frac{(z_i\,\mathbb{E}(S_i))^2}{\text{Var}\,(S_i)}\,.$$

Note that the retained proportion $1 - a_i$ depends in a crucial way on the value of the dispersion of S_i (the greater the dispersion of S_i, the smaller the retained proportion $1 - a_i$). Also, the retained proportion is higher if reinsurance is more expensive.

This result has an interesting consequence: interpret each risk X_i in a portfolio as a subportfolio on its own (containing only one risk). In that case, of course the claim history will not be sufficient to estimate the first two moments of each such claim size. However, as discussed in Section 7.4.1.2, for certain lines of business (like property lines) it is reasonable to assume that the loss degree V with respect to the sum insured (or PML) Q_i is identically distributed (here it is in fact sufficient that the first two moments coincide), i.e. $\mathbb{E}(V) = \mathbb{E}(X_i/Q_i) = \mu$ and $\mathrm{Var}(V) = \sigma^2, 1 \le i \le I$. It is also reasonable to assume that all safety loadings z_i are equal, and then (8.2.24) simplifies to

$$1 - a_i = \sqrt{\frac{c}{\sigma^2 I}} \frac{1}{Q_i}, \quad 1 \le i \le I.$$

That is, the resulting proportionality factors are a constant $M := \sqrt{c/I}/\sigma$ divided by Q_i (and $a_i = 0$ if $Q_i < M$), but this is the structure of a *surplus reinsurance* contract! Hence the present setting suggests a situation and criterion within QS strategies for which a surplus treaty is optimal, and one may in fact use the above reasoning as a guideline for the choice of the retention M. For an extension of this result to include cost-of-capital considerations, refer to Section 8.3.

If the reinsurance premium principle (8.2.23) is extended to include a variance component

$$P_{R_i} = (1+z)\mathbb{E}(R_i) + \beta \mathrm{Var}(R_i), \quad 1 \le i \le I \tag{8.2.25}$$

for some $\beta > 0$, then a similar calculation yields the adaptation

$$1 - a_i = \frac{z}{2(\lambda+\beta)} \frac{\mathbb{E}(S_i)}{\mathrm{Var}(S_i)} + \frac{\beta}{\lambda+\beta}, \quad 1 \le i \le I,$$

so the variance term in the premium principle adds a fixed proportion for all subportfolios.

8.2.6.2 Excess-of-loss Reinsurance

Let us now look for the best choice of retention a_i of an XL treaty for subportfolio i with total risk S_i, which is assumed compound Poisson distributed with rate λ_i and claim sizes $X_1^{(i)}, \dots, X_{N_i}^{(i)}$ and c.d.f. F_i (in contrast to above, here the result will not only depend on the first two moments, but on the entire distribution). The deducted part of subportfolio i then is

$$D_i = \sum_{j=1}^{N_i} \min(X_j^{(i)}, a_i), \quad 1 \le i \le I.$$

Let us consider a reinsurance premium principle of the general form (8.2.25). Then under the same variance condition as above we have to maximize the expression

$$\sum_{i=1}^{I} \left(P_i - \mathbb{E}(S_i) - z\mathbb{E}(R_i) - \beta \mathrm{Var}(R_i)\right) + \lambda \left(c - \sum_{i=1}^{I} \mathrm{Var}(D_i) \right).$$

By the Poisson assumption we see that for $1 \le i \le I$ (cf. also Section 2.3.1),

$$\frac{\partial}{\partial a_i} \mathbb{E}(R_i) = -\lambda_i (1 - F_i(a_i)),$$

$$\frac{\partial}{\partial a_i} \operatorname{Var}(R_i) = -2\lambda_i \mathbb{E}(X_i - a_i)_+,$$

$$\frac{\partial}{\partial a_i} \operatorname{Var}(D_i) = 2\lambda_i a_i (1 - F_i(a_i)).$$

The equation for a_i is therefore given by

$$z(1 - F_i(a_i)) + 2\beta \mathbb{E}(X^{(i)} - a_i)_+ - 2\lambda a_i \, (1 - F_i(a_i)) = 0 \, .$$

This leads quickly to the solution

$$2\lambda a_i = z + 2\beta \mathbb{E}(X^{(i)} - a_i | X^{(i)} > a_i), \; 1 \le i \le I \tag{8.2.26}$$

where λ is determined by the condition

$$c = \sum_{i=1}^{I} \lambda_i \mathbb{E}(\min(X^{(i)}, a_i(\lambda)))^2 \, .$$

Equation (8.2.26) can be rewritten as

$$\mathbb{E}(X^{(i)} - a_i | X^{(i)} > a_i) = \frac{2\lambda a_i - z}{2\beta},$$

which is an equation for a_i in terms of the mean excess function of the individual claim sizes. The existence and the value of the solution depend heavily on the form of this mean excess function. The right-hand side is a straight line in the variable a_i, starting at the value $-z/(2\beta)$ and increasing with slope λ/β. The function on the left starts at the positive value $\mathbb{E}(X^{(i)})$ but behaves differently depending on the tail behavior of the distribution. If the tail is exponentially bounded, then a unique solution exists. If, however, the distribution is strict Pareto with index α, then a solution only exists if $\alpha > 1 + \beta/\lambda$.

It is interesting to see that an increase in $\mathbb{E}(X^{(i)})$ results in a similar increase of the retention a_i. Also note that in the absence of a variance component in the premium principle (8.2.25), that is, $\beta = 0$, it follows from (8.2.26) that there is a constant retention $a_i = z/(2\lambda)$ across the subportfolios. Indeed, in practice it seems rather common not to vary the retention of XL treaties across different subportfolios.

8.3 Solvency Constraints and Cost of Capital

In Section 7.2.2 we discussed how the safety loading in insurance premiums may be determined in order to meet capital costs arising from regulatory solvency constraints. At the same time, for the final first-line insurance premium $P(1)$ further factors will play a role, (prominently) including market competition. Let us therefore now assume that the premium $P = P(1)$ is already given and fixed, and we look for reinsurance to improve the overall situation.

Recall that the solvency capital requirement for the annual loss is determined by some risk measure ρ (typically VaR or CTE). In addition to the viewpoint of Section 7.2.2, let us now assume that the capital costs increase the annual loss (e.g., because they are paid to external investors). Then, instead of $S - P$, the annual loss is

$$\text{Loss} = S - P + r_{CoC} \cdot \rho(\text{Loss}),$$

and the final necessary solvency capital will be higher than $\rho(S - P)$, leading to the recursive relationship

$$\rho(\text{Loss}) = \rho(S) - P + r_{CoC} \cdot \rho(\text{Loss}),$$

which yields

$$\rho(\text{Loss}) = \frac{\rho(S) - P}{1 - r_{CoC}}$$

(we use here that ρ is positively homogeneous and translation-invariant, cf. Section 7.1). The resulting gain (i.e., negative loss) at the end of the year (without reinsurance) then is

$$\frac{P}{1 - r_{CoC}} - S - \frac{r_{CoC}}{1 - r_{CoC}} \cdot \rho(S). \tag{8.3.27}$$

If a reinsurance treaty is entered for a premium P_R, this changes to

$$\frac{P - P_R}{1 - r_{CoC}} - (S - R) - \frac{r_{CoC}}{1 - r_{CoC}} \cdot \rho(S - R), \tag{8.3.28}$$

and one can now again try to find optimal reinsurance forms R according to the criteria discussed in the previous sections, now with the correction factor for cost of capital (and hence the solvency constraint) included.

For instance, if P is fixed and the goal is to maximize the expected income after reinsurance, one obtains from (8.3.28) that the goal is to identify R for which

$$\min_R \left(P_R - (1 - r_{CoC})\mathbb{E}(R) + r_{CoC} \cdot \rho(S - R)\right) \tag{8.3.29}$$

is attained. This approach of looking at the capital cost problem can be found, for example, in Kull [514], who gives a variety of optimal reinsurance results under this setup. For an expected utility approach of (8.3.28) in the spirit of (8.2.4), see Haas [417].

It may also frequently happen that the amount of available risk capital is fixed (a *risk budget* is available), and then one has to identify the reinsurance form that maximizes some objective (e.g., expected profit) given this capital constraint. Also, sometimes the overall risk capital of a company is fixed (by some general considerations) and then the question is how to most efficiently allocate this capital to subportfolios, that is, to design reinsurance arrangements on the individual subportfolios whose risk capital implications aggregate to the overall target. In this context, an extension of the problem of optimal proportions of subportfolios of Section 8.2.6.1 to the case with capital costs was developed by Kull [514]. Capital and its allocation have recently become crucial in the light of risk-based management of insurance companies, so the criteria discussed in the present section can be regarded as particularly relevant for applications (cf. Dacorogna [243]).

A variant of the above criterion is to focus on the expected profit relative to the required solvency capital. The goal then is to maximize the *return on risk-adjusted capital (RORAC)*, which in the notation of Section 7.2.2 translates into

$$\max_D \frac{P_D - \mathrm{BE}_D}{\mathrm{SCR}_D} = \max_R \frac{P - P_R - \mathbb{E}(S - R)}{\rho(S - R) - P + P_R} \tag{8.3.30}$$

for given first-line premium P, aggregate claim size S, reinsurance premium rule P_R and risk measure ρ. That is, here r_{CoC} is not prespecified, but the quantity to be maximized. One immediately sees that under this criterion a QS treaty $R = aS$ with $P_R = aP$ is not of interest at all, as it does not change the ratio (only $P_R < aP$ would be of interest).

Lampaert and Walhin [526] use the RORAC criterion (8.3.30) to compare quota share and surplus treaties with both fixed and variable proportions (lines, respectively). Performing an empirical study on a large data set from reinsurance practice, they conclude under some additional assumptions (including expected value premium principles for both P and P_R, with fixed safety loading for all risk sizes) that it is not beneficial to use *tables of lines* constructed by an inverse rate method (cf. Section 2.2), which is contrary to what one may intuitively expect. It is, however, beneficial for the RORAC criterion to construct a table of lines according to the method of Section 8.2.6.1.

It will be useful finally to adapt the obtained results to include claims reserves, market risk, counterparty risk etc. in the calculation of the solvency capital requirement and respective cost of capital. However, the resulting optimization problems then quickly become complex, and their solutions heavily depend on the respective posed assumptions.

8.4 Minimizing Other Risk Measures

In recent years many variants of the optimization problem

$$\min_R \rho(P_R + S - R) \tag{8.4.31}$$

have been studied in the literature. For instance, Cai and Tan [178] studied the subproblem of optimizing retention levels within SL contracts for (8.4.31) when ρ is the VaR or CTE and P_R is determined by an expected value principle. Cai et al. [179] then extended this analysis to a larger class of contracts, indicating that change-loss contracts (and QS and SL as special cases) are optimal. For a comparison across reinsurance premium principles, see Tan et al. [727]. An extension to optimal treaties in the presence of several reinsurers can be found in Asimit et al. [52].

Balbas et al. [73] provide an in-depth study on characterizing optimal reinsurance forms under (8.4.31) for a general class of risk measures ρ by exploiting duality theory in functional analysis. It is shown that a SL treaty is often optimal when P_R is an expected value principle. For investigations on the stability of optimality results when switching from one risk measure to another, see Balbás et al. [74]. Also, Balbás et al. [72] identified optimal reinsurance forms when there is uncertainty about the involved claim distribution (see also Asimit et al. [53] and Bernard et al. [123]).

Another risk measure that has recently gained considerable attention due to discussions on backtesting issues for risk measures is the *expectile*, see, for example, Ziegel [811], Bellini et al. [107], and Bellini and Bignozzi [106]. Cai and Weng [175] show that for certain reinsurance premium rules a superposition of two limited XL treaties is then optimal, where the reinsurer takes over two disjoint layers.

The criterion (8.4.31) is not likely to directly be a driving criterion in practice, as then (akin to minimizing the ruin probability in Section 8.2.5) it is optimal to stay out of the insurance business altogether, resulting in a zero value for ρ. However, if one takes the viewpoint that the primary business is already written and reinsurance premiums are more expensive than premiums in primary markets, then the identification of reinsurance forms minimizing such risk measures leads to interesting mathematical problems. Moreover, minor modifications can embed the solutions into the cost-of-capital framework of Section 8.3. To see this, note that by translation-invariance of ρ, the optimization problem (8.3.29) can be rewritten as

$$\min_R \rho\big(P_R - (1 - r_{CoC})\mathbb{E}(R) + r_{CoC} \cdot (S - R)\big), \tag{8.4.32}$$

and so the goal of maximizing the expected gain under regulatory solvency constraints translates into minimizing the risk measure ρ of a weighted sum of the safety loading $P_R - \mathbb{E}(R)$ in the reinsurance premium, the retained risk $S - R$ and $\mathbb{E}(R)$. If, for instance, $P_R = (1 + \theta_R)\mathbb{E}(R)$, then (8.4.32) simplifies to the problem

$$\min_R \rho\big((1 + \theta_R/r_{CoC})\mathbb{E}(R) + S - R\big),$$

so that results for (8.4.31) can be interpreted within the framework of (8.4.32) for a modified value of θ_R.

For optimal risk-sharing rules for convex risk measures expressed in the framework of monetary utility functions, see Jouini et al. [471], Barrieu and El Karoui [85], as well as Acciaio [7], who studied optimal risk sharing under non-monotone risk measures. A general solution to this problem is provided by Filipović and Svindland [351], who showed that for law-invariant convex risk measures on the model space L^p, where $1 \leq p \leq \infty$, the optimal risk allocation always exists and is composed of increasing Lipschitz continuous functions of the aggregate risk. For a study based on spectral risk measures, see Brandtner and Kürsten [160]. Barrieu and Scandolo [87] extend the Pareto-optimal allocation analysis to a multi-period situation by dealing with preference functionals on general vector spaces.

8.5 Combining Reinsurance Treaties

We have seen above that in some cases a change-loss contract turns out to be optimal, where a SL (or XL) feature is put on top of a QS arrangement. Apart from defining a contract with such a structure, the first-line insurer can also apply different reinsurance treaties simultaneously, which together lead to (or approximate) the desired coverage structure. For instance, a reinsurer may offer only a certain lower layer in an XL treaty, and another reinsurer specializing in catastrophe reinsurance may offer a treaty for a higher layer. Sometimes also the premium offer from a reinsurer may only be attractive for a certain part of the coverage, and a combination of two separate treaties is then preferable. Such combinations are frequently applied in reinsurance practice. Naturally, however, it is difficult (if not impossible) to formalize such a demand/supply situation together with the premium rules and objectives in order to identify a certain combination as optimal. One may rather be restricted to choosing an optimal parameter of an already pre-specified reinsurance form. For a study of this kind, see Verlaak [759] and Verlaak and Beirlant [760], where a variety of different combinations are tried out under a mean-variance optimality condition. We list here a few further examples:

- Suppose that there are n independent portfolios and that the claims in the ith portfolio are given by $\{X_{i,j}, 1 \leq j \leq N_i\}$. A QS treaty with factor a_i is applied to this portfolio together with an XL treaty with retention M_i. The overall retained risk for the first-line insurer then is

$$\sum_{i=1}^{n} \sum_{j=1}^{N_i} \min(a_i X_{i,j}, M_i) .$$

  In Centeno and Simões [195], the determination of the factors $a_i, (1 \leq i \leq n)$ and the retentions $M_i, (1 \leq i \leq n)$ under a minimization of the adjustment coefficient is treated (for an early reference, see also [187]).
- For a combination of surplus and XL treaties, see Benktander and Ohlin [114].
- Combinations of proportional and SL contracts are treated by, for example, Schmitter [679]. As another form to combine these two reinsurance forms, Hürlimann [453] uses a mixture $qX + (1-q)(X-C)_+$. Under some reasonable optimization criteria, one of the two reinsurance types prevails.

8.6 Reinsurance Chains

Another way of introducing more than one insurer is to deal with the risk in a hierarchical cascade (*retrocession*). Let the original risk be denoted by X. The first company faces the risk X, gets a premium P_1 and buys reinsurance for a part X_2 from a first reinsurer for a premium P_2. This company itself accepts X_2 as its own risk but in turn takes reinsurance for a part X_3 for a premium P_3. Ultimately, the nth reinsurer keeps the remaining part X_n. If one studies the question of optimal sharing in a utility framework, then the ith partner with utility function u_i will try to maximize

$$\mathbb{E}\left(u_i((P_i - P_{i+1}) + (X_{i+1} - X_i))\right),$$

where $X_1 = X$ and $P_{n+1} = 0$. Note that the Pareto-optimal risk sharing discussed in Section 8.2.1 has a similar flavor, but here the sequential character of the decision process leads to different results. For an early contribution on this problem for proportional treaties, certain choices for utility functions and premium rules, see Gerber [384].

Substantial generalizations of this problem have been investigated in d' Ursel et al. [312], Lemaire et al. [537], and Heijnen et al. [430, 432].

A critical question in the distribution of reinsurance is the optimal number of reinsurers and their respective responsibilities (e.g., see Powers et al. [626]). It also needs to be ensured that through repackaging of risks, retrocession does not lead to a cycle. One unfortunate such case in history is the so-called *LMX spiral* in the early 1990s (see Baluch et al. [76]).

8.7 Dynamic Reinsurance

In Section 8.2.5.2 we discussed the criterion of minimizing the ruin probability for a discrete-time or continuous-time surplus process through an optimal form of reinsurance. There the assumption was that the concrete form of the reinsurance treaty has to be chosen in the beginning, and then does not change over time. In the academic literature it has also been studied by how much the overall target (of minimal ruin probability) could be improved if one were allowed to adapt the reinsurance form along the way. When the surplus process is of Markovian type, the optimal strategy is typically a *feedback* strategy, which for the ruin probability criterion at each time point only depends on the current value of the surplus process.

For a discrete-time risk process, the respective optimal adaptation at each time point can be obtained by stochastic dynamic programming. In most cases one cannot obtain explicit results, but approximate the optimal solution numerically (see Schäl [669] for a theoretically sound basis of respective numerical algorithms). In the setting of classical continuous-time risk models (like the compound Poisson model), from a mathematical point of view somewhat more explicit results can be obtained if the times at which one can adapt the reinsurance form are not fixed (e.g., annual) time points, but are the claim occurrence times, as then the underlying process is a random walk with much simpler increment distribution. However, this little advantage comes at the expense of a less realistic model setup, as adapting the contract at every single claim occurrence will not be feasible in applications.

If, instead, one switches to the possibility of continuous-time adaptation of the contract, the analysis becomes more transparent. Mathematically, the minimal ruin probability (*value function*) can then often be determined by solving a Hamilton–Jacobi–Bellman (HJB) equation and in a number of cases also the corresponding optimal dynamic reinsurance strategy can be identified. It should be noted that an actual implementation of such a continuous-time adaptive strategy is even less "realistic", but this approach allows the potential improvement through continuous-time action (compared to a static strategy) to be quantified, and also gives some intuition for the suitability (and long-term consequences) of parameter choices within certain types of contracts.

When trying to solve such a stochastic control problem, a number of challenges arise. The solution of the HJB equation usually can only be obtained numerically and is just a possible candidate for the value function, and a separate verification step is needed. Also, the solution of the HJB equation may not be unique, and the value function may not be as regular as is needed for the equation. Correspondingly the actual calculations can be highly involved, and it is beyond the scope of this book to provide details on these aspects (see Schmidli [676] and Azcue and Muler [70] for profound surveys on all the mathematical aspects in such an approach). Instead, for illustration we simply state here the results one obtains for the classical risk model.

Consider the Cramér–Lundberg model (7.2.8) with a dynamic reinsurance strategy $u_t = u_t(X_i)$, where $u_t(X_i)$ is the retained amount of claim X_i for the cedent when the claim occurs at time t. Let $c_R(u_t)$ be the reinsurance premium intensity for strategy u_t and assume that more reinsurance is more expensive as well as $c_R(0) > c$ (otherwise it would be optimal to reinsure the entire portfolio, see also Section 8.2.5). Then the optimal control problem is to identify u_t such that the ruin probability for the process

$$C(t) = w + \int_0^t (c - c_R(u_s))\,ds - \sum_{i=1}^{N(t)} u_{T_i}(X_i)$$

is minimized, where T_i is the occurrence time of the ith claim.

- *Dynamic proportional reinsurance*: $u(y) = u\,y$ with $0 \le u \le 1$
  If one can dynamically adapt the proportionality constant in a QS contract, then (under the weak condition $\liminf_{u\uparrow 1} c_R(u)/(1-u) > 0$) it turns out to be optimal for the cedent not to purchase any proportional reinsurance as long as the current surplus is below a threshold value w_0 (i.e., $u = 1$ for any $w < w_0$). This may at first sight not look intuitive, as particularly for small surplus values reinsurance can be useful, but since reinsurance is expensive, for the long-term goal of minimal ruin probability it is then preferable to not pay reinsurance premiums and use the premium income from primary insurance to more quickly decrease the ruin probability before one can "afford" reinsurance again. Furthermore, if a solution $\gamma_R > 0$ to the equation

$$\inf_{u\in[0,1]} \left\{\lambda \mathbb{E}(e^{r\cdot u\cdot X}) - \lambda - \left(c - c_R(u)\right)r\right\} = 0 \tag{8.7.33}$$

  exists, then (under mild assumptions) the asymptotic behavior of the ruin probability improves (from the Cramér–Lundberg approximation (7.2.14) without reinsurance) to

$$\psi_R(w) \sim C\,e^{-\gamma_R w}, \quad w \to \infty \tag{8.7.34}$$

  for some constant $C > 0$. If the optimal value u^* for which the infimum in (8.7.33) is attained is moreover unique, then $\lim_{w\to\infty} u(w) = u^*$, that is, the optimal proportionality factor converges to a constant for increasing surplus.

  Note that due to the continuity and convexity properties of the involved functions, γ_R defined through (8.7.33) is also the supremum across all fixed u of the maximal solution of $\lambda\mathbb{E}(e^{r\cdot u\cdot X}) - \lambda - \left(c - c_R(u)\right)r = 0$. Hence the adjustment coefficient obtained by optimal dynamic proportional reinsurance is the same as the one obtained from determining the static u for a QS treaty that maximizes the adjustment coefficient (cf. Section 8.2.5.2). Correspondingly, dynamic reinsurance cannot improve the optimal adjustment coefficient among static QS treaties. However, for moderately sized surplus values, dynamic reinsurance can of course outperform the best static counterpart.

- *Dynamic XL reinsurance*: $u(y) = \min(y, u)$
  If it is possible to dynamically adapt the retention level u, then for exponentially bounded claims one can show (under some mild additional conditions) that also here an asymptotic behavior of the form (8.7.34) holds, if a solution $\gamma_R > 0$ to

$$\inf_{u\in[0,\infty]} \left\{ \lambda \int_0^u \left(1 - F_X(z)\right)e^{rz}\, dz - \left(c - c_R(u)\right) \right\} = 0$$

  exists. If the minimizer u^* is unique, then $\lim_{x\to\infty} u(x) = u^*$, that is, the optimal retention again converges to a constant value for increasing surplus. As for the proportional case, the resulting adjustment coefficient of the best static strategy cannot be improved through dynamic adaptation.

Several variants of the stochastic control problems of the above type have been studied extensively in the literature, in particular also in connection with optimal investment of parts of the surplus in the financial market and optimal dividend payments (see the Notes at the end of the chapter).

8.8 Beyond Piecewise Linear Contracts

One observes that almost all contracts identified as optimal in this chapter exhibit a piecewise linear shape of $R(X)$ (or $R(S)$) and likewise the corresponding retained amount. This is nicely in line with the fact that most of the reinsurance forms implemented in practice are of such a form. However, the question arises whether there could not exist more efficient ways for risk sharing, particularly since one may be inclined to think that smoother transitions between two regions are preferable to non-differentiable kinks (at junction points of linear pieces, like at the retention). This line of reasoning of course hinges on the underlying objectives and constraints for the optimization procedure. When using first and second moments of the retained risk as criteria, then piecewise linear solutions are no surprise. Note that for some other criteria in this chapter, such as in Sections 8.2.5 and 8.4, one sometimes also received optimal shapes of piecewise linear form because – in loose terms – on from a certain point (determined by some constraint) maximal coverage was needed until the available reinsurance premium is used up. It is natural to expect that in more general models or under different criteria non-linear shapes will be optimal. Among the cases for which this was already explicitly studied and shown, are:

- in the presence of default risk of the reinsurer (e.g., see Bernard and Ludkovki [125])
- in some cases when maximizing a linear combination of insurer's and reinsurer's utility (e.g., see Albrecher and Haas [20])
- taking into account cost-of-capital considerations (cf. Section 8.3)
- maximizing the adjustment coefficient under a variance principle or standard deviation principle for the reinsurance premium (cf. Section 8.2.5.2).

Even if non-linear retention functions are clearly non-standard in practice in an explicit form, they do appear implicitly as a result of certain combinations of reinsurance contracts and profit-sharing mechanisms.

Using non-linear shapes is also quite conceivable from a general viewpoint: reinsurance is about reshaping the insurer's loss distribution (often referred to as *tapering* in engineering circles), and if one has a target distribution for the retained amount, then one can define a respective retention function that leads to the desired target. For instance, if the goal after reinsurance of X is an exponential distribution with parameter λ for D, then

$$D(X) = \log(1 - F_X(X))^{-1/\lambda} \tag{8.8.35}$$

serves the purpose, since

$$\mathbb{P}(D(X) < x) = \mathbb{P}(F_X(X) \leq 1 - e^{-\lambda x}) = 1 - e^{-\lambda x}, \quad x > 0,$$

and $F_X(X)$ is uniformly distributed on $[0, 1]$. Note that (8.8.35) ensures $0 \leq D(X) \leq X$ as long as we have $F_X(x) \leq 1 - e^{-\lambda x}$ for $x \geq 0$ (which, for large x, is the case of interest). For instance, if X is a strict Pareto random variable with $F_X(x) = 1 - (x/x_0)^{-\alpha}$, $x \geq x_0 > 0$, then

$$D(X) = \log(X/x_0)^{\alpha/\lambda}$$

has the desired exponential distribution. Intuitively, rather than accepting the entire tail of X like in unlimited XL, the reinsurer accepts a larger proportion of the claim, the larger its size is. In a practical implementation, both the XL contract and such a contract (8.8.35) would only be written with an upper limit, but the latter can result in a considerably cheaper reinsurance form for the cedent than passing on an entire layer. Depending on the employed objectives and constraints, this may also be the formal solution of an optimization problem, in particular when involving capital costs in the criteria.

Finally, we mention another non-standard reinsurance form that is not implemented, but which could have nice properties in terms of reshaping the tail. In a *randomized reinsurance* contract, the reinsured amount $R(X)$ is not a deterministic function of X, but the mapping R is a random variable itself, that is, once X is known, there is an additional random mechanism to determine the reinsured amount according to some given distribution (e.g., a lottery under the supervision of a notary). While there may be several practical (and psychological) arguments against such a type of treaty, it can serve as an interesting thought experiment, since randomization is another effective tool for reshaping a loss distribution (in a possibly cost-effective way) (see Albrecher and Cani [17]). For the criteria of Section 8.2.5.1, Gajek and Zagrodny [366] showed that for certain discrete loss variables, a randomized reinsurance form in fact outperforms the deterministic ones.

A final general comment is in order. Even if one is able to formalize the objectives and constraints, and identify a corresponding optimal reinsurance form, in practice this particular contract may not be available or negotiable. In that case, one will look at all available reinsurance forms, determine the parameters (like the retention) such as to match the safety criterion and then choose the cheapest among those. In fact, some of the approaches discussed above dealt with identifying the best parameters within a given reinsurance form. It should be mentioned, however, that in such a procedure one should also be careful with model risk and estimation risk. At the same time, from the viewpoint of capital costs, whenever the resulting reduction of capital costs exceeds the reinsurance premium, the reinsurance contract will be of interest, even if the underlying model may have a certain degree of inaccuracy.

8.9 Notes and Bibliography

The modern economic theory of efficient risk sharing can be traced back to Borch [148, 151, 152] and Wilson [791]. Later, Arrow [47, 48] and in a more general framework Raviv [641] showed that the safety loading in insurance contracts leads to optimal insurance contracts containing deductibles. For an overview, see also Eeckhoudt et al. [313]. Gerber [385] extends Borch's equilibrium theorem (see also Pressacco [630] and Taylor [732, 733]). An alternative approach to Borch's result based on comonotonicity is Cheung et al. [206] and an extension to the more general notion of essential margins is Flåm [355]. Criterion (8.2.1) is also known as the *weighted-sum* or *scalarization* method in multi-objective optimization (cf. Ehrgott [314]). Note that the constants k_i in (8.2.1) can alternatively be interpreted as Lagrange multipliers of a marginal expected utility problem formulation with lower bounds on the expected utilities of the other companies. Knispel et al. [494] look into the effects of enlarging the number of (re)insurers and the implied diversification with respect to Pareto-optimal allocations. Recently, Malamud et al. [560] extended the results of Raviv to the situation where n reinsurers are available and the optimal reinsurance contracts $R_1(S), \dots, R_n(S)$ with $0 \leq R(S) = \sum_{i=1}^{n} R_i(S) \leq S$ are determined in an expected utility framework with given risk-averse utility functions for all participants, leading to different deductibles and proportionality constants for each reinsurance contract. A related study of the role of the leading reinsurer is Boonen et al. [147]. For a nice recent survey on optimal reinsurance from an economical point of view, refer to Bernard [120].

Cummins et al. [240] use the proportionality result of Section 8.2.1 to assess the capacity of the (re)insurance market for catastrophic losses in view of limited liabilities. Ludkovksi and Rüschendorf [550] provide deeper insight into the comonotonicity property of Pareto-optimal risk allocations. An attempt to bridge the actuarial and economic view of reinsurance is Aase [2], who gives an economic equilibrium analysis of a reinsurance market (see also Bühlmann [166] for an important early reference on the topic). Note that in an ideal frictionless environment there would be no demand for reinsurance; demand is created by, for example, financial distress cost, solvency cost, corporate tax (e.g., see Doherty [297] and Carneiro and Sherris [184]). Major [559] implements financial friction concepts from financial economics into actuarial optimal dividend models to valuate firm value (as originally proposed by de Finetti [256]) and illustrates by numerical solutions for some case studies the emergence of some stylized facts about reinsurance in the steering of firm value (see also Panning [603]).

Early contributions to the optimality of SL contracts include Vajda [749], Ammeter [37], Ohlin [594], Lemaire [535], Taylor [730], de Vylder et al. [264], and Heijnen [431]. Upper bounds for the variance for a general reinsurance treaty can be found in Birkel [137]. For some desirable theoretical properties of XL treaties, see Bühlmann [165]. Taylor [734] deals with reserving issues of XL layers.

The results in Section 8.2.3.3 have been extended in various directions. Bounds on the reinsurance cover lead to a truncated change-loss contract (Kaluszka [479]). Gajek and Zagrodny [365] show that for the standard deviation principle, (8.2.17) remains optimal if the objective is to minimize the truncated variance $E\big([(D - \mathbb{E}(D))^+]^2\big)$, whereas the absolute deviation measure $E|(D - \mathbb{E}(D))^+|$ is minimized for a truncated SL contract of the form $R(S) = \min((S-b)_+, L-b)$. For more general convex risk measures and premium principles, see Kaluszka [479, 481]. For a general survey, see Centeno and Simões [196].

In this chapter we always assumed $0 \le R(S) \le S$, which seems natural, as otherwise the insurer may be tempted to report the claim size differently (e.g., by delaying the settlement of claims or hiding losses), leading to moral hazard problems. Likewise, $R(S)$ should be a non-decreasing function (a property that the optimal reinsurance forms discussed in this chapter fulfilled in any case). If also $D(S)$ is non-decreasing (which is plausible for the same reasons, for example see Pesonen [614]), then the resulting reinsurance form $R(S)$ is necessarily continuous in S (e.g., see Carlier and Dana [183]). If there is a longer-term contract between the insurer and reinsurer, it may, however, also be preferable to drop the positivity constraint on $R(S)$ in favor of a profit-sharing arrangement (cf. Section 10.2). One way to deal with moral hazard in a multiperiod reinsurance framework is retrospective rating (see Doherty and Smetters [299]; for modelling audit costs in this connection see Picard [617]).

The results of Deprez and Gerber [279] discussed in Section 8.2.4 were generalized by Young [800] to the case of the risk-adjusted premium principle, and in a more general context in Promislow and Young [634].

Optimal retention levels for XL treaties with respect to minimizing the ruin probability in infinite and finite time are studied in Dickson and Waters [290, 291, 293], see also Hesselager [440]. For the Sparre Andersen model see Centeno [192, 193]. In Steenackers et al. [709] and De Longueville [259] the stability criterion is also the finite-time ruin probability. The effect of the initial reserve on the retention is studied in Centeno [190]. Centeno [191] also studied an upper bound on the ruin probability in finite time as a function of the retention limit in an XL arrangement. The allocation of the premiums is discussed by Sundt [719]. Amsler [41] proposed a reinsurance arrangement under which the reinsurer provides a full protection against ruin during a contractual period. This can be related to models for the expected present value of capital injections, see, for example, Dickson and Waters [292] and Eisenberg and Schmidli [318].

Heilmann [433] deals with the retention problem from the point of view of claim experience. Maximum retentions under an expected utility criterion have been determined by Zecchin [804]. One of the first attempts to come up with a decision analysis of reinsurance was given by Samson et al. [664]. Gollier [395] proposed a partially disappearing deductible when dealing with nuisance claims.

For a fixed reinsurance premium principle, the goal to minimize the VaR of the sum of retained risk and the reinsurance premium is closely related to minimizing the one-period ruin probability, as discussed in Section 8.2.5.1. It is, however, not identical, as there the goal was to minimize the probability of a loss larger than the current capital position, whereas with a VaR criterion of level α, one looks into optimally reshaping the loss distribution so that the specific α-quantile is minimal. Cheung [204] provides a geometric argument that simplifies the identification of the optimal reinsurance form in the setting of Cai and Tan [178] (see also Cheung et al. [208]). Rather general treatments of the topic are Cui et al. [236] and Cheung and Lo [205]. For multivariate risks see Cheung et al. [207]. If for the minimization of VaR and CTE additional constraints on the shape of the retention function are introduced, the optimal treaty turns out to be less robust for the VaR criterion (see Chi and Tan [209]). Chi and Weng [210] extend some of the analysis to more general premium principles. Tan et al. [728] identify non-intuitive optimal solutions when minimizing CTE. For an approach based directly on empirical loss data rather than on distributional assumptions for the risks, see Tan and Weng [726]. Cerqueti et al. [197] provide estimates for aggregate claims under a change-loss contract from empirical data. Another, more implicit way to exploit the Neyman–Pearson lemma for minimizing the one-period ruin probability than [366] is described in Weng and Zhuang [778]. Embrechts et al. [330] studied optimal risk sharing and its robustness for the so-called *range-VaR*, which is a variant of the Tail-VaR (cf. Section 7.2.2) for which the averaging of VaR values is restricted to a smaller interval (α, β) with $\alpha < \beta < 1$).

For optimal reinsurance forms in the presence of default risk of the reinsurer, see Cai et al. [175] and Asimit et al. [50]. Related earlier work includes Doherty and Schlesinger [301] and Mahul and Wright [557]. Optimal reinsurance shapes will then strongly depend on the type of involved dependence (systemic risks) and the stochastic nature of default in general, see, for example, Dana and Scarsini [245], Bernard and Ludkovki [125] and Biffis and Millossovich [133] as well as Cong and Tan [223]. For a low credit rating, the reinsurer may also be asked for a deposit, and such features need to be considered in the respective models. For an empirical study of the connectedness of the US insurance and reinsurance market and the implied potential systemic risk in case of defaults, see Park and Xi [605].

For an early contribution on considering capital costs for optimal reinsurance contracts, see Krvavych and Sherris [513]. A recent approach linked to an empirical study in this direction is Boyer [159], as well as Mango et al. [561] and Wacek [763]. Gurenko and Itigin [416] also empirically investigated reinsurance as a capital optimization tool in a regulatory environment. Asimit et al. [54] formulated and solved a concrete optimal reinsurance problem within the Solvency II framework, see also Clemente et al. [218] for a discussion on practical implications of reinsurance within Solvency II. For related studies of optimal intra-group risk transfers and resulting risk capital allocation in the framework of Solvency II, see Asimit et al. [51], whereas Asimit et al. [52] determine optimal risk transfers between legal entities in a group that are subject to different regulatory requirements. Group level diversification and possible limits of capital fungibility between respective entities is an important general topic in this context (cf. Keller [485] and Filipović and Kupper [350]). Gatzert and Schmeiser [377] study the impact of group diversification on shareholder value, see also Schlütter and Gründl [671] for effects on policyholder welfare. Another recent paper dealing with the cost of capital when determining optimal reinsurance forms is Chi and Weng [210]. Zhu et al. [809] studied the simultaneous optimization of various subportfolios in such a context, whereas Zhang [807] studied the overall capital cost reduction of both insurer and reinsurer.

Some problems in identifying optimal reinsurance forms may be modelled as cooperative games with stochastic payoffs. Such connections are exploited in Lemaire [536] and Baton and Lemaire [88]. For a good introduction, see Suijs et al. [715]. Game-theoretic approaches can also serve as arguments for randomized reinsurance forms, as respective solutions are often measure-valued. The ideas of Aase [4] are also relevant for the reinsurance context. A more recent contribution on Nash bargaining solutions in a reinsurance syndicate is Aase [5], see also Bensoussan et al. [117] and Boonen et al. [145, 146]. For an alternative general approach to pick a particular Pareto-optimal risk exchange based on actuarial fairness, see Bühlmann and Jewell [170].

The general role of minimal elements with respect to stop-loss order for optimal reinsurance forms was identified in Van Heerwaarden et al. [752]. For sequels, see Denuit et al. [277, 278], where a stochastic ordering is defined by means of pointwise comparison of the Lorenz curves of the risks. For an early attempt in this direction, see Heilmann [434]. Hu et al. [452] derived extremal choices for SL retentions based on stochastic ordering results. For an application of the combination of exponential utility and stop-loss to the reinsurance problem for inflation compensation for current old age annuities, see Lüthy [552]. There are still other optimality criteria for a single insurance risk, for example see the paper by Young [800], which has interpretations within a reinsurance context as well.

For an early reference on dynamic programming to identify optimal reinsurance schemes, see Pechlivanides [609]. Schmidli [674] first solved the continuous-time stochastic control problem with variable QS, and the XL case was tackled for the first time in Hipp and Vogt [448]. For a combination with an optimal dividend problem see Azcue and Muler [69]. Using dynamic reinsurance to minimize capital injections over time was studied by Eisenberg and Schmidli [317, 318], see also Schäl [668] and Bulinskaya et al. [172] for a discrete-time model. For modifications of optimal reinsurance in the presence of investments, see Schmidli [675], Asmussen et al. [61], Højgaard and Taksar [450], Irgens and Paulsen [458], and Hipp and Taksar [447], as well as Romera and Runggaldier [653] for a discrete-time approach. Since then many extensions of this problem have been studied. For a recent treatment of a two-dimensional case with dependence see Liang and Yam [542]. Meng et al. [571] allow for several reinsurers. Optimal reinsurance in a Markov-modulated environment when the states are not observable, but have to be filtered from observations is studied in Liang and Bayraktar [541]. Cani and Thonhauser [182] identified the dynamic reinsurance strategy that maximizes the surplus level at an exogeneously determined future exponential time. In addition to the concise general texts Schmidli [676] and Azcue and Muler [70] on the topic of stochastic control, we also refer to Hipp [444, 446] and Asmussen and Albrecher [57, Ch. XIV] for short surveys providing the intuition behind the concepts.

A dynamic approach to worst-case reinsurance is proposed in Korn et al. [498]. As an alternative to minimizing the probability of ruin, for an upper bound on number and sizes of claims the authors maximize the expected utility of surplus at a fixed time horizon for the worst-case claim scenario by a dynamic QS contract and treat the problem as a differential game, where the insurance company plays against mother nature. The resulting strategy is remarkably robust with respect to number and sizes of claims as well as the choice of the utility function.

In many cases discussed in this chapter the optimality was considered from the viewpoint of the insurer, and the reinsurer's perspective was reflected through the offered premium principle. Borch's result on Pareto-optimal risk exchanges, which in contrast considers to maximize a linear combination of expected utilities of insurer and reinsurer (cf. Section 8.2.1) can be refined to include asymmetries with respect to premium rules etc., see, for example, Albrecher and Haas [20], leading to non-linear retention functions. Cai et al. [176] considered the minimization of a convex combination of risk measures of the insurer and reinsurer (see also Jiang et al. [466]). Another formulation of combining the views of market participants is given in Assa [68], see also Zhuang et al. [810]. An approach considering joint survival probabilities of insurer and reinsurer can be found in Ignatov et al. [457], Kaishev and Dimitrova [295, 477], and Cai et al. [174], as well as Kchouk and Mailhot [484]. The setting of Ivanovs and Boxma [459] can also be interpreted in this context.

9

Simulation

In many practical situations, particularly also in reinsurance, the model assumptions or the number of model ingredients and their interactions are too complex to allow for an explicit calculation (or direct approximation) of quantities of interest. Many such quantities can, however, be expressed as expectations of a random variable X,[1] for which the distribution is specified through the model assumptions (albeit typically as a complicated interaction of other random variables, so that the direct calculation is impossible). For such situations, one can generate independent samples of that random variable and estimate the expectation by the arithmetic mean of those sample values. This is the core idea of Monte Carlo (MC) simulation. In view of its importance in reinsurance applications, we will discuss some of the main ideas in this chapter. For links to more detailed surveys on the topic see the Notes at the end of the chapter.

9.1 The Monte Carlo Method

Assume that for some random variable X we want to estimate $\mathbb{E}(X)$, and we know that this value is finite. If we are able to generate n independent sample values X_i, then the strong law of large numbers guarantees that, with probability 1,

$$\widehat{\mu}_n := \frac{1}{n}\sum_{i=1}^{n} X_i \stackrel{n\to\infty}{\longrightarrow} \mathbb{E}(X).$$

The estimator $\widehat{\mu}_n$ is unbiased and strongly consistent. If also $\mathrm{Var}(X) < \infty$, then by the Central Limit Theorem, in distribution,

$$\frac{\sqrt{n}}{\sqrt{\mathrm{Var}(X)}}\left(\frac{1}{n}\sum_{i=1}^{n} X_i - \mathbb{E}(X)\right) \stackrel{n\to\infty}{\longrightarrow} \mathcal{N}(0,1).$$

1 In this chapter X does not necessarily refer to a claim size. In fact, X may even be a function of hundreds of other random variables.

Reinsurance: Actuarial and Statistical Aspects, First Edition.
Hansjörg Albrecher, Jan Beirlant and Jozef L. Teugels.

That is, for large n the sample mean has the approximate distribution

$$\frac{1}{n}\sum_{i=1}^{n} X_i \sim \mathcal{N}\left(\mathbb{E}(X), \frac{1}{n}\mathrm{Var}\,(X)\right),$$

respectively

$$\frac{1}{n}\sum_{i=1}^{n} X_i \approx \mathbb{E}(X) + Z \quad \text{with } Z \sim \mathcal{N}\left(0, \frac{1}{n}\mathrm{Var}\,(X)\right).$$

This shows that the convergence rate of the Monte Carlo method is of the order $O\left(\sqrt{\frac{\mathrm{Var}\,(X)}{n}}\right) = O\left(\frac{1}{\sqrt{n}}\right)$.[2] Clearly, the error bounds for the Monte Carlo estimator will by construction always be probabilistic (i.e., not certain). From $\Phi(1.96) = 0.975$ for the standard normal c.d.f. one then receives the approximate 95% confidence interval for $\mathbb{E}(X)$:

$$\mathbb{P}\left(\hat{\mu}_n - 1.96\sqrt{\frac{\mathrm{Var}\,(X)}{n}} \leq \mathbb{E}(X) \leq \hat{\mu}_n + 1.96\sqrt{\frac{\mathrm{Var}\,(X)}{n}}\right) \approx 0.95, \tag{9.1.1}$$

and this confidence interval is the usual way Monte Carlo estimates are reported. In order to state this interval, one therefore needs to estimate the variance, and this is done by the (unbiased) sample variance of the generated replications:

$$\mathrm{Var}\,(X) \approx \hat{\sigma}_n^2 = \frac{1}{n-1}\left(\sum_{i=1}^{n} X_i^2 - n\,\hat{\mu}_n^2\right).$$

Note that adding one decimal place of precision requires 100 times as many replications, so for a high accuracy of the estimate very large sample sizes are needed, which can be very time-consuming. One alternative to reduce the size of the confidence interval is to reduce the variance of the estimator, which is sometimes possible with some smart ideas and workarounds. We will discuss such possibilities in Section 9.2.

In many cases relevant for reinsurance purposes, one in fact wants to estimate a probability for a certain event A (e.g., the event that the aggregate loss exceeds some threshold). However, such a probability can also be expressed as the expectation over the indicator function of the event A, that is, $\mathbb{P}(A) = \mathbb{E}(1_A)$. The corresponding Monte Carlo estimate for $\mathbb{P}(A)$ is then simply the relative frequency of the occurrence of A among n independent experiments

$$\mathrm{rf}_n(A) := \frac{1}{n}\sum_{i=1}^{n} 1_{A_i}.$$

---

2 If X is a function of s other random variables, this error bound is still independent of s, unlike classical numerical integration methods, which makes MC a popular alternative also in numerical integration, particularly for dimensions $s \geq 5$.

Since the variance of a Bernoulli random variable is given by

$$\mathrm{Var}\,(1_A) = \mathbb{P}(A)(1 - \mathbb{P}(A)) \tag{9.1.2}$$

the sample variance is $\widehat{\sigma}_n^2 = \mathrm{rf}_n(A)(1-\mathrm{rf}_n(A))$ and we then obtain the approximate 95% confidence interval for $\mathbb{P}(A)$ in the form

$$\left[\mathrm{rf}_n(A) - \frac{1.96}{\sqrt{n}}\widehat{\sigma}_n,\ \mathrm{rf}_n(A) + \frac{1.96}{\sqrt{n}}\widehat{\sigma}_n\right]. \tag{9.1.3}$$

For the implementation of the Monte Carlo method, one needs to produce random numbers through a deterministic algorithm ("pseudorandom numbers") which imitates randomness to such a degree that the result is practically indistinguishable from truly random numbers, that is, that the respective statistical tests for randomness are passed. Since $F_X(X)$ is uniform(0,1) distributed, the random number $F_X^{-1}(U)$ has distribution F_X and so it typically suffices to focus on the generation of *uniform* pseudorandom numbers (even if in many circumstances there are more efficient algorithms than this inversion method available to generate a random sample with distribution F_X, particularly when the inverse function F_X^{-1} is cumbersome to work with).[3] Since for the generation of one sample value of X one may in fact need many pseudorandom numbers (either X can be generated more efficiently by combining several variables or, more typically, X itself depends on many further random variables which all need to be generated for one realization of X), it is also important that there are no significant deterministic patterns in a produced pseudorandom sequence if one looks at certain blocks of numbers. Developing pseudorandom number generators and studying their properties is a classical topic of mathematical research, and nowadays there are very efficient and quick pseudorandom number generators available, so that we can generally take it for granted that the generation of "sufficiently" random and independent samples is possible (see Korn et al. [497] for an overview). The computational bottleneck in this context is hence typically not the generation of the pseudorandom sequence, but the evaluation of the functions that need to be applied to these numbers.

Another interpretation of the MC method is as follows: if the distribution of X (e.g., as a function of many other, possibly dependent, random variables) is not available explicitly, one can simulate n realizations $X_1, \ldots, X_n$ and build up an empirical c.d.f. $\hat{F}_n$ from these n points by assigning a weight of $1/n$ to each (cf. (4.2.7)). Using $\hat{F}_n$ as an approximation of the true F_X in calculations (e.g., for quantiles, expectations of functions of X, etc.) exactly corresponds to the MC estimator of the respective quantity.

Example 9.1 *(Estimation of VaR)* In Section 4.6 we discussed how to estimate quantiles from data. In applications, many risk factors can contribute to the distribution of the final random variable X of interest for which the high quantile $Q(\alpha) = \mathrm{VaR}_\alpha(X)$

3 An alternative interpretation is that the domain of the integration in the expression for $\mathbb{E}(X)$ can be transformed into the unit interval (or unit cube) by a suitable substitution, and any non-uniform density function is considered as a part of the function of which the expectation is calculated, that is, one can sample from a uniform distribution.

needs to be approximated. The crude MC estimator is obtained by simulating n replicates $X_1, \dots, X_n$ with c.d.f. F_X, and reading off the quantile

$$\hat{Q}_n(\alpha) = \inf\{x : \hat{F}_n(x) \geq \alpha\}$$

from the respective empirical c.d.f. $\hat{F}_n(x)$ (cf. (4.2.7)). Assume that the density function f_X of X exists. Applying the central limit theorem for quantiles, one gets the 95% confidence interval for the true value of $Q(\alpha)$ by

$$\left[\hat{Q}_n(\alpha) - 1.96\,\frac{\sqrt{\alpha(1-\alpha)}}{f_X(Q(\alpha))\sqrt{n}}, \hat{Q}_n(\alpha) + 1.96\,\frac{\sqrt{\alpha(1-\alpha)}}{f_X(Q(\alpha))\sqrt{n}}\right].$$

However, the expression $f_X(Q(\alpha))$ is hard to approximate (f_X is not available explicitly and $Q(\alpha)$ needs to be approximated by $\hat{Q}_n(\alpha)$). Even if f_X were available, the resulting interval length can be huge (e.g., if $X \sim \mathcal{N}(0,1)$, the simulation of $Q(0.995) = \mathrm{VaR}_{0.995}(X) = 2.576$ needs more than 366 million replications, if we want the confidence interval to be of length 0.001).

9.2 Variance Reduction Techniques

If Var (X) (respectively its estimate) in the confidence interval (9.1.1) is large, it can take a large number of replications to arrive at a satisfactory confidence interval. If we can replace the original estimate by another one with the same expectation, but smaller variance, then we can achieve an efficiency gain.[4]

Example 9.2 An example where the need for variance reduction becomes particularly obvious is *rare event sampling*, where one needs to estimate the probability $z = \mathbb{P}(A)$ of an event A which occurs with very low probability (a situation that frequently occurs in reinsurance applications). From (9.1.2), the variance of the crude MC estimator $Z = 1_A$ is $\sigma_Z^2 = z(1-z)$, and clearly $\sigma_Z^2 \sim z \to 0$ as $z \searrow 0$. However, what matters is the relative error (coefficient of variation)

$$\frac{\sigma_Z}{z} = \frac{\sqrt{z(1-z)}}{z} \sim \frac{1}{\sqrt{z}},$$

and the latter grows beyond any bound when $z \searrow 0$. If one wants to guarantee a certain relative precision of the MC estimate (in terms of the width of the confidence interval (9.1.3)), the sample size n needs to grow according to $n \sim const.\, z^{-1}$ as $z \searrow 0$, and this can be infeasible for small z. The techniques illustrated in the examples below will lead to estimators Z' for z with *bounded relative error* $\sigma_{Z'}/z < \infty$ or (slightly weaker) *logarithmic efficiency* $\limsup_{z\to 0} \sigma_{Z'}/z^{1-\epsilon} < \infty$ for any $\epsilon > 0$.

4 However, the development and implementation of such improvements also takes effort, and for each application one finally needs to find the appropriate tradeoff between developing such techniques or just running the crude MC algorithm with more replications.

In the following we briefly discuss three popular variants of variance reduction.

9.2.1 Conditional Monte Carlo

Let Y be another random variable and $\mathbb{E}(X|Y)$ can be calculated. Then

$$Y_1 := \mathbb{E}(X|Y)$$

is an unbiased estimator of $\mathbb{E}(X)$, and from

$$\mathrm{Var}\,(X) = \mathrm{Var}\,(\mathbb{E}(X|Y)) + \mathbb{E}(\mathrm{Var}\,(X|Y))$$

it follows that $\mathrm{Var}\,(Y_1) \leq \mathrm{Var}\,(X)$. Consequently, this always leads to a reduction of variance, but it will not always be easy to find an appropriate Y to serve the purpose.

We illustrate this approach with an impressive example for the simulation of tail probabilities of compound sums of heavy-tailed risks:

Example 9.3 (Asmussen–Kroese estimator)
Assume that one wants to simulate an estimate of $z(u) = \mathbb{P}\,(S(t) > u)$, where $S(t) = \sum_{i=1}^{N(t)} X_i$ and X_i are i.i.d. and heavy tailed. Crude MC suggests simulation of n replicates of $S(t)$ and reporting the fraction of the cases where the result exceeded u. This is clearly not efficient, particularly for large values of u (which, however, is often the region of interest in practical applications). If the claims X_i are subexponential, then we know from Chapter 3 that for large u the largest summand dominates the sum and is the crucial ingredient in the calculation. The ingenious idea is now to condition on the fact that the last term $X_{N(t)}$ is the largest, simulate all but that largest term, and estimate $z(u)$ from those terms, "calculating" rather than simulating that largest (crucial) term. If we denote $S_k := \sum_{i=1}^{k} X_i$ and $M_k := \max\{X_1, \ldots, X_k\}$, then by symmetry

$$\mathbb{P}(S_k > u) = k \cdot \mathbb{P}(S_k > u, M_k = X_k)$$

and one can formulate the conditional Monte Carlo estimator

$$\begin{aligned} Z(u) &= N(t) \cdot \mathbb{P}(S(t) > u, M_{N(t)} = X_{N(t)} | N(t), X_1, \ldots, X_{N(t)-1}) \\ &= N(t) \cdot \overline{F}_X\left(\max\left\{M_{N(t)-1}, u - \sum_{i=1}^{N(t)-1} X_i\right\}\right). \end{aligned}$$

Case study: Dutch fire insurance data. Figure 9.1 illustrates how this conditional MC estimator improves the 95% confidence intervals of crude MC for the simulated tail probability of the compound Poisson random variable $S(365) = \sum_{i=1}^{N(365)} X_i$ beyond 919,100,000 (which was calculated explicitly to be the $\mathrm{VaR}_{0.995}(S(365))$ in Table 6.2). □

9.2.2 Importance Sampling

One of the problems of crude MC estimation is that many generated sample points will not really be in the relevant region for the quantity to be approximated. For instance,

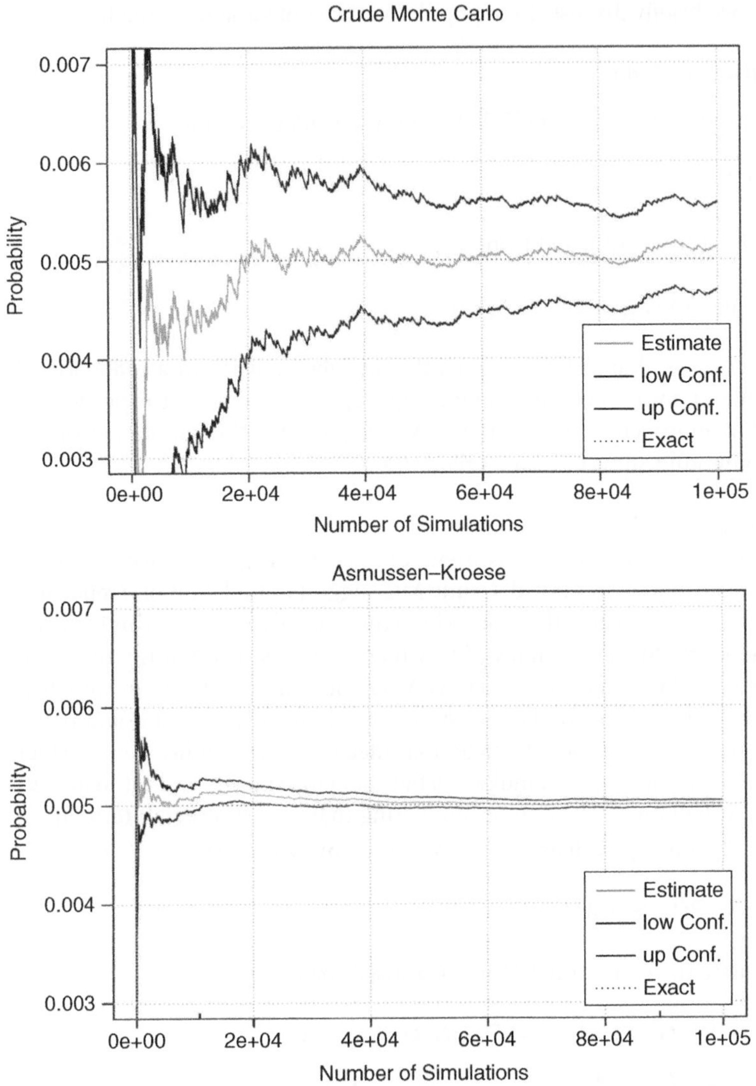

Figure 9.1 Dutch fire insurance data: confidence intervals for the simulation of $\mathbb{P}(S(365) \geq 919, 100, 000))$ for the compound Poisson case of Figure 6.2 as a function of n for crude MC and the Asmussen–Kroese estimator.

in Example 9.1 for the estimation of the quantile, for very small or very large α, there will not be enough sample points close to the quantile (it is inefficient to determine the exact value of a replication if it is in any case far to one side of the quantile; only the fact that it lands on that side is relevant), and this adds to the variance of the estimator.

Assume again that F_X allows for a density f_X. The idea of *importance sampling* is now to switch from f_X to another density $f_{\widetilde{X}}$ that concentrates more strongly on the region of interest (in Example 9.1 this would mean that $f_{\widetilde{X}}$ has a lot of probability mass around the suspected value of $Q(\alpha)$). Such a new density $f_{\widetilde{X}}$ can be obtained from f_X by shifting, rescaling, twisting etc. The quantity $f_X(x)/f_{\widetilde{X}}(x)$ is called the *likelihood ratio function.* We then have

$$\mathbb{E}(X) = \int x f_X(x)dx = \int \left(x\frac{f_X(x)}{f_{\widetilde{X}}(x)} \right) f_{\widetilde{X}}(x)dx = \mathbb{E}\left(\widetilde{X}\frac{f_X(\widetilde{X})}{f_{\widetilde{X}}(\widetilde{X})} \right).$$

We instead simulate n independent replicates $\widetilde{X}_1, \ldots, \widetilde{X}_n$ from the new random variable $\widetilde{X}$ (with density $f_{\widetilde{X}}$) and use the importance sampling estimator

$$\widehat{\mu}_n^I = \frac{1}{n}\sum_{i=1}^{n} \widetilde{X}_i \frac{f_X(\widetilde{X}_i)}{f_{\widetilde{X}}(\widetilde{X}_i)}. \tag{9.2.4}$$

This represents a weighted MC estimator for $\mathbb{E}(X)$ with weights according to the likelihood ratio function, in order to "correct" for using the new density $f_{\widetilde{X}}$. The variance of this estimator is

$$\text{Var}\,(\widehat{\mu}_n^I) = \frac{1}{n}\left(\int \frac{x^2 f_X(x)}{f_{\widetilde{X}}(x)} f_X(x)dx - \mathbb{E}^2(X) \right),$$

so that we achieve a variance reduction whenever $\mathbb{E}[X^2 f_X(X)/f_{\widetilde{X}}(X)] < \mathbb{E}(X^2)$. Correspondingly, importance sampling will be most efficient, if heuristically

- $f_{\widetilde{X}}(x)$ is large whenever $x^2 f_X(x)$ is large,
- $f_{\widetilde{X}}(x)$ is small whenever $x^2 f_X(x)$ is small.

Furthermore, $f_{\widetilde{X}}(x)$ should be easy to evaluate and $\widetilde{X}$ should be easy to simulate. The following example illustrates the efficiency gain.

Illustration: VaR of a gamma-distributed random variable. Consider the simulation of $\text{VaR}_{0.995}(X)$ for the case where X has a gamma(3.3,0.9) density. If one has an a priori guess for that value, then one can use importance sampling to concentrate the probability mass into that region. For instance, a large deviation argument may suggest that the final value is 14.2. If one decides to use *exponential twisting*

$$f_{\widetilde{X}}(x) = e^{\theta x} f_X(x)/\mathbb{E}(e^{\theta X})$$

(which is in essence an Esscher transform of X, cf. Chapter 7), then one can identify the value of θ for which $\mathbb{E}(\widetilde{X}) = 14.2$, which turns out to be $\theta = 0.88$. Figure 9.2 illustrates the efficiency gain of using the importance sampling estimate (9.2.4) compared to crude MC as a function of number of simulations. Note that although the first guess 14.2 on the true value was actually quite bad, the resulting twist of X is still much better than using the original distribution. □

Let us now turn to estimating the tail probability $z(u) = \mathbb{P}(S(t) > u)$ of an aggregate sum $S(t) = \sum_{i=1}^{N(t)} X_i$ for large u and light-tailed claims X_i. Then one can apply the exponential twisting idea described in the above illustration to the entire random variable $S(t)$

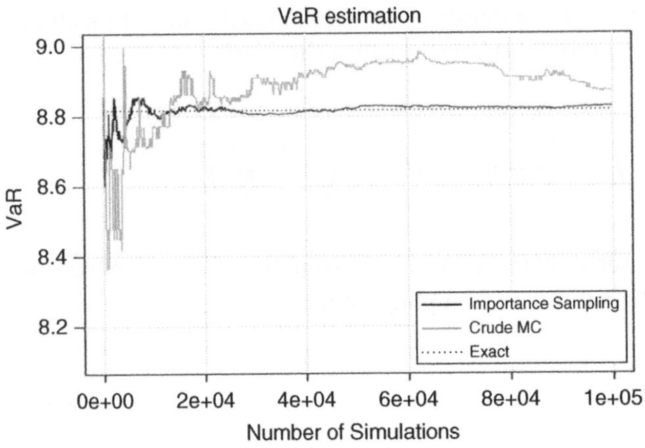

Figure 9.2 Estimation of $\mathrm{VaR}_{0.995}(X)$ for a gamma(3.3,0.9)-distributed X as a function of n for crude MC and an importance sampling estimator.

(which modifies the distribution of both $N(t)$ and X_i), to get the importance sampling estimate

$$Z^*(u) = \exp\left\{ -\theta(u) \sum_{i=1}^{N(t)} X_i \right\} e^{\kappa_S(\theta(u))} \, 1_{\{\sum_{i=1}^{N(t)} X_i > u\}},$$

where $\kappa_S(\theta(u)) = \log \mathbb{E}(e^{\theta(u)S(t)})$ is the cumulant-generating function of $S(t)$. We can now choose $\theta(u)$ (i.e., define the amount of tilting) in such a way that under the new measure we expect $S(t)$ to be equal to the threshold value u, so that[5]

$$\mathbb{E}_{\tilde{X}}(S(t)) = \kappa_S'(\theta(u)) \stackrel{!}{=} u.$$

Illustration: VaR of a compound Poisson sum of gamma-distributed claims. If $N(t)$ is a homogeneous Poisson process with rate λ, then $\kappa_S(\theta(u)) = \lambda t(M_{X_i}(\theta(u)) - 1)$, where $M_{X_i}(\theta(u)) = \mathbb{E}(e^{\theta(u)X_i})$ denotes the moment-generating function of the individual claims X_i. Fix now $t = 1$. Correspondingly we can choose $\theta(u)$ as the solution of $\lambda M_{X_i}'(\theta(u)) = u$. Due to $\mathbb{E}(e^{r+\theta(u)S(1)})/\mathbb{E}(e^{\theta(u)S(1)}) = e^{\lambda M_{X_i}(\theta(u))(M_{X_i}(r+\theta(u))/M_{X_i}(\theta(u))-1)}$, we thus simulate a compound Poisson process with rate $\lambda \cdot M_{X_i}(\theta(u))$ and individual claims with density $e^{\theta(u)x} f_{X_i}(x)/\mathbb{E}(e^{\theta(u)X_i})$ and calculate $Z^*(u)$. Figure 9.3 illustrates the remarkable efficiency improvement for a particular compound Poisson sum of gamma-distributed random variables. □

For heavy-tailed claim sizes, exponential twisting is not applicable because the moment-generating function $M_{X_i}(r)$ does not exist for any $r > 0$. For such cases it is popular to

5 The rationale behind this choice can also be seen in terms of a saddlepoint approximation for $u \to \infty$, since $S(t)$ under the new measure is then increasingly well approximated by a normal distribution with mean $\kappa_S'(\theta(u))$ and variance $\kappa_S''(\theta(u))$.

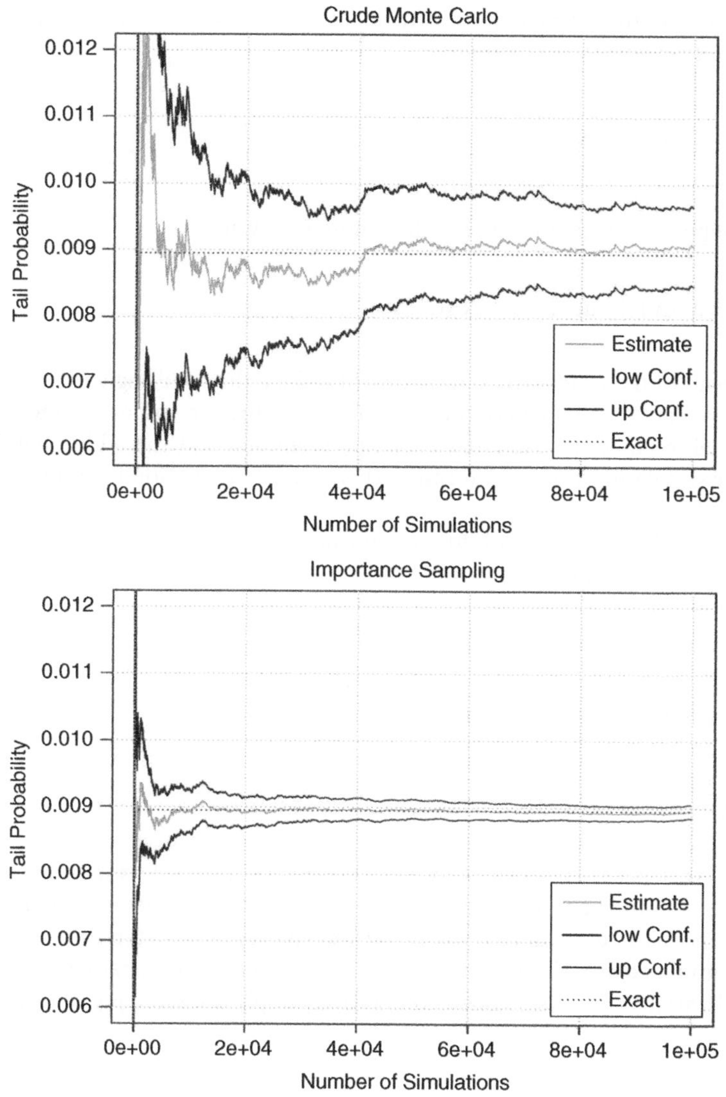

Figure 9.3 Confidence intervals for the simulation of $\mathbb{P}(\sum_{i=1}^{N_{10}(1)} X_i \geq 130)$, where X_i are gamma(3.3,2) distributed, as a function of n for crude MC and the importance sampling estimator (9.2.4).

twist the hazard rate $f_X(x)/\overline{F}_X(x)$ instead (further refinements are so-called *asymmetric hazard rate twisting* and *delayed hazard rate twisting*, see Juneja and Shahabuddin [472]).

9.2.3 Control Variates

Assume that a random variable Y is available for which $\mathbb{E}(Y)$ is known exactly and which is positively correlated with X. Then the deviation of the simulated from the exact value of Y may be used to correct the estimate for X. The identity $\mathbb{E}(X) = \mathbb{E}(X - Y) + \mathbb{E}(Y)$ indeed suggests the *control variate MC estimator*

$$\overline{X}_Y = \frac{1}{n}\sum_{i=1}^{n}(X_i - Y_i) + \mathbb{E}(Y),$$

where (X_i, Y_i) are independent copies of (X, Y). One immediately gets

$$\text{Var}\,(\overline{X}_Y) = \frac{1}{n}\,(\text{Var}\,(X) + \text{Var}\,(Y) - 2\,\text{Cov}(X, Y)),$$

so that using X_Y instead of the crude MC estimator will reduce the resulting variance whenever $2\,\text{Cov}(X, Y) - \text{Var}\,(Y) \geq 0$. In particular, the closer Y is to X, the more variance of crude MC can be eliminated. If a suitable Y can be identified and the additional computation time for simulating Y is not excessive, this variance reduction technique can perform remarkably well.

Note that if Y is a control variate, so is $a\,Y$ for any constant $a > 0$ and this suggests to choose a in the most favorable way, namely such that

$$n\,\text{Var}\,(\overline{X}_{aY}) = \text{Var}\,(X) + a^2\text{Var}\,(Y) - 2a\text{Cov}(X, Y)$$

is minimized. This leads to the optimal constant

$$a^* = \frac{\text{Cov}(X, Y)}{\text{Var}\,(Y)}$$

and a resulting overall variance reduction of $\text{Cov}^2(X, Y)/(n\text{Var}\,(Y))$. The corresponding relative variance reduction then amounts to $\rho^2_{X,Y}$, where $\rho_{X,Y}$ is the correlation coefficient. In most cases, $\text{Cov}(X, Y)$ and $\text{Var}\,(Y)$ will not be known explicitly, so typically the latter values will be simulated as well (in a smaller pre-run simulation or in parallel).

If a further control variate is available, one can iteratively apply the same procedure again and use

$$\overline{X}_{Y,Z} = \overline{X}_Y - \frac{1}{n}\sum_{i=1}^{n} Z_i + \mathbb{E}(Z),$$

which will lead to a further reduced variance if

$$n\,\text{Var}\,(\overline{X}_{Y,Z}) = \text{Var}\,(X_Y) + \text{Var}\,(Z) - 2\text{Cov}(X_Y, Z) < \text{Var}\,(X_Y).$$

Example 9.4 Let us reconsider Example 9.3, but now consider the simulation of the stop-loss premium $\mathbb{E}(S(t) - u)_+$. Then for subexponential F_X one has the first-order asymptotic approximation

$$\mathbb{E}(S(t) - u)_+ \sim \mathbb{E}(N(t))e_X(u)\overline{F}_X(u), \;\; u \to \infty,$$

where $e_X(u)$ denotes the mean excess function of X. In the spirit of Example 9.3, one then can formulate the Asmussen–Kroese estimator

$$\begin{aligned} Z_{SL}(u) &= N(t) \cdot \mathbb{E}((S(t) - u)_+ 1_{\{M_{N(t)} = X_{N(t)}\}} | N(t), X_1, \dots, X_{N(t)-1}) \\ &= N(t) \left[\left(M_{N(t)-1} + \sum_{i=1}^{N(t)-1} X_i - u \right)_+ + e_X(W(u)) \right] \overline{F}_X(W(u)) \end{aligned}$$

with $W(u) = \max \left\{ M_{N(t)-1}, u - \sum_{i=1}^{N(t)-1} X_i \right\}$. In order to control the variability of N, one can now introduce the control variate $N(t) e_X(u) \overline{F}_X(u)$. This leads to the improved estimator

$$Z_{SL}^c(u) = Z_{SL}(u) - N(t) e_X(u) \overline{F}_X(u) + \mathbb{E}(N(t))\, e_X(u) \overline{F}_X(u).$$

Case study: Dutch fire insurance data. Figure 9.4 illustrates the stop-loss premium $\mathbb{E}((S(365) - C)_+)$, where the retention $C = 919{,}100{,}000$ is the $\mathrm{VaR}_{0.995}(S(365))$ calculated in this model. One sees that the Asmussen–Kroese estimator is an impressive improvement over the crude MC estimator, but the control variate is a very minor additional improvement in this case. The reason for this is that the variability of $N(t)$, which the control estimate $Z_{SL}(u)$ improves upon, only makes up a small part of the overall variance, so that the efficiency gain is not considerable. □

Illustration: SL premium for another compound Poisson sum. In order to show how the control variate estimate $Z_{SL}^c(u)$ can improve the Asmussen–Kroese estimator substantially, Figure 9.5 gives the the Asmussen–Kroese estimator for the SL premium $\mathbb{E}((\sum_{i=1}^{N(1)} X_i - 482.96)_+)$, where the individual claims are Pareto-distributed with tail $1 - F_X(x) = (1 + x)^{-1.5}$ and $N(1)$ is Poisson-distributed with $\lambda = 10$. One sees that in this case the importance of the variance of $N(1)$ is considerable, and correspondingly the control variate improvement is pronounced. □

For further details on the above methods and further variance reduction techniques like antithetic variables and stratified sampling, see, for example, Asmussen and Glynn [59].

9.3 Quasi-Monte Carlo Techniques

The simulation of one replicate of $\sum_{i=1}^{k} X_i$ in fact involves k random variables (or even more if X_i is generated with the help of several pseudo-random numbers). If the number of summands is itself random (as for $S(t) = \sum_{i=1}^{N(t)} X_i$), then, for each replicate, $N(t)$ needs to be simulated first and the outcome determines the resulting number of needed pseudo-random numbers for this replicate. If s such pseudo-random numbers are needed, one can interpret that each replication needs an s-dimensional pseudorandom number, that is, (after suitable transformation) one uses a uniformly distributed point in the s-dimensional unit cube $[0, 1]^s$. One advantage of MC methods (and partly the reason why they are so popular) is that the error bound does not depend on s. However, the error bound is (by construction) only probabilistic and the convergence rate of $1/\sqrt{n}$ is not overly fast.

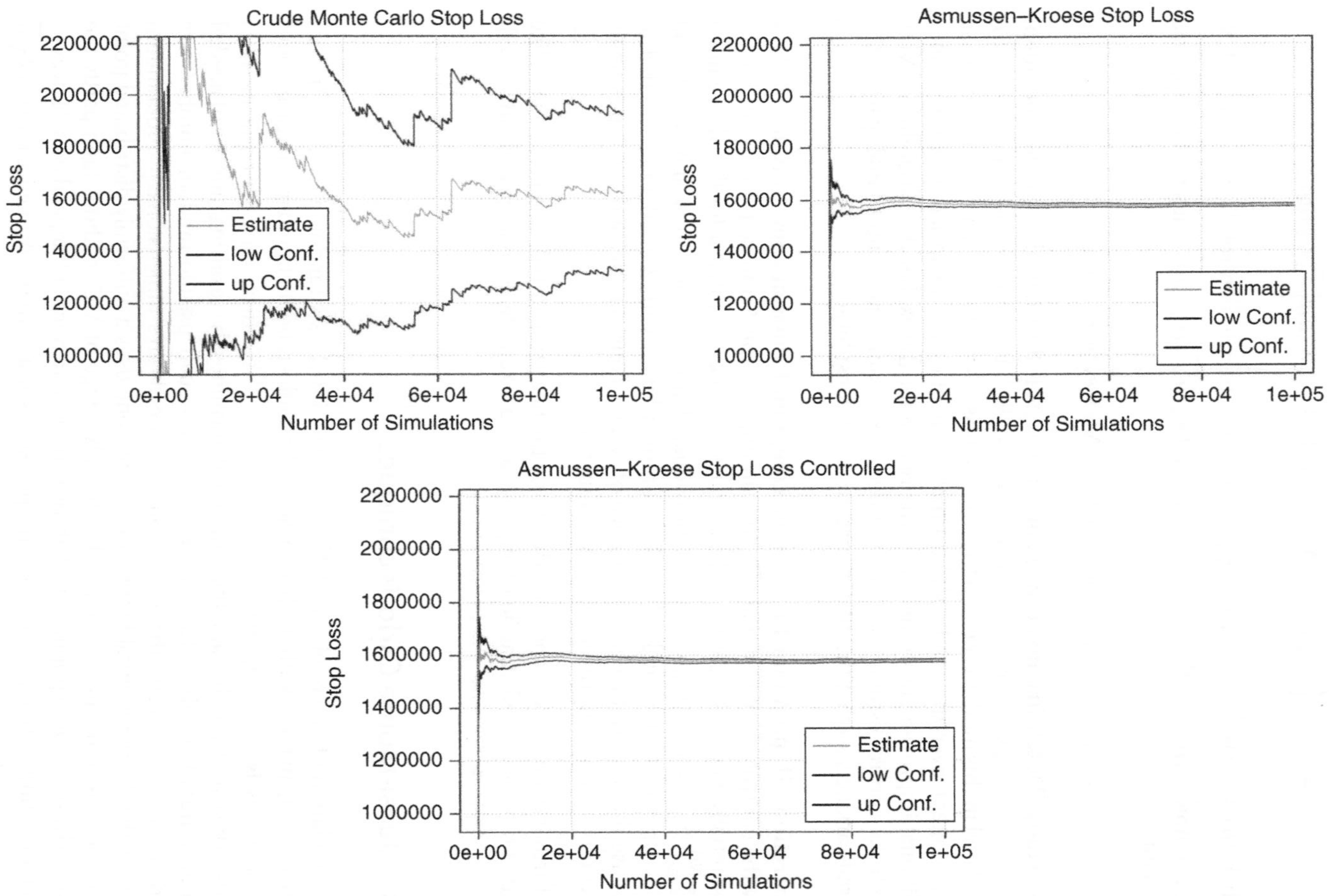

Figure 9.4 Dutch fire insurance data: confidence intervals for the simulation of $\mathbb{E}((S(365) - 919,100,000)_+)$ for the compound Poisson case of Figure 6.2 as a function of n for crude MC (top left), Asmussen–Kroese estimator $Z_{SL}(u)$ (top right), and its control variate improvement $Z^c_{SL}(u)$ (bottom).

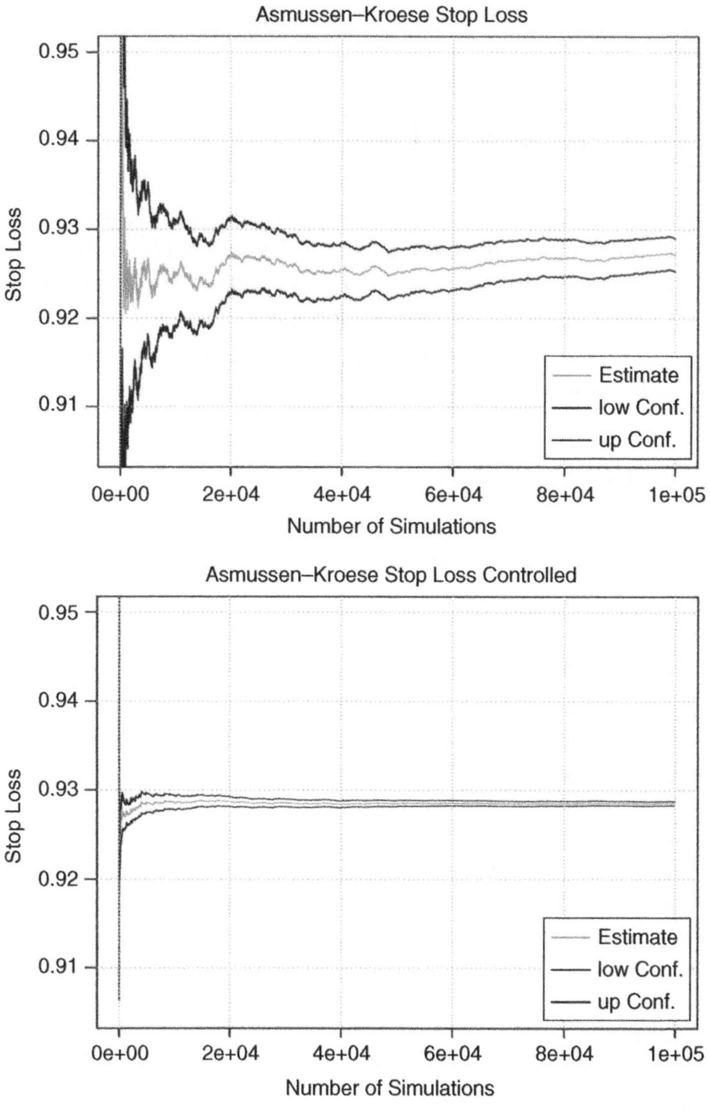

Figure 9.5 Confidence intervals for the simulation of $\mathbb{E}((\sum_{i=1}^{N(1)} X_i - 482.96)_+)$ for $1 - F_X(x) = (1+x)^{-1.5}$ and $\lambda = 10$ as a function of n for $Z_{SL}(u)$ and its control variate improvement $Z_{SL}^c(u)$.

To improve the performance of the MC method, an alternative to reducing the variance is to improve the convergence speed of the method. This can be achieved by replacing the pseudorandom numbers with deterministic point sequences in $[0,1]^s$ which imitate the properties of the uniform distribution in this unit cube well. The construction of such point sequences with good distribution properties is a classical topic in mathematics and nowadays there is a plethora of available algorithms to quickly generate such *quasi-Monte Carlo (QMC) sequences* $(\mathbf{x}_j)_{j=1,\ldots,n}$ in $[0,1]^s$, for example see Dick and Pillichshammer [288] for a recent survey.

One way to measure the distribution properties of such a deterministic sequence is the *star discrepancy*

$$D_n^*(\mathbf{x}_1, \dots, \mathbf{x}_n) = \sup_{\mathbf{y} \in [0,1]^s} \left| \frac{1}{n} \sum_{j=1}^{n} 1_{[0,\mathbf{y})}(\mathbf{x}_j) - \prod_{i=1}^{s} y_i \right|,$$

where $[0, \mathbf{y}) = [0, y_1) \times \dots \times [0, y_s)$. That is, one considers the worst-case deviation of the empirical fraction of points from the theoretical fraction $\prod_{i=1}^{n} y_i$ over all intervals $[0, \mathbf{y}) \in [0, 1]^s$. Correspondingly, $(\mathbf{x}_j)_{j=1,2,\dots,}$ is called *uniformly distributed* if $\lim_{n\to\infty} D_n^*(\mathbf{x}_1, \dots, \mathbf{x}_n) = 0$. If such a sequence is now used for approximating $\int_{[0,1]^s} f(\mathbf{u})\, d\mathbf{u}$ (which in our context represents the expected value of a random variable), then an upper bound for the approximation error is given by the *Koksma–Hlawka inequality*

$$\left| \frac{1}{n} \sum_{j=1}^{n} f(\mathbf{x}_j) - \int_{[0,1]^s} f(\mathbf{u})\, d\mathbf{u} \right| \leq V(f)\, D_n^*(\mathbf{x}_1, \dots, \mathbf{x}_n),$$

where $V(f)$ is the variation of the function f.[6] Hence, one can split the error into a term that only depends on the integrand and a second term that only depends on the quality of the used point sequence. The star discrepancy of the best known sequences has an asymptotic order of $O\left(\frac{\log^s n}{n}\right)$, so that for not too large values of s the convergence rate of QMC integration can considerably outperform MC. The actual performance can then even be better than this upper bound, but it also depends on the constant involved in the O term for the concrete QMC sequence. Typically, it can be a significant advantage to use QMC sequences for up to 20 or 30 dimensions.[7]

Example 9.5 A simple example for a QMC sequence is the Halton sequence. Choose an integer b and define for each integer j the sequence $a_k(j)$ as the digits of j w.r.t. to base b, that is, $j = \sum_{j=0}^{\infty} a_k(j) b^k$. Then the radical inverse function $\phi_b(j) = \sum_{k=0}^{\infty} a_k(j) b^{-k-1}$ is the reflection of this representation at the decimal point. If one repeats this procedure now for relatively prime integers $b_1, \dots, b_s \geq 2$, one receives the s-dimensional Halton sequence with bases $b_1, \dots, b_s$

$$\mathbf{x}_j = \left(\phi_{b_1}(j), \dots, \phi_{b_s}(j)\right) \in [0, 1]^s, \quad j \geq 1.$$

This sequence is uniformly distributed and satisfies $D_n^*(\mathbf{x}_1, \dots, \mathbf{x}_n) = O\left(\frac{\log^s n}{n}\right)$. Figure 9.6 depicts (for dimension $s = 2$) 250 MC points and the first 250 points from a Halton sequence with bases $b_1 = 2$ and $b_2 = 3$. One sees how the QMC sequence fills up the space much more regularly.

While the construction of Halton sequences is extremely simple, their properties in higher dimensions are not favorable (the constant in the upper discrepancy bound

6 Concretely, the total variation in the sense of Hardy and Krause.

7 If more dimensions are needed, the remaining dimensions can, for instance, be filled with regular MC points, leading to a so-called *hybrid MC* algorithm.

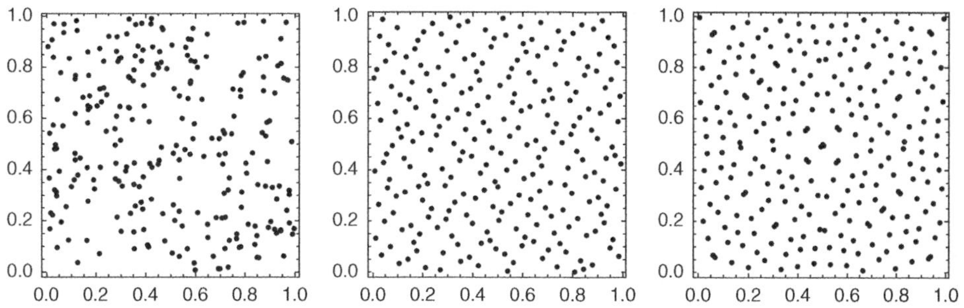

Figure 9.6 Two-dimensional sequence of 250 pseudorandom numbers (left), Halton numbers with $b_1 = 2$ and $b_2 = 3$ (middle) and Sobol numbers (right).

increases fast as s grows). Better alternatives are, for example, *(t,s)* sequences (with *Sobol sequences* [705] as particular examples, cf. Figure 9.6 (right) for a plot of the first 250 points in the first two dimensions). Their construction is more involved, but respective codes for a quick generation are nowadays available in many computer packages (see the next section for references).

Case study: Dutch fire insurance data. Figure 9.7 shows the performance of a QMC implementation of the Asmussen–Kroese estimator of Figure 9.1 (right) for a Halton and a Sobol sequence. As can be observed, the performance of the Halton sequence (in the plot the bases $b_1, \ldots, b_s$ are the first s prime numbers) is somewhat disappointing for this particular application. Note that each individual simulation run is generated from one high-dimensional point, where the first dimension of the point is used to determine the number $N(365)$ of claims to be generated for that run; in the present case the largest dimension finally needed is $s = 149$. But as mentioned above, in such high dimensions the uniformity properties of Halton sequences deteriorate, and it takes many simulation runs to approximate the exact value well. At the same time, one sees that the estimate based on a Sobol sequence is excellent already for a low number of simulation runs. In both cases one can achieve further improvements using hybrid MC methods, as mentioned above. This illustrates that a judicious choice of an appropriate sequence is crucial (and will depend on the quantity to be approximated), yet the Sobol plot shows that the potential of QMC sequences to speed up computations is very promising. □

Since the first dimensions of a QMC sequence typically have better properties (an effect known as *curse of dimensionality*), one strategy to increase the efficiency of QMC algorithms is to use the low dimensions of the sequence for the generation of realizations of those random variables that are particularly important for the outcome (like $N(t)$ in the simulation of $\sum_{i=1}^{N(t)} X_i$). One way to formalize this is by means of a refined version of the Koksma–Hlawka inequality:

$$\left| \frac{1}{n} \sum_{j=1}^{n} f(\mathbf{x}_j) - \int_{[0,1)^s} f(\mathbf{u}) d\mathbf{u} \right| \leq \sum_{l=0}^{s-1} \sum_{F_l} D_n^* \left(\mathbf{x}_n^{(F_l)} \right) V^{(s-l)}(f^{(F_l)}),$$

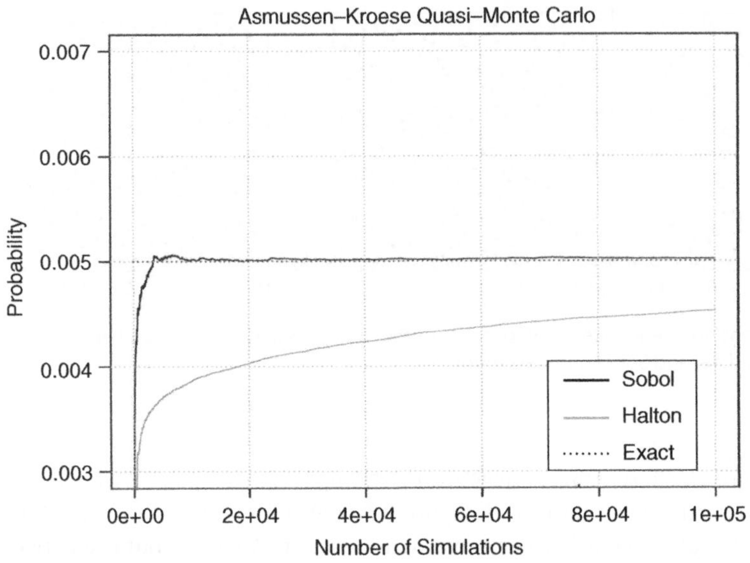

Figure 9.7 Dutch fire insurance data: QMC simulation of $\mathbb{P}(S(365) \geq 919,100,000))$ for the compound Poisson case of Figure 6.2 as a function of n for the Asmussen–Kroese estimator.

where F_l are $(s-l)$-dimensional faces, $\mathbf{x}_j^{(F_l)}$ are points in those faces with $x_{i_1} = \cdots = x_{i_l} = 1$ elsewhere, and $V^{(s-l)}(f^{(F_l)})$ are the respective lower-dimensional variations of the function f. Hence, one can try to keep the overall error bound low by assigning the "good" dimensions (with lower discrepancy) of the sequence to those variables that have a higher (low-dimensional) variation and hence contribute most to the overall variation of f. In particular, it is often observed that only some dimensions of the function f are really crucial for the total variability (leading to the notion of the *effective dimension* of the integrand, cf. Wang and Fang [774]).

9.4 Notes and Bibliography

Excellent surveys on stochastic simulation techniques are Asmussen and Glynn [59], Korn et al. [497], Glasserman [390]. See also Asmussen and Albrecher [57, Ch. XV]. The first logarithmically efficient rare event simulation algorithm can be found in Asmussen and Binswanger [58], and the Asmussen–Kroese estimator was proposed in [64]. For the study of rare event simulation algorithms for stop-loss premiums, see also Hartinger and Kortschak [422] and Asmussen and Kortschak [62, 63] for further performance improvements. An important contribution for rare event sampling is Blanchet and Glynn [141].

Classical texts on the theory of QMC methods are Niederreiter [590] and Drmota and Tichy [309]. For more recent accounts see, for example, Dick and Pillichshammer [288] and Lemieux [538]. Since, by its deterministic nature, the QMC method does not provide a confidence interval for the obtained estimate, it is sometimes common to sequentially use randomized QMC sequences (e.g., choosing different starting values) and then providing a statistical error estimate, as for MC (e.g., see L'Ecuyer and Lemieux [553]). QMC techniques have been used in various application areas in insurance (for classical risk models see Albrecher and Kainhofer [24] and Preischl et al. [629], and for an application to CAT bond pricing see Albrecher et al. [22]). One further advantage of QMC methods is that due to its deterministic construction an entire simulation run can easily be replicated, whereas this can be more tricky for a MC implementation.

In recent years there also has been a lot of research activity on simulation techniques for dependent risks (see Mai and Scherer [558] for a survey). QMC techniques for dependent scenarios are now being studied in more detail (see Cambou et al. [181] and Preischl [628] for some first contributions).

Whenever a model admits an explicit expression (or an explicit approximation) in terms of the parameters, model sensitivities and respective tuning can be done in a very efficient way, and then such an approach is preferable to simulation, since the latter only gives one number and changing parameters entails an entirely new simulation exercise. However, when it comes to aggregating the entire portfolio (which, for instance, is needed for the determination of the solvency capital of a company), the number and type of involved risks will be too complex to allow for explicit expressions, and simulations are the essential tool to assess the resulting profit and loss distribution, and particularly its tail. For sensitivity analysis in connection with simulation techniques in semi-explicit situations see, for example, Glasserman [390]. It is also conceptually easy to add particular additional scenarios in a simulation approach (see Mack [555] for early suggestions in that direction). The regulator in fact often asks to store all the simulated values for potential control purposes, which in view of the enormous number of data can be a challenge. In this connection Arbenz and Guevara [45] recently proposed a data compression technique that allows such empirical c.d.f. to be stored much more efficiently, keeping in mind the reproducability of the concrete implemented risk measures.

10

Further Topics

10.1 More on Large Claim Reinsurance

While large claim treaties as discussed in Section 2.5 are not popular in practice (even if in particular on the facultative basis such contracts can occasionally be found), we want to illustrate here that from a mathematical perspective there are a number of elegant results on the quantities of interest available. We first deal with larger order statistics and subsequently look into large claims reinsurance and ECOMOR.

10.1.1 The Ordered Claims

The sizes of the largest claims also depend on the random number of claims. Recall our notation for the order statistics

$$X_{1,N(t)} \leq \cdots \leq X_{N(t)-s-1,N(t)} \leq X_{N(t)-s,N(t)} \leq \cdots \leq X_{N(t),N(t)}. \tag{10.1.1}$$

Distribution. Consider an XL treaty with unbounded layer and retention x, and $N_R(t)$ denoting the number of claims for the reinsurer. The number of claims overshooting a retention x is at least equal to r if and only if the r largest order statistics overshoot the level x. Consequently

$$\mathbb{P}\{X_{N(t)-r+1,N(t)} \leq x\} = \sum_{n=0}^{r-1} \mathbb{P}(N_R(t) = n)$$

and

$$\mathbb{P}\{X_{N(t)-r,N(t)} \leq x < X_{N(t)-r+1,N(t)}\} = \mathbb{P}(N_R(t) = r).$$

(which can already be found in Galambos [367]). Correspondingly, if $N(t)$ comes from a Poisson process or Pascal process, so does $N_R(t)$ and these formulas lead to simple expressions for the c.d.f. of order statistics in terms of the respective sums (see Franckx [358] and Benktander [112] for the Poisson case and Kupper [519] as well as Ciminelli [214] for the Pascal case).

Reinsurance: Actuarial and Statistical Aspects, First Edition.
Hansjörg Albrecher, Jan Beirlant and Jozef L. Teugels.

Let us consider the general case with independent claim number and claim size processes, and use the notation

$$\Pi_r(t) := \mathbb{P}\{N(t) \le r-1\} = \sum_{n=0}^{r-1} p_n(t) \tag{10.1.2}$$

for the probability of having fewer than r claims up to time t. Then from (5.6.37) we can write

$$\mathbb{P}\{X_{N(t)-r+1,N(t)} \le x\} = \sum_{n=0}^{r-1}\sum_{k=n}^{\infty} p_k(t) \binom{k}{n} (1-F_X(x))^n F_X^{k-n}(x).$$

Let us denote $q = 1-p = F_X(x)$. Then by reversing the order of summation we easily find that

$$\mathbb{P}\{X_{N(t)-r+1,N(t)} \le x\} = \Pi_r(t) + \sum_{k=r}^{\infty} p_k(t) \sum_{n=0}^{r-1} \binom{k}{n} p^n q^{k-n}$$

where the first term corresponds to the fact that not even r claims have turned up by time t. For the second term we use the well-known equality

$$\frac{1}{r}\sum_{n=0}^{r-1} \binom{k}{n} p^n q^{k-n} = \binom{k}{r} \int_p^1 v^{r-1}(1-v)^{k-r} dv$$

together with the definition (5.1.1) to obtain

$$\mathbb{P}\{X_{N(t)-r+1,N(t)} \le x\} = \Pi_r(t) + \frac{1}{(r-1)!} \int_{1-F_X(x)}^1 Q_t^{(r)}(1-v) v^{r-1} dv, \tag{10.1.3}$$

where

$$Q_t^{(r)}(z) = \sum_{k=r}^{\infty} \frac{k!}{(k-r)!} p_k(t)\, z^{k-r}.$$

If X is a continuous random variable with density f_X, the corresponding density part (apart from a jump of size $\Pi_r(t)$ at the origin) is

$$f_{X_{N(t)-r+1,N(t)}}(x) = \frac{1}{(r-1)!}[1-F_X(x)]^{r-1} Q_t^{(r)}(F_X(x)) f_X(x).$$

Limit distribution. We now look into the one-dimensional limit distribution of the rth largest order statistics (where r is kept fixed). Assume that there exist deterministic functions $d(t) > 0$ and $c(t)$ such that when $t \uparrow \infty$

$$d^{-1}(t)\{X_{N(t)-r+1,N(t)} - c(t)\} \xrightarrow{D} V_r \tag{10.1.4}$$

where V_r is a non-degenerate random variable. It is natural to assume here that $F_X \in C_\gamma(a)$. In terms of the tail quantile function $U(x) = Q_X\left(1-\frac{1}{x}\right)$ this means that for an extreme value index $\gamma \in \mathbb{R}$ and some auxiliary function $a(x) > 0$,

$$\frac{U(vu)-U(v)}{a(v)} \to h_\gamma(u) := \int_1^u w^{\gamma-1}\, dw,$$

cf. Definition 3.1. Take $x \in \mathbb{R}$ fixed. From (10.1.3),

$$\begin{aligned}&\mathbb{P}\{X_{N(t)-r+1,N(t)} \le c(t) + x\, d(t)\}\\ &\quad = \Pi_r(t) + \frac{1}{(r-1)!}\int_{\psi_t(x)}^t w^{r-1}\left(\frac{1}{t^r}Q_t^{(r)}\left(1-\frac{w}{t}\right)\right) dw\end{aligned}$$

where

$$\psi_t(x) = t\left\{1 - F_X(c(t) + xd(t))\right\}.$$

When $t \uparrow \infty$ both the integrand and the lower limit should tend to a reasonable limit. For the integrand it therefore seems natural to assume that the claim number process is nearly mixed Poisson, as defined in Section 5.3.1. Note, however, that the integrand is actually independent of the quantity t if the claim number process is mixed Poisson.

The remaining convergence condition can then be formulated in the form that when $t \uparrow \infty$

$$\psi_t(x) = t\left(1 - F_X(c(t) + xd(t))\right) \to \psi(x)$$

for a function $\psi(x)$ to be determined. The latter condition can be identified with the extremal condition in terms of U if we choose $\psi_t(0) = 1$, $c(t) = U(t)$ and $d(t) = a(t)$ as in the condition on U. Under the two conditions we see that when $t \to \infty$

$$\mathbb{P}\{X_{N(t)-r+1,N(t)} - U(t) \le x\, a(t)\} \to \frac{1}{(r-1)!}\int_{(1+\gamma x)^{-\gamma^{-1}}}^{\infty} w^{r-1} q_r(w)dw.$$

This result follows since for any fixed r, $\Pi_r(t) \to 0$ with $t \uparrow \infty$ while $q_r(w) = \mathbb{E}(\Lambda^r e^{-\Lambda w})$. Rewriting the extremal laws by an affine transformation, the result can be expressed in the following easier form:

For $X \in C_\gamma(a)$ and $\{N(t); t \ge 0\}$ near mixed Poisson, one has

$$\mathbb{P}\{X_{N(t)-r+1,N(t)} - c(t) \le y\, d(t)\} \to \frac{1}{(r-1)!}\int_{\psi(y)}^{\infty} w^{r-1} q_r(w)dw,$$

where one of the three following cases emerges necessarily:

(i) $\gamma > 0,\ c(t) = 0,\ d(t) = U(t)$ and $\psi(y) = y^{-1/\gamma}, y \ge 0$
(ii) $\gamma = 0,\ c(t) = U(t),\ d(t) = a(t)$ and $\psi(y) = e^{-y}$
(iii) $\gamma < 0,\ c(t) = x_+ := U(\infty),\ d(t) = x_+ - U(t)$ and $\psi(y) = |y|^{1/|\gamma|}, y \le 0$.

Notice that for a deterministic $N(t)$ this result reduces to the classical extreme value laws discussed in Chapter 3. For the case $r = 1$ and $N(t)$ discrete, see Galambos [367]. In this form the result is a special case of a more general weak convergence result in Silvestrov et al. [701] where even the independence condition between claim number and claim size processes is weakened considerably. It seems possible to derive more explicit statements in the case where F_X is assumed to belong to C_γ with remainder conditions. In the mixed Poisson case particularly, no further complications should show up.

Moments. Of course a full treatment of the behavior of the moments under a max-domain of attraction is possible. We restrict our attention to a few direct results. Equation (10.1.3) can also be used to derive the consecutive moments of the order statistics. For $\beta \in \mathbb{N}$ we have

$$\mathbb{E}\{X_{N(t)-r+1,N(t)}\}^{\beta} = \frac{1}{(r-1)!}\int_0^t U^{\beta}\left(\frac{t}{w}\right) w^{r-1}\frac{1}{t^r}Q_t^{(r)}\left(1-\frac{w}{t}\right) dw. \qquad (10.1.5)$$

The easiest way to prove this is by noticing that for a non-negative random variable W and $\beta > 0$,

$$\mathbb{E}(W^{\beta}) = \beta \int_0^{\infty} x^{\beta-1}\,\mathbb{P}(W > x)\,dx$$

and that by (10.1.3)

$$\mathbb{P}\{X_{N(t)-r+1,N(t)} > x\} = \frac{1}{(r-1)!}\int_0^{1-F_X(x)} Q_t^{(r)}(1-v)v^{r-1}dv.$$

The appearance of the quantile function is of course pleasing. It indicates precisely what information on the underlying two processes is needed. For the claim number process we need $\frac{1}{t^r}Q_t^{(r)}(1-\frac{w}{t})$ while the quantity $U^{\beta}\left(\frac{t}{w}\right)$ incorporates the information requested on the claim size distribution.

The asymptotic expression of the moments under the domain of attraction condition is then easy if the claim number process is also near mixed Poisson. Indeed if $F_X \in C_\gamma$ with $\gamma > 0$, then for $\beta \geq 0$ and $\frac{r}{\beta} > \gamma$

$$\mathbb{E}(X_{N(t)-r+1,N(t)})^{\beta} \sim U^{\beta}(t)\frac{\Gamma(r-\gamma\beta)}{(r-1)!}\mathbb{E}(\Lambda^{\gamma\beta}).$$

When $\gamma = 0$ one can sharpen this result while using the full strength of the domain of attraction condition.

Combinations of order statistics. In this section we state a versatile formula that will allow us later to derive almost all desired expressions as special cases. In (10.1.1) we see that the first part of this ordered sample contains the small claims while on the right-hand side of the scale we find the large claims; they are separated by intermediate claims. Let us use the following two abbreviations:

$$\Sigma_s(t) = \sum_{j=1}^{N(t)-s-1} X_{j,N(t)}$$

and

$$\Lambda_s(t) = \sum_{j=N(t)-s+1}^{N(t)} X_{j,N(t)}.$$

Here Σ refers to *small* while Λ refers to *large*. We easily arrive at properties of the claim fragments through the joint Laplace transform

$$\Omega_s(u, v, w; t) = \mathbb{E}(\exp[-u\,\Lambda_s(t) - v\,X_{N(t)-s,N(t)} - w\,\Sigma_s(t)])$$

where $u, v, w \geq 0$. Conditioning on the number of claims at the time epoch t and interpreting $X_{r,N(t)} = 0$ whenever $r \leq 0$, one can derive

$$\Omega_s(u, v, w; t) = \sum_{n=0}^{s} \widehat{F}_X^n(u) p_n(t)$$
$$+ \frac{1}{s!} \int_0^{\infty} e^{-vy} \{ \int_y^{\infty} e^{-uz} dF_X(z) \}^s Q_t^{(s+1)}(\int_0^y e^{-wx} dF_X(x)) dF_X(y). \quad (10.1.6)$$

Many special cases can be derived from this formula by properly choosing s, u, v and w. We give a couple of examples.

- For *one order statistic*. Choose $u = w = 0$ and take $s = r - 1$. Then we fall back on the Laplace transform of the order statistic $X_{N(t)-r+1,N(t)}$ as treated in (10.1.3).
- For *all but the largest claims*. Take now $u = v = 0$ and $s = r - 1$. Then

$$\mathbb{E}(\exp\{-w \sum_{i=1}^{N(t)-r} X_{i,N(t)}\})$$
$$= \Pi_r(t) + \frac{1}{(r-1)!} \int_0^{\infty} [1 - F_X(y)]^{r-1} Q_t^{(r)}(\int_0^y e^{-wx} dF_X(x)) dF_X(y).$$

For the special case where $r = 1$ we find the formula

$$\mathbb{E}(\exp\{-w(S_{N(t)} - X_{N(t),N(t)})\}) = p_0(t) + \int_0^1 Q_t'(\int_0^v e^{-wQ_X(b)} db) dv$$

for the transform of the sum of all but the largest claim. In the case of the Poisson claim number process this result goes back to Ammeter [38].

- For the *smallest claims*. Take $u = 0$ and again $s = r - 1$. Then

$$\mathbb{E}\left(\exp\left\{-vX_{N(t)-r+1,N(t)} - w\sum_{j=1}^{N(t)-r} X_{j,N(t)}\right\}\right)$$
$$= \Pi_r(t) + \frac{1}{(r-1)!}\int_0^\infty e^{-vy}[1-F_X(y)]^{r-1}Q_t^{(r)}\left(\int_0^y e^{-wx}dF_X(x)\right)dF_X(y).$$

If we invert this with respect to v we obtain the hybrid expression containing the distribution of $X_{N(t)-r+1}$, $N(t)$ and the transform of $\Sigma_{r-1}(t)$. From the latter relation one can, for example, derive a further relation that gives information on the ratio between any of the claims and all the smaller ones. The resulting formula might be used to normalize that portion of the portfolio that corresponds to the smaller claims.

10.1.2 Large Claim Reinsurance

We can now use the above formulae to describe some general properties of quantities relevant in large claim reinsurance.

Distributional aspects. Denote the reinsured amount by

$$L_r(t) := \sum_{i=1}^{r} X_{N(t)-i+1,N(t)}$$

on the set $1_{\{N(t)\geq r\}}$. By putting $u = \theta$, $s = r$, and $v = w = 0$ in (10.1.6) we deduce

$$\mathbb{E}(\exp[-\theta L_r(t)1_{\{N(t)\geq r\}}])$$
$$= \Pi_r(t) + \frac{1}{r!}\int_0^1 Q_t^{(r+1)}(1-v)\left(\int_0^v e^{-\theta U(1/y)}dy\right)^r dv.$$

For the case of exponential claim sizes more explicit results can be obtained, as shown by Kremer in [505].

Weak limit. The above formula can be used to obtain the general form for the limit in distribution for the appropriately normalized expression $L_r(t)$ when $t \uparrow \infty$. As is the case of the limit behavior for the extreme order statistics we will assume that $F_X \in C_\gamma(a)$ for $\gamma \in \mathbb{R}$ and that the counting process is mixed Poisson. A little reduction yields

$$r!\,\mathbb{E}\left(\exp\left[-\theta\frac{L_r(t)-c(t)}{d(t)}1_{\{N(t)\geq r\}}\right] - \Pi_r(t)\right)$$
$$= \int_0^t q_{r+1}(w)\left(\int_0^w e^{-\theta\chi_t(z)}\,dz\right)^r dw$$

where

$$\chi_t(z) := \frac{1}{d(t)}\left[U\left(\frac{t}{z}\right) - \frac{c(t)}{r}\right].$$

It is now quite obvious what we should do. We choose $c(t) = rU(t)$ and $d(t) = a(t)$; then $\chi_t(z) \to h_\gamma(1/z)$. Therefore the right-hand side tends to the limit

$$\int_0^\infty q_{r+1}(w)\left(\int_0^w e^{-\theta h_\gamma(1/z)}dz\right)^r dw.$$

Let us specialize a bit.

(i) If $\gamma > 0$, then one can replace $a(t)/U(t)$ by its limit γ, yielding the somewhat simpler result

$$\exp\left[-\theta\frac{L_r(t)}{U(t)}\right] \to \frac{1}{r!}\int_0^\infty q_{r+1}(w)\left(\int_0^w e^{-\theta z^{-\gamma}}dz\right)^r dw.$$

For $r = 1$ the right-hand side can also easily be written as a Laplace transform. The resulting expression for the limit in distribution coincides with that of the maximum from the previous section.

(ii) For $\gamma = 0$, the inner integral reduces to

$$\int_0^w z^\theta\, dz = \frac{w^{\theta+1}}{\theta+1}.$$

Using the structure variable we find for $t \uparrow \infty$

$$\mathbb{E}\left\{\exp\left[-\theta\frac{L_r(t) - rU(t)}{a(t)}1_{\{N(t)\geq r\}}\right]\right\} \to \frac{\Gamma(r(\theta+1)+1)}{\Gamma(r+1)}\frac{\mathbb{E}(\Lambda^{-r\theta})}{(1+\theta)^r}.$$

The mean. From the above Laplace transform we can immediately deduce the first few moments. We restrict attention to the mean. An easy deduction yields

$$\mathbb{E}(L_r(t)) = \frac{1}{(r-1)!}\int_0^1 Q_t^{(r+1)}(1-v)v^{r-1}\int_0^v U\left(\frac{1}{z}\right)\, dzdv. \tag{10.1.7}$$

We link the above expression with our knowledge about the classes C_γ used in the treatment of one order statistic. First make the change of variable $v = w/t$. Then replace $z = \frac{w}{t}y$. One easily finds that

$$\frac{\mathbb{E}(L_r(t))}{U(t)} = \frac{1}{(r-1)!}\int_0^t \frac{w^r}{t^{r+1}}Q_t^{(r+1)}\left(1-\frac{w}{t}\right)\int_0^1 \frac{U\left(\frac{t}{wy}\right)}{U(t)}\, dydw.$$

Taking limits for $t \uparrow \infty$ shows that we need to assume that $\gamma < 1$. It then follows that

$$\frac{\mathbb{E}(L_r(t))}{U(t)} \to \frac{1}{(r-1)!(1-\gamma)} \int_0^\infty w^{r-\gamma} q_{r+1}(w)\, dw.$$

Further comments.

- As can be expected, the calculation of premiums quickly runs into mathematically intractable formulas. This has been recognized by Benktander [112], who deals with a relation between XL and the largest claim situations. For the calculation of the pure premium, see Berglund [118] and references therein.
- Kupper [520] compares the pure premium for the XL cover at retention M with that of the largest claims cover at retention r. He specifically deals with the case where the claim size distribution is strict Pareto. For example, the effect of truncation on the largest claims depends strongly on the index of this claim, as shown in [519].
- Berliner [119] considers a set of interesting problems connected with the largest claims covers. He assumes the claim number process to be Poisson and derives the joint distribution of two large claims $X_{N(t)-r+1,N(t)}$ and $X_{N(t)-s+1,N(t)}$, and computes their covariance for the case of a strict Pareto law, as well as $\text{Cov}(L_r(t), S(t))$.
- Kremer [501] gives crude upper bounds for the pure premium under a Pareto claim distribution. The asymptotic efficiency of the largest claims reinsurance treaty is discussed in Kremer [502].
- Some practical aspects of large claim distributions have been treated by Schnieper [682]. Albrecher et al. [26] studied the joint distribution of larger and smaller claims for regularly varying claim distributions (see also Ladoucette and Teugels [525] for an overview).

10.1.3 ECOMOR

In some sense the ECOMOR treaty rephrases the largest claims treaty by giving it an additional SL character. However, one can also consider ECOMOR as an XL treaty with a random retention that follows the oscillations showing up, for example, by inflation.

Distributional aspects. As defined in Chapter 2, the reinsured amount in an ECOMOR treaty is

$$E_r(t) = \sum_{i=1}^{r} X_{N(t)-i+1,N(t)} - rX_{N(t)-r,N(t)}.$$

Again the expression of the reinsured amount equals 0 if $N(t) \le r$. The special choice $s = r$, $w = 0$ and $v = -ru$ in (10.1.6) gives us the following expression for the Laplace transform of the reinsured amount:

$$\mathbb{E}\{\exp[-\theta E_r(t)]\} = \Pi_{r+1}(t) + \frac{1}{r!} \int_0^1 Q_t^{(r+1)}(1-v) \left(\int_0^v \exp\{-\theta(U(1/w) - U(1/v))\} dw \right)^r dv.$$

Weak limit. We use a procedure similar to the one for the largest claims reinsurance. So we start from the expression above where we normalize by the auxiliary function $a(t)$ from the max-domain of attraction. We have

$$r!\mathbb{E}\left\{\exp\left[-\theta\frac{E_r(t)}{a(t)}\right]-\Pi_{r+1}(t)\right\}$$
$$=\int_0^t \frac{1}{t^{r+1}}Q_t^{(r+1)}(1-\frac{w}{t})\left(\int_0^w e^{-\theta\phi_t(z,w)}\,dz\right)^r\,dw,$$

where

$$\phi_t(z,w):=\frac{U(t/z)\;-\;U(t/w)}{a(t)}.$$

Note that the very definition of $E_r(t)$ makes further centering unnecessary. We again assume that $F_X\in C_\gamma$ and that the counting process is mixed Poisson. It then easily follows that

$$\mathbb{E}\left\{\exp\left[-\theta\frac{E_r(t)}{a(t)}\right]\right\}\to\frac{1}{r!}\int_0^\infty w^r\,q_{r+1}(w)\left(\int_0^1 e^{\left\{-\theta\int_{1/w}^{1/wb}z^{\gamma-1}dz\right\}}db\right)^r\,dw.$$

In general the resulting limit distribution seems very hard to recover. However, in the integrand we find for a fixed w a power of a Laplace transform

$$\left(\int_0^1 e^{\left\{-\theta\int_{1/w}^{1/wb}z^{\gamma-1}dz\right\}}db\right)=:\int_0^\infty e^{-\theta v}\,dG_w(v)\,.$$

For $0<\gamma\le 1$ it is not difficult to show that then

$$G_w(v)=1-(1+\gamma v w^\gamma)^{-1/\gamma}\,,$$

which looks very much like a generalized Pareto distribution.

Still two values of γ seem to give something special.

(i) When $\gamma=0$ then we get a simple expression in that then

$$\mathbb{E}\left\{\exp\left[-\theta\frac{E_r(t)}{a(t)}\right]\right\}\to\left(\frac{1}{\theta+1}\right)^r,$$

which can be directly interpreted as the Laplace transform of a gamma distribution.

(ii) When $\gamma=-1$ (as for the uniform distribution on $[0,1]$) a simple calculation yields

$$\mathbb{E}\left\{\exp\left[-\theta\frac{E_r(t)}{a(t)}\right]\right\}\to\mathbb{E}\left\{\prod_{j=0}^r\frac{\Lambda}{\Lambda+j\theta}\right\}.$$

For degenerate Λ the limit distribution is a product of independent exponentials.

The mean. From the above relation we derive the expression for the first moment.

$$\mathbb{E}(E_r(t)) = \frac{1}{(r-1)!}\int_0^1 Q_t^{(r+1)}(1-v)v^{r-1}\int_0^v (U(1/y)-U(1/v))^r\,dy\,dv.$$

We again indicate what happens when the C_γ-classes are in force. With the notation introduced for one order statistic, we see that we again need $\gamma < 1$. Then, as before,

$$\frac{\mathbb{E}(E_r(t))}{a(t)} \to \frac{1}{(r-1)!(1-\gamma)}\int_0^\infty q_{r+1}(w)w^{r-\gamma}dw = \frac{\Gamma(r-\gamma+1)}{(r-1)!(1-\gamma)}\mathbb{E}(\Lambda^\gamma),$$

illustrating the role played by the structure variable Λ.

Further comments. As for the ECOMOR treaty, a few more explicit results can be found in the actuarial literature.

- Ammeter [39] points out how the exclusion of one or more of the largest claims has the result of reducing the expected amount of the remaining aggregate claim amount. In some cases even an infinite expectation becomes finite after such a reduction. He also points out how in a portfolio with Pareto distributed claim sizes there is a preponderance of small claims. See Albrecher et al. [26] for generalizations.
- For a general study of ECOMOR treaties, see Ladoucette and Teugels [523]. Crude bounds for the pure premium go back to Kremer [501]. For a study of LCR and ECOMOR treaties for more general claim number processes, see also Asimit and Jones [55].
- Here is another possibility for a large claims reinsurance treaty that imitates an ECOMOR treaty but that has a different kind of retention. Accumulating information on the largest claims, one can first get an estimate for the number of claims $\mathcal{K}_t(a)$ falling within a fixed distance a from the observed record claim within the time slot $[0,t]$. The average over this number of claims yields an alternative large claim reinsurance treaty. Information on the quantity $\mathcal{K}_t(a)$ can be found in Li et al. [540] and Hashorva [423]. For studies on the asymptotic behavior of the tail probability of $E_r(t)$ for exponentially bounded claim sizes, see Jiang and Tang [465] and Hashorva and Li [426]. Peng [610] studied the joint tail of the reinsured amounts in ECOMOR and large claim treaties.
- For a bivariate version of ECOMOR, see Hashorva [424, 425].

10.2 Alternative Risk Transfer

There are several alternatives to traditional reinsurance available in the market, summarized under the term *alternative risk transfer* (ART). They increase the efficiency of the marketplace and can also be particularly helpful at times when traditional reinsurance capacity is limited (e.g., after major natural catastrophes such as Hurricane Andrew in 1992 or Hurricane Katrina in 2005). Some ART solutions take a more integrated approach to reduce the risk over time (rather than treating the various types of risks separately), providing additional diversification over time in connection with underwriting cycles, counterparty risk, and coverage of large catastrophes. Others provide alternatives in the nature of the relationship between insurer and reinsurer by creating a secondary market where one can enter or leave coverage in a more flexible way than in classical reinsurance treaties. In terms of volume, ART nowadays constitutes about 15% of the total reinsurance business, measured in terms of dedicated capital (Source: JLT Re).

In this section we briefly discuss some of these alternative possibilities, and for more details see the references at the end of the section. In general, the risk transfer can be via alternative *risk carriers* and alternative *products*.

Alternative carriers. Next to *self-insurance* (which can be both regulated and non-regulated and is particularly popular in the USA), *reinsurance pools* act like a mutual, where each insurance company can cede its risk and its premiums of a specific risk class, for instance for very large risks. *Risk-retention groups* are based on a similar concept of pooling for companies to get access to certain types of liability insurance. As another example, *captives* are popular vehicles to lower insurance premiums as well as transaction costs. Typically located in a tax-friendly environment, captives are insurance or reinsurance companies which insure the risks of their parent company in a cost-effective manner. In that way, access to the global reinsurance market can be obtained, which due to the larger diversification possibilities of reinsurers may lower premiums through reduced capital cost. In addition, this allows a certain degree of time diversification (captives are usually allowed to hold equalization reserves, whereas according to the current accounting standards insurance companies are not). For smaller captives, it is also popular to operate as a *reinsurance captive*, which insures the risk of its parent as a reinsurer of a first-line insurance company, to which the risk was first transferred. A particular advantage of such a construction is that from a regulatory perspective it will then be treated as a reinsurer, for which different rules apply.

Particularly in times when traditional reinsurance premiums are very high or coverage is not available at all (which happened after Hurricane Katrina), an alternative are so-called *reinsurance side-cars*, where investors deposit funds and in turn participate in the premiums and claims of the insurance company (typically in the form of a QS-type treaty for a line of business, with a retention for the company that ensures alignment of interests of the two parties). The deposit equals the reinsurance contract limit (so it is a fully *collateralized* reinsurance form where counterparty risk is eliminated), so the investors' liability is limited to the amount of the deposit.

Sidecars are only one way for capital market investors to gain exposure to insurance risk. In general, the *capital market* is a major alternative risk carrier, particularly through insurance-linked securities (see below).

Finite risk reinsurance. If in addition to insurance risk transfer there is also a significant weight on other goals in a product, one speaks of *structured reinsurance*. A particular example is *finite risk reinsurance*, which is a combination of risk transfer and risk financing between an insurer and a reinsurer in a form that is tailored towards the concrete needs of the insurer. While a substantial goal of the transfer is to enhance the insurer's financial results, for tax reasons it has to contain a (limited) amount of insurance risk transferred to the reinsurer in order to be classified as reinsurance. Such contracts have a duration of several years and combine loss experience and investment returns. They formalize a longer-term relationship between the two parties which has a time diversification component, as the reinsurer can count on incoming premiums and the insurer on agreed coverage to known conditions over a longer time horizon.

There are *retrospective* and *prospective* variants. An example of the former is a *loss portfolio transfer*, where the insurer transfers outstanding claims from some long-tailed business of previous years to the reinsurer, and in turn pays a premium consisting of the net present value of these claims plus fees. In that way he passes on risks related to the timing and amount of loss development. In *adverse development covers* the IBNR losses are also included. Here the claims reserves are not transferred to the reinsurer, but the reinsurer only covers losses that exceed the reserves which the insurer has already built up, and for this transfer a premium is paid (this can be set up like an XL or SL contract on the adverse loss development and is also very convenient in cases of mergers or take-overs of a company).

Prospective variants of finite risk reinsurance include *spread loss treaties*, where for the transfer of specified losses (with annual and overall limits) the insurer pays premiums to the reinsurer onto an "experience account" and these premiums (minus fees and expenses) are then invested. At the end of the contract period, the balance is settled with the insurer, exposing the reinsurer to the counterparty risk that the insurer may not be able to pay a potential negative balance. Finally, *finite quota share* arrangements can include over- or undercompensation of claims over prespecified periods of time.

Integrated products. In *multi-year/multi-line products* several business lines are bundled together and/or over a longer time horizon, which leads to a smoothing of the aggregate risk and hence to lower premiums. A disadvantage is counterparty risk for the insurer, and it can also be non-trivial to cooperate across business lines within the insurance company.

Multi-trigger products. Here the reinsurer pays losses only contingent on a second event, which typically is correlated to the insurer's financial result. For instance, the reinsurer will only pay if the losses exceed a certain threshold and at the same time a stock index, commodity price, exchange rate etc. is below a prespecified level. Such an arrangement will lead to considerably lower premiums, but still may serve the overall financial result of the insurer well. At the same time, the reinsurer can also benefit from this variant, as resulting capital needs will decrease, particularly if there are several such contracts with independent triggers in the portfolio.

Contingent capital. This refers to the option for the insurance company to raise debt or equity capital for prespecified conditions, in case there is a severe aggregate insurance loss experience or another prespecified event occurs. It can be seen as a put option on the own shares with a predefined strike value. By setting up the conditions before financial distress, fresh capital can in such a case be acquired in a much cheaper way than on the market. In fact, this is a means of financing rather than a transfer of insurance risk. As a variant of this idea, in recent years *contingent convertible bonds* ("CoCo" bonds) have become increasingly popular (particularly since in many countries this is now considered as regulatory-efficient capital), and recently products have been issued that as a trigger combine the occurrence of natural catastrophes and the solvency ratio of the company.

Industry loss warranties (ILW). These contracts resemble reinsurance contracts, but the loss under consideration is the loss of the entire insurance industry arising from an event (measured through some index[1]) rather than the individual loss experience. The market for such contracts has considerably grown over the last years, with reinsurance companies and hedge funds being typical protection providers. A disadvantage for the insurer is that the reinsured amount is not based on their own loss experience, even if the industry index will usually be reasonably correlated. However, the resulting discrepancy (referred to as *basis risk*) may lead to inefficiencies.

Insurance-linked securities (ILS). In broad terms, these are financial instruments whose values are driven by insurance loss events. They enable insurance risk to be (directly) placed on the capital market (*insurance securitization*). Among the most important examples are *catastrophe bonds* (*CAT bonds*), which are bonds with the additional feature that the investor will not receive the coupon (or even not the principal) if a certain trigger related to the occurrence of natural catastrophes is hit. In turn, the coupon in the absence of that trigger is substantially higher. Such bonds are traded over the counter and can increase the insurance capacity for risks for which it is difficult to find traditional reinsurance. The initiator of the CAT bond is the insurance company that seeks this protection. The trigger can be the individual loss experience of the issuer due to a catastrophe or (more often) an index representing the average catastrophe loss experienced by the insurance sector in a prespecified time interval and business line. Finally, it is very common nowadays to have *parametric triggers*, that is, a physical measurement (such as wind speed, magnitude of an earthquake etc.), as this is a reliable, "objective", and usually easily accessible trigger for investors (whereas for a trigger linked to the individual loss experience of the issuer there are obvious moral hazard issues and the final settlement can take much longer). One then speaks of an *index transaction* (instead of an *indemnity transaction*). However, triggers based on an index or parametric triggers again introduce basis risk for the issuer (on the plus side, the insurer does not have to pass on the actual claim data to the outside in this case).

In practice, there is usually a special-purpose vehicle (SPV) that acts as an intermediary (located in a tax-friendly environment), which then issues a conventional reinsurance policy to the insurance company. The SPV uses the premium payments of the insurance company to pay the coupons to the investor, and if the event is triggered (or maturity of the bond is reached without the trigger), the amount in the SPV is paid out to the insurer and the investors according to the bond specifications. One particular advantage of the CAT bond is that – in contrast to traditional reinsurance – there is no counterparty risk for the insurer (unless the construction involves further parties like swap providers), since the reinsured amount is already available in the SPV (or the trust account). At the same time, for investors this can be an attractive product, since the underlying trigger event will typically be independent of other investments and the excess coupon can be considerable (in the long run, it will of course be determined by demand and supply, as well as the concrete loss experience of previous years). Conceptually, CAT bonds represent a secondary reinsurance market, where the investor (who takes the reinsurer's role here) has more flexibility to leave or enter the "contract"

---

1 The most common indices are the Property Claims Service (PCS) index for US perils and the PERILS index for European ones (cf. Lane [528] for details).

along the way. In recent years the CAT bond market has increased considerably in size. Further ILS products (with, however, a much smaller market) are, for example, longevity swaps and products related to embedded-value securitization and extreme mortality securitization. Altogether, the global ILS risk capital outstanding in 2016 exceeded 25 billion US$ (Source: Artemis).

10.2.1 Notes and Bibliography

The first implementation of a contract that resembles a CAT bond can be traced back to about 3000 BC (and hence long before the first insurance policy was issued!), when the Babylonians issued maritime loans in which the borrower did not have to repay the loan in case of a loss due to certain accidents (see Holland [451]). CAT bonds and other insurance derivatives in their modern forms started to be traded in the USA after Hurricane Andrew in 1992. In Europe, the concept became popular when the WinCAT coupon (a convertible bond with a trigger related to hail or storm) was issued by Winterthur Insurance in 1997. The model risk related to that product was studied by Schmock [680]. For early descriptions of the developments and discussions, refer to Doherty [296, 300], Gorvett [403], Swiss Re [723], Munich Re [584], and Doherty and Richter [298].

Excellent general surveys on alternative risk transfer are Lane [528], Culp [237], Liebwein [543], and the handbook edited by Barrieu and Albertini [84]. See also Gastel [375, Ch.8], Mürmann [586], and Banks [78]. For a rich source on convertible bonds, refer to De Spiegeleer and Schoutens [260]. Niedrig and Gründl [591] studied the effects of CoCo bonds on the solvency capital situation of insurance companies. Gibson et al. [389] investigated the choice between reinsurance and securitization of natural catastrophes from the viewpoint of information flow.

The ILS market did not grow as fast as anticipated in the beginning. For an attempt to explain this from a behavioral economics perspective, see Bantwal and Kunreuther [80]. Barrieu and Loubergé [86] suggested possible modifications of the product structure. For general reflections about the tradeoff between traditional reinsurance and securitization see Doherty and Schlesinger [302] and Cummins and Trainar [239]. The valuation of ILS products naturally is an interesting academic topic, since actuarial and financial pricing techniques have to be merged in a meaningful way. An early paper in this direction was Embrechts and Meister [333]. Time change techniques in catastrophe option pricing can be found in Geman et al. [378]. Later discussions include Cox et al. [230], Cox and Pedersen [231, 232], Embrechts [324], Dassios and Jang [251], Jaimungal and Wang [461], Lee and Yu [532], and Mürmann [582, 583] as well as Haslip and Kaishev [427] and Gatzert et al. [376]. For adaptations to models based on Cox processes, see Lin et al. [544]. A game-theoretic approach is given in Subramanian and Wang [714]. An empirical analysis of pricing practice of CAT bonds over a long time horizon can be found in Braun [161]. For computational issues in relation to the pricing of CAT bonds, see, for example, Vaugirard [753], and for QMC simulation refer to Albrecher et al. [21, 22]. Dieckmann [294] discusses a consumption-based equilibrium model for pricing CAT bonds. Bäuerle [90] studied the stochastic control problem to dynamically mix reinsurance and CAT bonds under basis risk. Securitization is also becoming an important instrument for risks in life insurance. For a recent study on the design and pricing of an inverse survivor bond for annuity securitization, see Lorson and Wagner [547].

An early discussion on finite risk reinsurance can be found in Hess [437], see also Von Dahlen [762]. Time diversification is an important element in these constructions, particularly since equalization reserves are nowadays typically not exempted from tax). Dacorogna et al. [244] studied the quantitative effect of time diversification for catastrophe risk from a shareholder perspective. For consideration of time diversification in terms of self-insurance on the individual level, see Gollier [397].

10.3 Reinsurance and Finance

As the title suggests, the focus of this book is on *actuarial* and *statistical* aspects of reinsurance. We want to emphasize that the concrete implementation of reinsurance in practice also has a considerable *financial* function for the ceding company. While some financial aspects of the risk transfer have entered the discussion at various places in the book, it is beyond its scope to provide a representative treatment of this element. Instead, we give some references and remarks here.

An excellent book for this topic is Liebwein [543]. Wilson [792] is a rich recent source concerning the formalization of the concepts of value and capital of financial institutions, and also contains sections that are specific to (re)insurance. Klaasen and Van Eeghen [489] give a detailed account of the concept of economic capital. The role of capital and capital management for insurers and reinsurers in view of risk-based regulation is discussed in Dacorogna [243]. For an introduction to *dynamic financial analysis* and its connections with reinsurance, see Kaufmann et al. [483], De Waegenaere et al. [266], and Eling and Parnitzke [321].

An early discussion of the influence of reinsurance on the stock value of an insurance company is given by Doherty and Tinic [303]. Zanjani [803] studied the effects of capital costs on catastrophe insurance markets, see also Harrington et al. [421] for a discussion of the extent to which insurance derivatives can reduce the need for equity capital. Froot [361] suggests a framework for capital structure decisions of (re)insurers. For a discussion of the valuation effects of reinsurance purchases in terms of firm leverage see Garven and Lamm-Tennant [374] and Garven [372]. Blazenko [142] provides an early study of reinsurance from an economic perspective. For an equilibrium model in reinsurance and capital markets in which professional reinsurers arise endogenously, see Plantin [624]. Upreti and Adams [748] investigated reinsurance as a strategic function in insurance markets through its impact on product-market outcomes, whereas Garven et al. [373] provide empirical evidence how the lengths of the relationship between insurer and reinsurer have positive effects on the insurer's profitability and credit quality. Altuntas et al. [34] investigated the capital structure of insurance companies and concluded that on the global level it is quite heterogeneous, as it also entails heterogeneous reinsurance demands. Rymaszewski et al. [662] discuss the benefits of pooling risks in the context of insurance guarantee funds, which can be interpreted as obligatory reinsurance (see also Schmeiser et al. [672]).

The impact of foreign exchange risks on reinsurance decisions is studied in Blum et al. [143] and Jacque et al. [460].

Financial pricing of (re)insurance contracts (i.e., applying financial asset pricing theory, empirical asset pricing, and mathematical finance tools to price insurance products) has become quite a prominent topic in the last two decades. Chang et al. [198] determined equilibrium reinsurance premiums within an option framework in terms of the underwritten risks by the ceding company and the first-line premiums. Incompleteness of the reinsurance market is the starting point of Kroll et al. [509]. A discussion on its effect on insurance pricing is given in Castagnolli et al. [186]. Financial pricing by line of business is discussed in Phillips and Cummins [616], as well as Gründl and Schmeiser [412]. It is quite natural to look for arbitrage-free pricing also in the context of reinsurance contracts. In loose terms, the Fundamental Theorem of Asset Pricing asserts that in the absence of arbitrage possibilities, the pricing functional ψ must be a positive and linear functional defined in the Hilbert space L^2. Then, by virtue of the Riesz representation theorem, one can express ψ as an expectation with respect to a modified (distorted) random variable (for some heuristic explanations in this application context, see Sherris [697]), that is, one looks for a corresponding risk-neutral probability measure. The reinsurance market is incomplete (illiquid), and so there is no unique choice for such an adjustment of the physical probability measure. There are, however, various justifications for certain choices of a risk-neutral probability measure, including the minimal martingale measure and the minimal entropy martingale measure, for example see Møller [580] and Jang and Krvavych [462], as well as Kreps [507], Sondermann [706], and Schweizer [685] for earlier work. Venter [754] and Venter et al. [756] provide further explanations from a practical perspective. For a combined model where trading can occur on financial as well as on reinsurance markets, see De Waegenaere et al. [265]. The pricing of CAT bonds also falls into this category, so we also refer to the references in the previous section. A unifying recent overview on the topic is Bauer et al. [89], see also Zweifel and Eisen [812]. For a survey on hybrid (re)insurance/financial instruments and their pricing, see Cummins and Weiss [242].

10.4 Catastrophic Risk

We have dealt with catastrophic risk already in earlier sections (particularly in the context of CAT bonds in Section 10.2), but due to its importance for reinsurance companies, we would like to finish with some more comments and references on the topic. There is first the difficult task of describing the concept itself in a quantitative way. It is not straightforward to agree on a concrete definition of a catastrophic claim; an early attempt can be found in Ajne et al. [10]. Clearly, a catastrophic claim falls into the category of large claims, where from the statistical side one may see an additional challenge in the fact that usually there are very few data points for a systematic study available, and the available ones are often only rough estimates of the true value, even long after the catastrophe has occurred. At the same time, it will often be difficult to make the data points comparable. For some general observations about links between catastrophes and insurability, see Schnieper [681], Zeckhauser [805], Gollier [396], Smith [703], and Punter [637]. Paudel et al. [608] is a comparative study on implemented public and private insurance systems for natural catastrophes.

A general survey on modelling and managing catastrophic risk is Banks [79], but see also Kozlowski et al. [499], Meyers [574], and more recently Woo [794] as well as Krvavych [511]. Pricing of financial products solely from knowing the aggregate amount of catastrophic claims is covered in Christensen and Schmidli [213]. For surveys see Epstein [339], Anderson and Dong [42], Aase [3], and d'Arcy et al. [250]. In

O'Brien [593] hedging strategies are introduced to deal with catastrophe insurance options. The appearance of such options on the financial market has given rise to interesting discussions. Securitization of catastrophe risks by capital, CAT options, and reinsurance has been dealt with by Krieter et al. [508], Albrecht et al. [33], Pentikäinen [612], Balford et al. [83], and Meyers et al. [575]. For a general approach, see Jones and Casti [469]. Forecasting using extreme value methods has been illustrated in Coles and Pericchi [222], and see also Lescourret and Robert [539].

Insurability of natural catastrophes can only be achieved in a sustainable way if there is an equilibrium between losses and premium income, over both time and in space. However, the geographical distribution of the claims is often difficult to assess. Insurance companies therefore typically use a *bottom-up* approach in which they use scientific/expert knowledge in connection with the time and size of a natural catastrophe (e.g., with high-resolution physical models for weather parameters), and calibrated in terms of risk exposure (this can involve very detailed information from engineering on building structures, see Heneka and Ruck [436] for an illustration). Missing data are often estimated by expert knowledge, and parameters in the model are sometimes hard to determine in the presence of sparse and inaccurate loss data. Nevertheless, in recent decades scientists have built up an impressive toolkit to quantify respective risks, and nowadays there exist several commercial firms who professionally assess the risk for certain natural catastrophes in specific regions and who offer their services to the (re)insurance industry. When studying the patterns of natural catastrophes and building models, one also needs to carefully consider and incorporate systematic changes in risks due to climate change (see Botzen [155] for a general discussion). When possible covariates for a partial explanation of systematic changes can be identified, extreme value techniques with covariates, as discussed in Chapter 4, can be very helpful in the analysis. Here there is also still a lot of potential for future research.

In view of the above, techniques from credibility theory can be appropriate, but from the point of view of heavy-tailed distributions. Also, alternative proposals to the credibility paradigm have to be developed. Moreover, many data from the realm of catastrophic risk are censored, so that techniques as discussed in Chapter 4 can be helpful. For some further statistical issues, including the influence of inflation, see Cozzolino et al. [233]. For USA-based natural disasters and their impact on reinsurance, see Patrat et al. [607]. Maccaferri et al. [554] is a survey of the relevance of the various natural disaster risks for European countries and the development of the respective insurance markets, particularly focusing on flood, storm, earthquake, and drought. A long-term empirical study of adaptive premium strategies for catastrophe insurance can be found in Born and Viscusi [153]. Niehaus [592] summarized research contributions on the question to what extent the allocation of catastrophe risk is consistent with notions of optimal risk sharing, and how respective efficiency could be increased.

We mention here some examples of models or collections of claim data for specific types of natural catastrophes.

- Catastrophic wind losses and connected XL covers have been considered in Sanders [665]. For Japan, see Mayuzumi [565] and for Europe refer to Matulla et al. [564]. For concrete models of country-wide storm losses see, for example, Dorland et al. [305], Donat et al. [304], Klawa and Ulbrich [490], and Prettenthaler et al. [633].

- Hurricanes are treated in Burger et al. [173], Watson and Johnson [776], Cole et al. [220], and Pita et al. [620].
- Hail insurance is dealt with by Benktander [113] and Brown et al. [164], and for a recent hail model, see Mohr et al. [579].
- Flood losses have been studied in Merz et al. [572, 573], Jongman et al. [470], and Prettenthaler et al. [631, 632]. For a study of the implications of climate change on flood risks in Europe, see Feyen et al. [348].
- For earthquakes, see Ryder et al. [661], Wakuri et al. [764], Bertogg et al. [130], Crowley et al. [235], Asprone et al. [67], and Chen et al. [202].
- Business interruptions are dealt with in Zajdenweber [802], and see also Rose and Lim [656] and Rose and Huyck [655].

Man-made and other types of catastrophes are equally challenging to deal with. We list a few examples below.

- For risk assessment of terrorism, see Monahan [581]. For the role of insurance in covering such risks see, for instance, Kunreuther [515], Ericson and Doyle [340], Thomas [743], and Swiss Re [725].
- A recent overview of the field of cyber risk is given in Eling and Wirfs [322]. For an empirical analysis of the insurability of cyber risk see Biener et al. [132]. While the size of some of such claims can be moderate, there is a considerable potential for catastrophic cyber losses, for example see Coburn et al. [219].
- Actuarial risks related to pandemic diseases also have potential to be disastrous, for examples see Swiss Re [724] and Van Broekhoven et al. [750]. For integration of pandemic risk into an internal model, see Planchet [623], and a general actuarial modelling approach to this topic can be found in Feng and Garrido [346].

Some reinsurance companies offer interesting illustrative material to catastrophes, see, for instance, the web-sites http://www.swissre.com and http://www.munichre.com, where information on recent catastrophes and their actuarial consequences is regularly updated. For recent general reflections on the development of reinsurance and its role in dealing with catastrophes, see Haueter and Jones [429].

References

1 Aas, K., Czado, C., Frigessi, A., and Bakken, H. (2009) Pair-copula constructions of multiple dependence. *Insurance Math. Econom.*, **44** (2), 182–198.

2 Aase, K. (1992) Dynamic equilibrium and the structure premium in a reinsurance market. *Geneva Papers on Risk and Insurance Theory*, **17**, 93–136.

3 Aase, K. (1999) An equilibrium model of catastrophe insurance futures and spreads. *Geneva Papers on Risk and Insurance Theory*, **24**, 69–96.

4 Aase, K.K. (2002) Perspectives of risk sharing. *Scand. Actuar. J.*, pp. 73–128.

5 Aase, K.K. (2009) The Nash bargaining solution vs. equilibrium in a reinsurance syndicate. *Scand. Actuar. J.*, pp. 219–238.

6 Aban, I.B., Meerschaert, M.M., and Panorska, A.K. (2006) Parameter estimation for the truncated Pareto distribution. *J. Am. Stat. Assoc.*, **101** (473), 270–277.

7 Acciaio, B. (2007) Optimal risk sharing with non-monotone monetary functionals. *Finance and Stochastics*, **11** (2), 267–289.

8 Acerbi, C. and Tasche, D. (2002) On the coherence of expected shortfall. *J. Banking & Finance*, **26** (7), 1487–1503.

9 Aebi, M., Embrechts, P., and Mikosch, T. (1992) A large claim index. *Mitt. Ver. Schweiz. Versich. Math.*, pp. 143–156.

10 Ajne, B. and Wide, H. (1987) On the definition of catastrophe claims and the calculation of their expected cost for the purpose of long range planning and profit centre control. *Astin Bull.*, **17**, 171–178.

11 Akritas, M. and Van Keilegom, I. (2003) Estimation of bivariate and marginal distributions with censored data. *J. R. Stat. Soc. Ser. B Stat. Methodol.*, **65** (2), 457–471.

12 Albers, W. (1999) Stop-loss premiums under dependence. *Insurance Math. Econom.*, **29**, 173–185.

13 Albrecher, H. (2010) Reinsurance, in *Encyclopedia of Quantitative Finance*, Wiley, Chichester, pp. 1539–1543.

14 Albrecher, H. and Asmussen, S. (2006) Ruin probabilities and aggregate claims distributions for shot-noise Cox processes. *Scand. Actuar. J.*, pp. 86–110.

15 Albrecher, H., Asmussen, S., and Kortschak, D. (2006) Tail asymptotics for the sum of two heavy-tailed dependent risks. *Extremes*, **9**, 107–130.

16 Albrecher, H. and Boxma, O. (2005) On the discounted penalty function in a Markov-dependent risk model. *Insurance Math. Econom.*, **37** (3), 650–672.

17 Albrecher, H. and Cani, A. (2017) *On randomized reinsurance contracts*, Preprint, University of Lausanne.

Reinsurance: Actuarial and Statistical Aspects, First Edition.
Hansjörg Albrecher, Jan Beirlant and Jozef L. Teugels.

18 Albrecher, H., Constantinescu, C., and Loisel, S. (2011) Explicit ruin formulas for models with dependence among risks. *Insurance Math. Econom.*, **48** (2), 265–270.

19 Albrecher, H., Eisele, K., Steffensen, M., and Wuethrich, M. (2016) *A Framework for Cost-of-Capital Rate Analysis in Insurance*, Preprint, University of Lausanne.

20 Albrecher, H. and Haas, S. (2016) *The Joint Perspective of Cedent and Reinsurer on the Optimality of Reinsurance Contracts*, Preprint, University of Lausanne.

21 Albrecher, H., Hartinger, J., and Tichy, R. (2003) Multivariate approximation methods for the pricing of catastrophe-linked bonds. *Internat. Ser. Numer. Math.*, **145**, 21–39.

22 Albrecher, H., Hartinger, J., and Tichy, R. (2004) Quasi-Monte Carlo techniques for CAT bond pricing. *Monte Carlo Methods & Appl.*, **10**, 197–212.

23 Albrecher, H., Hipp, C., and Kortschak, D. (2010) Higher-order expansions for compound distributions and ruin probabilities with subexponential claims. *Scand. Actuar. J.*, pp. 105–135.

24 Albrecher, H. and Kainhofer, R. (2002) Risk theory with a non-linear dividend barrier. *Computing*, **68**, 289–311.

25 Albrecher, H., Ladoucette, S.A., and Teugels, J.L. (2010) Asymptotics of the sample coefficient of variation and the sample dispersion. *J. Statist. Plann. Inference*, **140** (2), 358–368.

26 Albrecher, H., Robert, C., and Teugels, J.L. (2014) Joint asymptotic distributions of smallest and largest insurance claims. *Risks*, **2**, 289–314.

27 Albrecher, H. and Teugels, J. (2006) Asymptotic analysis of a measure of variation. *Theory of Probability and Mathematical Statistics*, **74**, 1–10.

28 Albrecher, H. and Teugels, J. (2006) Exponential behavior in the presence of dependence in risk theory. *J. Appl. Probability*, **43** (1), 257–273.

29 Albrecher, H. and Teugels, J.L. (2008) On excess-of-loss reinsurance. *Teor. Ǐmovīr. Mat. Stat.*, (79), 5–19.

30 Albrecher, H., Teugels, J.L., and Tichy, R.F. (2001) On a gamma series expansion for the time-dependent probability of collective ruin. *Insurance Math. Econom.*, **29**, 345–355.

31 Albrecht, P. (1984) Laplace transforms, Mellin transforms and mixed Poisson processes. *Scand. Actuar. J.*, pp. 58–64.

32 Albrecht, P. (1984) Summary report on large claims, in *Proceedings of the 4 countries Astin Coll.*, pp. 153–164.

33 Albrecht, P. and König, A. (1995) Risikomanagement von Rückversicherungsunternehmen. *Trans. 25th Intern. Congress Actuaries*, **II**, 1–22.

34 Altuntas, M., Berry-Stölzle, T.R., and Wende, S. (2015) Does one size fit all? determinants of insurer capital structure around the globe. *J. Banking & Finance*, **61**, 251–271.

35 Alves, M.F., Gomes, M.I., and De Haan, L. (2003) A new class of semi-parametric estimators of the second order parameter. *Portugaliae Mathematica*, **60** (2), 193–214.

36 Ammeter, H. (1948) A generalization of the collective theory of risk in regard to fluctuating basic probabilities. *Skand. Akt. Tidskr.*, pp. 171–198.

37 Ammeter, H. (1963) Spreading of exceptional claims by means of an internal stop-loss cover. *Astin Bull.*, **2**, 380–386.

38 Ammeter, H. (1964) Note concerning the distribution function of the total loss excluding the largest individual claims. *Astin Bull.*, **3**, 132–143.

39 Ammeter, H. (1964) The rating of "largest claim" reinsurance covers. *Quarterly Allgem. Reinsur. Comp. Jubilee*, **2**.

40 Ammeter, H. (1971) Grösstschaden-Verteilungen und ihre Anwendungen. *Mitt. Ver. Schweiz. Versich. Math.*, **71**, 35–62.

41 Amsler, M. (1993) Reassurance du risque de ruine. *Bulletin of the Swiss Association of Actuaries*, **1**, 33–39.

42 Anderson, R.R. and Dong, W. (1998) Pricing catastrophe reinsurance with reinstatement provisions using a catastrophe model, in *CAS Forum*, Summer, pp. 303–322.

43 Antal, P. (2009) Mathematical methods in reinsurance. *Lecture Notes, ETH Zurich.*

44 Applebaum, D. (2009) *Lévy Processes and Stochastic Calculus*, Cambridge University Press.

45 Arbenz, P. and Guevara-Alarcón, W. (2016) Risk measure preserving piecewise linear approximation of empirical distributions. *Eur. Actuarial J.*, **6** (1), 113–148.

46 Arrow, K.J. (1963) Uncertainty and the welfare economics of medical care. *The American Economic Review*, **53** (5), 941–973.

47 Arrow, K.J. (1971) Insurance, risk and resource allocation. *Essays in the theory of risk-bearing*, pp. 134–143.

48 Arrow, K.J. (1974) Optimal insurance and generalized deductibles. *Scand. Actuar. J.*, pp. 1–42.

49 Artzner, P., Delbaen, F., Eber, J.M., and Heath, D. (1999) Coherent measures of risk. *Mathematical Finance*, **9**, 203–228.

50 Asimit, A.V., Badescu, A.M., and Cheung, K.C. (2013) Optimal reinsurance in the presence of counterparty default risk. *Insurance Math. Econom.*, **53** (3), 690–697.

51 Asimit, A.V., Badescu, A.M., Haberman, S., and Kim, E.S. (2016) Efficient risk allocation within a non-life insurance group under solvency II regime. *Insurance Math. Econom.*, **66**, 69–76.

52 Asimit, A.V., Badescu, A.M., and Tsanakas, A. (2013) Optimal risk transfers in insurance groups. *Eur. Actuarial J.*, **3** (1), 159–190.

53 Asimit, A.V., Bignozzi, V., Cheung, K.C., Hu, J., and Kim, E.S. (2017) Robust and Pareto optimality of insurance contract. *European Journal of Operational Research*. In press.

54 Asimit, A.V., Chi, Y., and Hu, J. (2015) Optimal non-life reinsurance under solvency II regime. *Insurance Math. Econom.*, **65**, 227–237.

55 Asimit, A.V. and Jones, B.L. (2008) Dependence and the asymptotic behavior of large claims reinsurance. *Insurance Math. Econom.*, **43** (3), 407–411.

56 Asmussen, S. (1989) Risk theory in a Markovian environment. *Scand. Actuar. J.*, pp. 66–100.

57 Asmussen, S. and Albrecher, H. (2010) *Ruin probabilities*, Advanced Series on Statistical Science & Applied Probability, 14, World Scientific Publishing Co. Pte. Ltd., Hackensack, NJ, 2nd edn.

58 Asmussen, S., Binswanger, K., and Højgaard, B. (2000) Rare events simulation for heavy-tailed distributions. *Bernoulli*, **6** (2), 303–322.

59 Asmussen, S. and Glynn, P.W. (2007) *Stochastic Simulation: Algorithms and Analysis*, Stochastic Modelling and Applied Probability, vol. 57, Springer, New York.

60 Asmussen, S., Goffard, P., and Laub, P. (2016) Orthonormal polynomial expansions and lognormal sum densities. *Festschrift for Ragnar Norberg, World Scientific.*

61 Asmussen, S., Højgaard, B., and Taksar, M. (2000) Optimal risk control and dividend distribution policies. Example of excess-of-loss reinsurance for an insurance corporation. *Finance and Stochastics*, **4** (3), 299–324.

62 Asmussen, S. and Kortschak, D. (2012) On error rates in rare event simulation with heavy tails, in *Proceedings of the Winter Simulation Conference*, pp. 38–48.
63 Asmussen, S. and Kortschak, D. (2015) Error rates and improved algorithms for rare event simulation with heavy Weibull tails. *Methodology and Computing in Applied Probability*, **17** (2), 441–461.
64 Asmussen, S. and Kroese, D.P. (2006) Improved algorithms for rare event simulation with heavy tails. *Adv. Appl. Probab.*, **38** (2), 545–558.
65 Asmussen, S., Nerman, O., and Olsson, M. (1996) Fitting phase-type distributions via the EM algorithm. *Scand. J. Stats*, **23** (4), 419–441.
66 Asmussen, S., Schmidli, H., and Schmidt, V. (1999) Tail probabilities for non-standard risk and queueing processes with subexponential jumps. *Adv. Appl. Probab.*, **31** (2), 422–447.
67 Asprone, D., Jalayer, F., Prota, A., Manfredi, G., Simonelli, S., and Acconcia, A. (2013) Earthquake loss analysis of the Italian building stock investigating the feasibility of an earthquake insurance system, in *Proceedings of the 11th Int. Conf. Structural Safety & Reliability*, Taylor & Francis Group, pp. 4077–4083.
68 Assa, H. (2015) On optimal reinsurance policy with distortion risk measures and premiums. *Insurance Math. Econom.*, **61**, 70–75.
69 Azcue, P. and Muler, N. (2005) Optimal reinsurance and dividend distribution policies in the Cramér-Lundberg model. *Mathematical Finance*, **15** (2), 261–308.
70 Azcue, P. and Muler, N. (2014) *Stochastic Optimization in Insurance: A Dynamic Programming Approach*, Springer.
71 Badescu, A.L., Lin, X.S., and Tang, D. (2016) A marked Cox model for the number of IBNR claims: Theory. *Insurance Math. Econom.*, **69**, 29–37.
72 Balbás, A., Balbás, B., Balbás, R., and Heras, A. (2015) Optimal reinsurance under risk and uncertainty. *Insurance Math. Econom.*, **60**, 61–74.
73 Balbás, A., Balbás, B., and Heras, A. (2009) Optimal reinsurance with general risk measures. *Insurance Math. Econom.*, **44** (3), 374–384.
74 Balbás, A., Balbás, B., and Heras, A. (2011) Stable solutions for optimal reinsurance problems involving risk measures. *Eur. J. Operation. Res.*, **214** (3), 796–804.
75 Balbás, A., Garrido, J., and Mayoral, S. (2009) Properties of distortion risk measures. *Methodology and Computing in Applied Probability*, **11** (3), 385–399.
76 Baluch, F., Mutenga, S., and Parsons, C. (2011) Insurance, systemic risk and the financial crisis. *The Geneva Papers on Risk and Insurance: Issues and Practice*, **36** (1), 126–163.
77 Bams, D. and Wielhouwer, J. (2001) Empirical issues in Value-at-Risk. *Astin Bull.*, **31**, 299–315.
78 Banks, E. (2004) *Alternative Risk Transfer: Integrated Risk Management through Insurance, Reinsurance, and the Capital Markets*, John Wiley & Sons.
79 Banks, E. (2005) *Catastrophic Risk: Analysis and Management*, John Wiley & Sons.
80 Bantwal, V. and Kunreuther, H. (2000) A Cat bond premium puzzle? *J. Psychol. Finan. Markets*, **1**, 76–91.
81 Barbe, P. and McCormick, W. (2008) Asymptotic expansions for infinite weighted convolutions of rapidly varying subexponential distributions. *Prob. Theory Relat. Fields*, **141**, 155–180.
82 Barbour, A.D. and Chen, L.H. (2005) *An Introduction to Stein's Method*, vol. 4, World Scientific.

83 Barfod, A. and Lando, D. (1996) On derivative contracts on catastrophic losses. *Colloq. XXVII Astin*, pp. 5–22.

84 Barrieu, P. and Albertini, L. (2010) *The Handbook of Insurance-Linked Securities*, John Wiley & Sons.

85 Barrieu, P. and El Karoui, N. (2005) Inf-convolution of risk measures and optimal risk transfer. *Finance and Stochastics*, **9** (2), 269–298.

86 Barrieu, P. and Loubergé, H. (2009) Hybrid cat bonds. *J. Risk and Insurance*, **76** (3), 547–578.

87 Barrieu, P. and Scandolo, G. (2008) General Pareto optimal allocations and applications to multi-period risks. *Astin Bull.*, **38** (1), 105–136.

88 Baton, B. and Lemaire, J. (1981) The bargaining set of reinsurance market. *Astin Bull.*, **12**, 101–114.

89 Bauer, D., Phillips, R.D., and Zanjani, G.H. (2013) Financial pricing of insurance, in *Handbook of Insurance*, Springer, pp. 627–645.

90 Bäuerle, N. (2004) Traditional versus non-traditional reinsurance in a dynamic setting. *Scand. Actuar. J.*, pp. 355–371.

91 Bäuerle, N. and Grübel, R. (2005) Multivariate counting processes: copulas and beyond. *Astin Bull.*, **35** (2), 379–408.

92 Beard, R. (1963) Some notes on the statistical theory of extreme values. *Astin Bull.*, **3**, 6–12.

93 Beard, R., Pentikäinen, T., and Pesonen, E. (1984) *Risk Theory*, Chapman & Hall, London., 3rd edn.

94 Beekman, J. (1974) *Two Stochastic Processes*, Almqvist & Wiksell International, Stockholm.

95 Beirlant, J., Alves, I.F., and Gomes, I. (2016) Tail fitting for truncated and non-truncated Pareto-type distributions. *Extremes*, **19** (3), 1–34.

96 Beirlant, J., Bardoutsos, A., de Wet, T., and Gijbels, I. (2016) Bias reduced tail estimation for censored Pareto type distributions. *Statistics & Probability Letters*, **109**, 78–88.

97 Beirlant, J., Dierckx, G., Goegebeur, Y., and Matthys, G. (1999) Tail index estimation and an exponential regression model. *Extremes*, **2** (2), 177–200.

98 Beirlant, J., Escobar-Bach, M., Goegebeur, Y., and Guillou, A. (2016) Bias-corrected estimation of stable tail dependence function. *J. Multivariate Anal.*, **143**, 453–466.

99 Beirlant, J. and Goegebeur, Y. (2003) Regression with response distributions of Pareto-type. *Computational Stat. Data Anal.*, **42** (4), 595–619.

100 Beirlant, J., Goegebeur, Y., Segers, J., and Teugels, J. (2004) *Statistics of Extremes*, Wiley Series in Probability and Statistics, John Wiley & Sons Ltd., Chichester.

101 Beirlant, J., Guillou, A., Dierckx, G., and Fils-Villetard, A. (2007) Estimation of the extreme value index and extreme quantiles under random censoring. *Extremes*, **10** (3), 151–174.

102 Beirlant, J., Joossens, E., and Segers, J. (2009) Second-order refined peaks-over-threshold modelling for heavy-tailed distributions. *J. Statist. Planning and Inference*, **139** (8), 2800–2815.

103 Beirlant, J., Matthys, G., and Dierckx, G. (2001) Heavy-tailed distributions and rating. *Astin Bull.*, **31**, 41–62.

104 Beirlant, J., Schoutens, W., De Spiegeleer, J., Reynkens, T., and Herrmann, K. (2016) Hunting for black swans in the European banking sector using extreme value analysis,

in *Advanced Modeling in Mathematical Finance, In Honour of E. Eberlein*, Springer.

105 Beirlant, J., Teugels, J., and Vynckier, P. (1996) *Practical Analysis of Extreme Values*, Leuven University Press, Belgium.

106 Bellini, F. and Bignozzi, V. (2015) On elicitable risk measures. *Quant. Finance*, **15** (5), 725–733.

107 Bellini, F., Klar, B., Müller, A., and Gianin, E.R. (2014) Generalized quantiles as risk measures. *Insurance Math. Econom.*, **54**, 41–48.

108 Benckert, L. (1962) The log-normal model for the distribution of one claim. *Astin Bull.*, **2**, 9–23.

109 Benckert, L. and Sternberg, I. (1957) An attempt to find an expression for the distribution of the fire damage amount. *Trans. 15th Intern. Congress Actuaries*, **II**, 288–296.

110 Benktander, G. (1962) A note on the most "dangerous" and skewest class of distributions. *Astin Bull.*, **2**, 387–390.

111 Benktander, G. (1971) Some aspects of reinsurance profits and loadings. *Astin Bull.*, **5**, 314–327.

112 Benktander, G. (1978) Largest claims reinsurance (LCR). a quick method to calculate LCR-risk rates from excess of loss risk rates. *Astin Bull.*, **10**, 54–58.

113 Benktander, G. (1993) A stop-loss rating formula. *XXIV Coll. Astin*, **2**, 1–11.

114 Benktander, G. and Ohlin, J. (1967) A combination of surplus and excess reinsurance of a fire portfolio. *Astin Bull.*, **4**, 177–190.

115 Benktander, G. and Segerdahl, C. (1960) On the analytical representation of claim distributions with special reference to excess of loss reinsurance. *Trans. 16th Intern. Congress Actuaries*, pp. 626–646.

116 Bennett, M. and Johnson, P. (1984) The treatment of large claims when deciding on a premium structure and on the relationship between the premiums of different groups. *Proc. 4 countries Astin Coll.*, pp. 175–194.

117 Bensoussan, A., Siu, C.C., Yam, S.C.P., and Yang, H. (2014) A class of non-zero-sum stochastic differential investment and reinsurance games. *Automatica*, **50** (8), 2025–2037.

118 Berglund, R. (1998) A note on the net premium for a generalized largest claims reinsurance cover. *Astin Bull.*, **28**, 153–162.

119 Berliner, B. (1972) Correlations between excess of loss reinsurance covers and reinsurance of the n largest claims. *Astin Bull.*, **6**, 260–275.

120 Bernard, C. (2013) Risk sharing and pricing in the reinsurance market, in *Handbook of Insurance*, Springer.

121 Bernard, C., Denuit, M., and Vanduffel, S. (2016) Measuring portfolio risk under partial dependence information. *J. Risk and Insurance*. In press.

122 Bernard, C., He, X., Yan, J.A., and Zhou, X.Y. (2015) Optimal insurance design under rank-dependent expected utility. *Mathematical Finance*, **25** (1), 154–186.

123 Bernard, C., Ji, S., and Tian, W. (2013) An optimal insurance design problem under Knightian uncertainty. *Decisions in Economics and Finance*, **36** (2), 99–124.

124 Bernard, C., Jiang, X., and Wang, R. (2014) Risk aggregation with dependence uncertainty. *Insurance Math. Econom.*, **54**, 93–108.

125 Bernard, C. and Ludkovski, M. (2012) Impact of counterparty risk on the reinsurance market. *North American Actuarial Journal*, **16** (1), 87–111.

126 Bernard, C., Rüschendorf, L., Vanduffel, S., and Yao, J. (2015) How robust is the value-at-risk of credit risk portfolios? *Eur. J. Finance*, **23** (6), 507–534.
127 Bernard, C. and Tian, W. (2009) Optimal reinsurance arrangements under tail risk measures. *J. Risk and Insurance*, **76** (3), 709–725.
128 Bernard, C. and Vanduffel, S. (2015) A new approach to assessing model risk in high dimensions. *J. Banking and Finance*, **58**, 166–178.
129 Bernegger, S. (1994) The Swiss Re exposure curves and the MBBEFD distribution. *Astin Bull.*, **27** (1), 99–111.
130 Bertogg, M., Schmid, E., and Kriesch, S. (2001) Earthquake modelling as a risk management tool for accumulation control in the insurance industry, in *El riesgo sísmico: prevención y seguro*, Consorcio de Compensación de Seguros, Madrid, pp. 233–242.
131 Bertram, J. (1981) Numerische Berechnung von Gesamtschadenverteilungen. *Blätter der DGVFM*, **15** (2), 175–194.
132 Biener, C., Eling, M., and Wirfs, J.H. (2015) Insurability of cyber risk: An empirical analysis. *The Geneva Papers on Risk and Insurance Issues and Practice*, **40** (1), 131–158.
133 Biffis, E. and Millossovich, P. (2012) Optimal insurance with counterparty default risk. *Available at SSRN 1634883*.
134 Bignozzi, V., Mao, T., Wang, B., and Wang, R. (2016) Diversification limit of quantiles under dependence uncertainty. *Extremes*, **19** (2), 143–170.
135 Bingham, N., Goldie, C., and Teugels, J. (1987) *Regular Variation, Encyclopedia of Mathematics and its Applications*, vol. 27, Cambridge University Press, Cambridge.
136 Bingham, N.H. and Teugels, J.L. (1981) Conditions implying domains of attraction, in *Proceedings of the Sixth Conference on Probability Theory, Braşov, 1979*, pp. 23–34.
137 Birkel, T. (1995) Elementary upper bounds for the variance of a general reinsurance treaty. *Blätter der DGVFM*, **21**, 309–312.
138 Björk, T. and Grandell, J. (1988) Exponential inequalities for ruin probabilities in the Cox case. *Scand. Actuar. J.*, pp. 77–111.
139 Bladt, M. and Nielsen, B.F. (2017) *Matrix–Exponential Distributions in Applied Probability*, Springer, forthcoming.
140 Bladt, M., Nielsen, B.F., and Samorodnitsky, G. (2015) Calculation of ruin probabilities for a dense class of heavy tailed distributions. *Scand. Actuar. J.*, pp. 573–591.
141 Blanchet, J. and Glynn, P. (2008) Efficient rare-event simulation for the maximum of heavy-tailed random walks. *Ann. Appl. Probab.*, **18** (4), 1351–1378.
142 Blazenko, G. (1986) The economics of reinsurance. *J. Risk and Insurance*, pp. 258–277.
143 Blum, P., Dacorogna, M., Embrechts, P., Neghaiwi, A., and Niggli, H. (2001) Using DFA for modelling the impact of foreign exchange risks on reinsurance decisions, in *CAS Forum*, Summer, pp. 49–94.
144 Bohman, H. (1975) Numerical inversions of characteristic functions. *Scand. Actuar. J.*, pp. 121–124.
145 Boonen, T.J. (2016) Nash equilibria of over-the-counter bargaining for insurance risk redistributions: The role of a regulator. *Eur. J. Operational Res.*, **250** (3), 955–965.
146 Boonen, T.J., Tan, K.S., and Zhuang, S.C. (2016) Pricing in reinsurance bargaining with comonotonic additive utility functions. *Astin Bull.*, **46** (2), 507–530.
147 Boonen, T.J., Tan, K.S., and Zhuang, S.C. (2016) The role of a representative reinsurer in optimal reinsurance. *Insurance Math. Econom.*, **70**, 196–204.
148 Borch, K. (1962) Equilibrium in a reinsurance market. *Econometrica*, **30**, 424–444.

149 Borch, K. (1974) *The Mathematical Theory of Insurance*, Lexington Books, Toronto.
150 Borch, K. (1975) Optimal insurance arrangements. *Astin Bull.*, **8**, 284–290.
151 Borch, K. (1978) Problems in the economic theory of insurance. *Astin Bull.*, **10**, 1–11.
152 Borch, K. (1990) *Economics of Insurance*, Elsevier, North Holland.
153 Born, P. and Viscusi, W.K. (2006) The catastrophic effects of natural disasters on insurance markets. *J. Risk and Uncertainty*, **33** (1-2), 55–72.
154 Borscheid, P., Gugerli, D., and Straumann, T. (2013) *The Value of Risk: Swiss Re and the History of Reinsurance*, Oxford University Press.
155 Botzen, W.W. (2013) *Managing Extreme Climate Change Risks through Insurance*, Cambridge University Press.
156 Bourne, S., Oates, S., van Elk, J., and Doornhof, D. (2014) A seismological model for earthquakes induced by fluid extraction from a subsurface reservoir. *J. Geophys. Res.: Solid Earth*, **119** (12), 8991–9015.
157 Bowers, N., Gerber, H., Hickman, J., Jones, D., and Nesbitt, C. (1997) *Actuarial Mathematics*, Society of Actuaries, Ithaca, Il., 2nd edn.
158 Box, G.E. and Jenkins, G.M. (1976) *Time Series Analysis: Forecasting and Control*, Holden-Day, revised edn.
159 Boyer, M.M. and Nyce, C.M. (2013) An industrial organization theory of risk sharing. *North American Actuarial Journal*, **17** (4), 283–296.
160 Brandtner, M. and Kürsten, W. (2014) Solvency II, regulatory capital, and optimal reinsurance: How good are conditional value-at-risk and spectral risk measures? *Insurance Math. Econom.*, **59**, 156–167.
161 Braun, A. (2016) Pricing in the primary market for cat bonds: new empirical evidence. *Journal of Risk and Insurance*, **83** (4), 811–847.
162 Brazauskas, V. (2003) Influence functions of empirical nonparametric estimators of net reinsurance premiums. *Insurance Math. Econom.*, **32** (1), 115–133.
163 Breiman, L. (1965) On some limit theorems similar to the arc sine law. *Theory Probab. Appl.*, **10**, 323–333.
164 Brown, T.M., Pogorzelski, W.H., and Giammanco, I.M. (2015) Evaluating hail damage using property insurance claims data. *Weather, Climate, and Society*, **7** (3), 197–210.
165 Bühlmann, H. (1970) *Mathematical Methods in Risk Theory*, Springer Verlag, Heidelberg.
166 Bühlmann, H. (1984) The general economic premium principle. *Astin Bull.*, **14** (1), 13–21.
167 Bühlmann, H. (1984) Numerical evaluation of the compound poisson distribution: recursion or fast fourier transform? *Scand. Actuar. J.*, pp. 116–126.
168 Bühlmann, H., Gagliardi, B., Gerber, H., and Straub, E. (1977) Some inequalities for stop-loss premiums. *Astin Bull.*, **9**, 75–83.
169 Bühlmann, H. and Gisler, A. (2006) *A Course in Credibility Theory and its Applications*, Springer Science & Business Media.
170 Bühlmann, H. and Jewell, W. (1979) Optimal risk exchanges. *Astin Bull.*, **10**, 243–262.
171 Bühlmann, H. and Jewell, W. (1982) Excess claims and data trimming in the context of credibility rating procedures. *Mitt. Ver. Schweiz. Versich. Math.*, pp. 117–147.
172 Bulinskaya, E., Gusak, J., and Muromskaya, A. (2015) Discrete-time insurance model with capital injections and reinsurance. *Methodology and Computing in Applied Probability*, **17** (4), 899–914.

173 Burger, G., Fitzgerald, B., White, J., and Woods, P. (1996) Incorporating a hurricane model into property ratemaking, in *CAS Forum*, Winter, pp. 129–190.

174 Cai, J., Fang, Y., Li, Z., and Willmot, G.E. (2013) Optimal reciprocal reinsurance treaties under the joint survival probability and the joint profitable probability. *J. Risk and Insurance*, **80** (1), 145–168.

175 Cai, J., Lemieux, C., and Liu, F. (2014) Optimal reinsurance with regulatory initial capital and default risk. *Insurance Math. Econom.*, **57**, 13–24.

176 Cai, J., Lemieux, C., and Liu, F. (2016) Optimal reinsurance from the perspectives of both an insurer and a reinsurer. *Astin Bull.*, pp. 1–35.

177 Cai, J., Liu, H., and Wang, R. (2016) Asymptotic equivalence of risk measures under dependence uncertainty. *Math. Finance*. In press.

178 Cai, J. and Tan, K.S. (2007) Optimal retention for a stop-loss reinsurance under the VaR and CTE risk measures. *Astin Bull.*, **37** (1), 93–112.

179 Cai, J., Tan, K.S., Weng, C., and Zhang, Y. (2008) Optimal reinsurance under VaR and CTE risk measures. *Insurance Math. Econom.*, **43** (1), 185–196.

180 Calderín-Ojeda, E. and Kwok, C.F. (2015) Modeling claims data with composite Stoppa models. *Scand. Actuar. J.*, pp. 817–836.

181 Cambou, M., Hofert, M., and Lemieux, C. (2016) A primer on quasi-random numbers for copula models. *Statistics and Computing*. In press.

182 Cani, A. and Thonhauser, S. (2016) An optimal reinsurance problem in the Cramér–Lundberg model. *Mathematical Methods of Operations Research*. In press.

183 Carlier, G. and Dana, R.A. (2003) Pareto efficient insurance contracts when the insurer's cost function is discontinuous. *Economic Theory*, **21** (4), 871–893.

184 Carneiro, L.A. and Sherris, M. (2005) Demand for reinsurance: Evidence from Australian insurers, in *Report*, Faculty of Commerce and Economics, Sydney.

185 Carter, R. (1979) *Reinsurance*, Kluwer, London.

186 Castagnoli, E., Maccheroni, F., and Marinacci, M. (2002) Insurance premia consistent with the market. *Insurance Math. Econom.*, **31**, 267 – 284.

187 Centeno, L. (1985) On combining quota-share and excess of loss. *Astin Bull.*, **15**, 49–63.

188 Centeno, L. (1986) Measuring the effects of reinsurance by the adjustment coefficient. *Insurance Math. Econom.*, **5**, 169–182.

189 Centeno, L. (1991) An insight into the excess of loss retention limit. *Scand. Actuar. J.*, pp. 97–102.

190 Centeno, L. (1995) The effect of the retention on the risk reserve. *Astin Bull.*, **25**, 67–74.

191 Centeno, L. (1997) Excess-of-loss reinsurance and the probability of ruin in finite horizon. *Astin Bull.*, **27**, 59–70.

192 Centeno, L. (2002) Excess of loss reinsurance and Gerber's inequality in the Sparre Andersen model. *Insurance Math. Econom.*, **31**, 415–427.

193 Centeno, L. (2002) Measuring the effects of reinsurance by the adjustment coefficient in the Sparre Anderson model. *Insurance Math. Econom.*, **30**, 37–49.

194 Centeno, L. and Guerra, M. (2010) The optimal reinsurance strategy–the individual claim case. *Insurance Math. Econom.*, **46** (3), 450–460.

195 Centeno, L. and Simões, O. (1991) Combining quota-share and excess-of-loss treaties on the reinsurance on *n* independent risks. *Astin Bull.*, **21**, 41–55.

196 Centeno, L. and Simões, O. (2009) Optimal reinsurance. *Rev. R. Acad. Cienc. Exactas Fís. Nat. Ser. A Math. RACSAM*, **103** (2), 387–404.

197 Cerqueti, R., Foschi, R., and Spizzichino, F. (2009) A spatial mixed Poisson framework for combination of excess-of-loss and proportional reinsurance contracts. *Insurance Math. Econom.*, **45** (1), 59–64.

198 Chang, J., Cheung, C., and Krinsky, I. (1989) On the derivation of reinsurance premiums. *Insurance Math. Econom.*, **8**, 137–144.

199 Chaubey, Y., Garrido, J., and Trudeau, S. (1998) On the computation of aggregate claims distributions: some new approximations. *Insurance Math. Econom.*, **23**, 215–230.

200 Chavez-Demoulin, V. and Davison, A. (2005) Generalized additive modelling of sample extremes. *J. R. Stat. Soc. Ser. C Appl. Stat.*, **54** (1), 207–222.

201 Chavez-Demoulin, V., Embrechts, P., and Hofert, M. (2016) An extreme value approach for modeling operational risk losses depending on covariates. *Journal of Risk and Insurance*, **83** (3), 735–776.

202 Chen, R., Jaiswal, K., Bausch, D., Seligson, H., and Wills, C. (2016) Annualized earthquake loss estimates for California and their sensitivity to site amplification. *Seismological Research Letters*, **87** (6), 1363–1372.

203 Cheridito, P., Delbaen, F., and Kupper, M. (2004) Coherent and convex monetary risk measures for bounded cadlag processes. *Stochastic Processes and their Applications*, **112** (1), 1–22.

204 Cheung, K.C. (2010) Optimal reinsurance revisited–a geometric approach. *Astin Bull.*, **40** (1), 221–239.

205 Cheung, K.C. and Lo, A. (2017) Characterizations of optimal reinsurance treaties: a cost-benefit approach. *Scand. Actuar. J.*, pp. 1–28.

206 Cheung, K.C., Rong, Y., and Yam, S. (2014) Borch's theorem from the perspective of comonotonicity. *Insurance Math. Econom.*, **54**, 144–151.

207 Cheung, K.C., Sung, K., and Yam, S. (2014) Risk-minimizing reinsurance protection for multivariate risks. *J. Risk and Insurance*, **81** (1), 219–236.

208 Cheung, K.C., Sung, K., Yam, S., and Yung, S. (2014) Optimal reinsurance under general law-invariant risk measures. *Scand. Actuar. J.*, pp. 72–91.

209 Chi, Y. and Tan, K.S. (2010) Optimal reinsurance under VaR and CVaR risk measures: a simplified approach. *Astin Bull.*, **41** (2), 487–509.

210 Chi, Y. and Weng, C. (2013) Optimal reinsurance subject to Vajda condition. *Insurance Math. Econom.*, **53** (1), 179–189.

211 Chistyakov, V. (1964) A theorem on sums of independent random variables and its application to branching processes. *Theory Probab. Appl.*, **9**, 640–648.

212 Chover, J., Ney, P., and Wainger, S. (1973) Functions of probability measures. *J. D'Anal. Math.*, **26**, 255–302.

213 Christensen, C. and Schmidli, H. (2000) Pricing catastrophic insurance products based on actually reported claims. *Insurance Math. Econom.*, **27**, 189–200.

214 Ciminelli, E. (1976) On the distribution of the highest claims and its application to the automobile insurance liability. *Trans. 20th Intern. Congress Actuaries*, pp. 501–517.

215 Clark, D. (1994) A simple tool for pricing loss sensitive features of reinsurance treaties. *Casualty Actuarial Society Call Papers.*

216 Clark, D. (1996) Basics of reinsurance pricing. *CAS Study Note.*

217 Clark, D. (2013) A note on the upper-truncated Pareto distribution, in *Proceedings of the Enterprise Risk Management Symposium, Chicago, April 22-24, 2013.*

218 Clemente, G.P., Savelli, N., and Zappa, D. (2015) The impact of reinsurance strategies on capital requirements for premium risk in insurance. *Risks*, **3** (2), 164–182.

219 Coburn, A., Ruffle, S., and Pryor, L. (2014) Cyber catastrophe. *The Actuary*, (December).

220 Cole, C.R., Macpherson, D.A., and McCullough, K.A. (2010) A comparison of hurricane loss models. *Journal of Insurance Issues*, **33** (1), 31–53.

221 Coles, S. (2001) *An Introduction to Statistical Modelling of Extreme Values*, Springer Verlag, Berlin.

222 Coles, S. and Pericchi, L. (2003) Anticipating catastrophes through extreme value modelling. *J. Roy. Statist. Soc. Series C*, **52**, 405–416.

223 Cong, J. and Tan, K.S. (2016) Optimal VaR-based risk management with reinsurance. *Annals of Operations Research*, **237** (1-2), 177–202.

224 Constantinescu, C., Hashorva, E., and Ji, L. (2011) Archimedean copulas in finite and infinite dimensions–with application to ruin problems. *Insurance Math. Econom.*, **49** (3), 487–495.

225 Cooley, J.W. and Tukey, J.W. (1965) An algorithm for the machine calculation of complex Fourier series. *Math. Comp.*, **19** (90), 297–301.

226 Cooray, K. and Ananda, M.M. (2005) Modeling actuarial data with a composite lognormal-Pareto model. *Scand. Actuar. J.*, pp. 321–334.

227 Corradin, S. and Verbrigghe, B. (2002) Economic risk capital and reinsurance: An application to fire claims of an insurance company, in *Proceedings of the 6th IME Conference, Lisbon.*

228 Corro, D. (2002) Fitting beta-densities to loss data, in *CAS Forum*, Summer, pp. 169–174.

229 Cox, D.R. (1955) Some statistical methods connected with series of events. *J. R. Stat. Soc. Ser. B Stat. Methodol.*, **17** (2), 129–164.

230 Cox, S.H., Fairchild, J.R., and Pedersen, H.W. (2000) Economic aspects of securitization of risk. *Astin Bull.*, **30** (1), 157–193.

231 Cox, S.H., Fairchild, J.R., and Pedersen, H.W. (2004) Valuation of structured risk management products. *Insurance Math. Econom.*, **34** (2), 259–272.

232 Cox, S.H. and Pedersen, H.W. (2000) Catastrophe risk bonds. *North American Actuarial Journal*, **4** (4), 56–82.

233 Cozzolino, J. and Gaydos, E. (1993) Measuring the probability of disastrous losses. *SCOR Notes: Rewarding Risk*, pp. 120–171.

234 Cramér, H. (1955) Collective risk theory: A survey of the theory from the point of view of stochastic processes. *The Jubilee Volume of Försäkringsbolaget Skandia, Stockholm.*

235 Crowley, H. and Bommer, J.J. (2006) Modelling seismic hazard in earthquake loss models with spatially distributed exposure. *Bulletin of Earthquake Engineering*, **4** (3), 249–273.

236 Cui, W., Yang, J., and Wu, L. (2013) Optimal reinsurance minimizing the distortion risk measure under general reinsurance premium principles. *Insurance Math. Econom.*, **53** (1), 74–85.

237 Culp, C.L. (2011) *Structured Finance and Insurance: The ART of Managing Capital and Risk*, John Wiley & Sons.

238 Cummins, J., Dionne, G., McDonald, J., and Pritchett, B. (1990) Applications of the GB2 family of distributions in modeling insurance loss processes. *Insurance Math. Econom.*, **9**, 257 – 272.

239 Cummins, J. and Trainar, P. (2009) Securitization, insurance, and reinsurance. *J. Risk and Insurance*, **76** (3), 463–492.

240 Cummins, J.D., Doherty, N.A., and Lo, A. (2002) Can insurers pay for the "big one"? Measuring the capacity of an insurance market to respond to catastrophic losses. *Journal of Banking and Finance*, **26**, 557–583.

241 Cummins, J.D. and Mahul, O. (2004) The demand for insurance with an upper limit on coverage. *J. Risk and Insurance*, **71** (2), 253–264.

242 Cummins, J.D. and Weiss, M.A. (2009) Convergence of insurance and financial markets: Hybrid and securitized risk-transfer solutions. *J. Risk and Insurance*, **76** (3), 493–545.

243 Dacorogna, M.M. (2015) A change of paradigm for the insurance industry. *SCOR Papers*, **34**.

244 Dacorogna, M.M., Albrecher, H., Moller, M., and Sahiti, S. (2013) Equalization reserves for natural catastrophes and shareholder value: a simulation study. *Eur. Actuar. J.*, **3** (1), 1–21.

245 Dana, R.A. and Scarsini, M. (2007) Optimal risk sharing with background risk. *J. Econ. Theory*, **133** (1), 152–176.

246 Danielsson, J. and de Vries, G. (2000) Value at risk and extreme returns. *Annals d'Economie et de Statistique*, **60**, 239–269.

247 Dannenburg, D., Kaas, R., and Goovaerts, M. (1996) *Practical Actuarial Credibility Models*, Institute of Actuarial Science, Amsterdam.

248 Daouia, A., Gardes, L., and Girard, S. (2013) On kernel smoothing for extremal quantile regression. *Bernoulli*, **19** (5B), 2557–2589.

249 Daouia, A., Gardes, L., Girard, S., and Lekina, A. (2011) Kernel estimators of extreme level curves. *Test*, **20** (2), 311–333.

250 D'Arcy, S., France, V., and Gorvett, R. (1999) Pricing catastrophic risk: could CAT-futures have coped with Andrew? *CAS Discussion Papers*, pp. 1–46.

251 Dassios, A. and Jang, J.W. (2003) Pricing of catastrophe reinsurance and derivatives using the Cox process with shot noise intensity. *Finance Stoch.*, **7** (1), 73–95.

252 Davison, A. and Ramesh, N. (2000) Local likelihood smoothing of sample extremes. *J. R. Stat. Soc. Ser. B Stat. Methodol.*, **62** (1), 191–208.

253 Davison, A. and Smith, R.L. (1990) Models for exceedances over high thresholds. *J. R. Stat. Soc. Ser. B Stat. Methodol.*, **52** (3), 393–442.

254 Daykin, C., Pentikäinen, T., and Pesonen, E. (1994) *Practical Risk Theory for Actuaries*, Chapman & Hall, London.

255 de Finetti, B. (1940) Il problema dei "Pieni.". *Giorn. Ist. Ital. Attuari*, **11**, 1–88.

256 de Finetti, B. (1957) Su un'impostazione alternativa della teoria collettiva del rischio. *Transactions of the 15th Int. Congress of Actuaries*, **2**, 433–443.

257 De Haan, L. (1970) *On Regular Variation and its Application to theWeak Convergence of Sample Extremes*, Mathematisch Centrum.

258 De Haan, L. and Ferreira, A. (2006) *Extreme Value Theory*, Springer Series in Operations Research and Financial Engineering, Springer, New York.

259 De Longueville, P. (1995) Optimal reinsurance from the point of view of the excess-of-loss reinsurer under the finite-time ruin criterion. *Trans. 25th Intern. Congress Actuaries*, pp. 121–140.

260 De Spiegeleer, J. and Schoutens, W. (2011) *The Handbook of Convertible Bonds: Pricing, Strategies and Risk Management*, vol. 581, John Wiley & Sons.

261 De Valk, C. (2016) Approximation and estimation of very small probabilities of multivariate extremes. *Extremes*, **19**, 687–717.

262 De Valk, C. and Cai, J. (2017) A high quantile estimator based on the log-generalised Weibull tail limit. *Econometrics and Statistics*. In press.

263 De Vylder, F. (1996) *Advanced Risk Theory, a Self-contained Introduction*, Editions de l'Université de Bruxelles, Brussels.

264 De Vylder, F. and Goovaerts, M. (1983) Best bounds on the stop loss premium in case of known range, expectation, variance and mode of the risk. *Insurance Math. Econom.*, **2**, 241–249.

265 De Waegenaere, A. (1994) Equilibria in a mixed financial-reinsurance market with constrained trading possibilities. *Insurance Math. Econom.*, **14**, 205–218.

266 De Waegenaere, A. and Delbaen, F. (1992) A dynamic reinsurance theory. *Insurance Math. Econom.*, **11**, 31–48.

267 Deelstra, G. and Plantin, G. (2014) *Risk Theory and Reinsurance*, EAA Series, Springer, Heidelberg.

268 Dekkers, A.L., Einmahl, J.H., and De Haan, L. (1989) A moment estimator for the index of an extreme-value distribution. *The Annals of Statistics*, **17** (4), 1833–1855.

269 Delaporte, P. (1959) Quelques problémes de statistique mathématique posés par l'assurance automobile et le bonus pour non sinistre. *Bull. Trimestriel de l' Institut des Actuaires Français*, **227**, 87–102.

270 Dell'Aquila, R. and Embrechts, P. (2006) Extremes and robustness: a contradiction? *Financial Markets and Portfolio Management*, **20** (1), 103–118.

271 Dempster, A.P., Laird, N.M., and Rubin, D.B. (1977) Maximum likelihood from incomplete data via the EM algorithm. *J. R. Stat. Soc. Ser. B Stat. Methodol.*, **39** (1), 1–38.

272 Den Isefer, P., Smith, M., and Dekker, R. (1997) Computing compound distributions faster! *Insurance Math. Econom.*, **20**, 23–34.

273 Denneberg, D. (1994) *Non-Additive Measure and Integral*, vol. 27, Springer.

274 Denuit, M., Dhaene, J., Goovaerts, M., and Kaas, R. (2005) *Actuarial Theory for Dependent Risks*, Wiley, Chichester.

275 Denuit, M., Genest, C., and Marceau, E. (1999) Stochastic bounds on sums of dependent risks. *Insurance Math. Econom.*, **25**, 85–104.

276 Denuit, M., Maréchal, X., Pitrebois, S., and Walhin, J. (2007) *Actuarial Modelling of Claim Counts. Risk Classification, Credibility and Bonus-Malus Systems*, John Wiley & Sons, Chichester.

277 Denuit, M. and Vermandele, C. (1998) Optimal reinsurance and stop-loss order. *Insurance Math. Econom.*, **22**, 229–233.

278 Denuit, M. and Vermandele, C. (1999) Lorenz and excess wealth orders, with applications in reinsurance theory. *Scand. Actuar. J.*, pp. 170–185.

279 Deprez, O. and Gerber, H.U. (1985) On convex principles of premium calculation. *Insurance Math. Econom.*, **4** (3), 179–189.

280 Desmedt, S., Snoussi, M., Chenut, X., and Walhin, J. (2012) Experience and exposure rating for property per risk excess of loss reinsurance revisited. *Astin Bull.*, **42** (1), 233–270.

281 Dhaene, J., Denuit, M., Goovaerts, M., Kaas, R., and D., V. (2002) The concept of comonotonicity in actuarial science and finance: Applications. *Insurance Math. Econom.*, **31** (2), 133–161.

282 Dhaene, J., Denuit, M., Goovaerts, M., Kaas, R., and Vyncke, D. (2002) The concept of comonotonicity in actuarial science and finance: Theory. *Insurance Math. Econom.*, **31** (1), 3–33.
283 Dhaene, J. and Goovaerts, M. (1996) Dependency of risks and stop-loss order. *Astin Bull.*, **26** (2), 201–212.
284 Dhaene, J., Goovaerts, M., and Kaas, R. (2003) Economic capital allocation derived from risk measures. *North American Actuarial Journal*, **7** (2), 44–56.
285 Dhaene, J., Laeven, R.J., Vanduffel, S., Darkiewicz, G., and Goovaerts, M. (2008) Can a coherent risk measure be too subadditive? *Journal of Risk and Insurance*, **75** (2), 365–386.
286 Dhaene, J., Wang, S., Young, V., and Goovaerts, M. (2000) Comonotonicity and maximal stop-loss premiums. *Mitt. Ver. Schweiz. Versich. Math.*, pp. 99–113.
287 Dhaene, J., Willmot, G., and Sundt, B. (1999) Recursions for distribution functions and stop-loss transforms. *Scand. Actuar. J.*, pp. 52–65.
288 Dick, J. and Pillichshammer, F. (2010) *Digital Nets and Sequences: Discrepancy Theory and Quasi–Monte Carlo Integration*, Cambridge University Press.
289 Dickmann, H. (1984) A contribution to the treatment of large claims in motor third party liability insurance. *Proc. 4 countries Astin Coll.*, pp. 195–202.
290 Dickson, D. and Waters, H. (1996) Reinsurance and ruin. *Insurance Math. Econom.*, **19** (1), 61–80.
291 Dickson, D. and Waters, H. (1997) Relative reinsurance retention levels. *Astin Bull.*, **27**, 207–227.
292 Dickson, D. and Waters, H. (2004) Some optimal dividends problems. *Astin Bull.*, **34** (1), 49–74.
293 Dickson, D. and Waters, H. (2006) Optimal dynamic reinsurance. *Astin Bull.*, **36** (2), 415–432.
294 Dieckmann, S. (2011) A consumption-based evaluation of the Cat bond market, *Tech. Rep.*, Working paper, University of Pennsylvania.
295 Dimitrova, D.S. and Kaishev, V.K. (2010) Optimal joint survival reinsurance: An efficient frontier approach. *Insurance Math. Econom.*, **47** (1), 27–35.
296 Doherty, N. (1997) Innovations in managing catastrophe risk. *Journal of Risk and Insurance*, **64** (4), 713–718.
297 Doherty, N. (2000) *Integrated Risk Management: Techniques and Strategies for Managing Corporate Risk: Techniques and Strategies for Managing Corporate Risk*, McGraw Hill.
298 Doherty, N. and Richter, A. (2002) Moral hazard, basis risk and gap insurance. *Journal of Risk and Insurance*, **69** (1), 9–24.
299 Doherty, N. and Smetters, K. (2005) Moral hazard in reinsurance markets. *Journal of Risk and Insurance*, **72** (3), 375–391.
300 Doherty, N.A. (1997) Financial innovation in the management of catastrophe risk. *J. Appl. Corporate Finance*, **10** (3), 84–95.
301 Doherty, N.A. and Schlesinger, H. (1991) Rational insurance purchasing: Consideration of contract non-performance, in *Managing the Insolvency Risk of Insurance Companies*, Springer, pp. 283–294.
302 Doherty, N.A. and Schlesinger, H. (2002) Insurance contracts and securitization. *J. Risk and Insurance*, **69** (1), 45–62.
303 Doherty, N.A. and Tinic, S.M. (1981) Reinsurance under conditions of capital market equilibrium: A note. *J. Finance*, **36** (4), 949–953.

304 Donat, M., Pardowitz, T., Leckebusch, G., Ulbrich, U., and Burghoff, O. (2011) High-resolution refinement of a storm loss model and estimation of return periods of loss-intensive storms over Germany. *Natural Hazards and Earth System Sciences*, **11** (10), 2821–2833.

305 Dorland, C., Tol, R.S., and Palutikof, J.P. (1999) Vulnerability of the Netherlands and Northwest Europe to storm damage under climate change. *Climatic Change*, **43** (3), 513–535.

306 Douglas, J. (1980) *Analysis of Standard Contagious Distributions*, Statistical distributions in scientific work, Vol. 4, International Cooperative Publishing House, Fairland, MD.

307 Douglas, J. (1980) *Analysis with Standard Contagious Distributions*, International Cooperative Publishing House.

308 Drieskens, D., Henry, M., Walhin, J.F., and Wielandts, J. (2012) Stochastic projection for large individual losses. *Scand. Actuar. J.*, pp. 1–39.

309 Drmota, M. and Tichy, R. (1997) *Sequences, Discrepancies and Applications, Lecture Notes in Mathematics*, vol. 1651, Springer, New York, Berlin, Heidelberg, Tokyo.

310 Dubourdieu, J. (1938) Remarques relatives à la théorie mathématique de l'assurance-accidents. *Bull. Trim. Inst. Actu. Français*, **44**, 79–146.

311 Duffie, D., Saita, L., and Wang, K. (2007) Multi-period corporate default prediction with stochastic covariates. *J. Financial Econom.*, **83** (3), 635–665.

312 d'Ursel, L. and Lauwers, M. (1985) Chains of reinsurance: non-cooperative equilibria and Pareto optimality. *Insurance Math. Econom.*, **4**, 279–285.

313 Eeckhoudt, L., G.C. and Schlesinger, H. (2005) *Economic and Financial Decisions under Risk*, Princeton University Press.

314 Ehrgott, M. (2005) *Multicriteria Optimization*, Springer, Heidelberg.

315 Einmahl, J., Fils-Villetard, A., and Guillou, A. (2008) Statistics of extremes under random censoring. *Bernoulli*, **14** (1), 207–227.

316 Einmahl, J.H., De Haan, L., and Zhou, C. (2016) Statistics of heteroscedastic extremes. *J. R. Stat. Soc. Ser. B Stat. Methodol.*, **78** (1), 31–51.

317 Eisenberg, J. and Schmidli, H. (2009) Optimal control of capital injections by reinsurance in a diffusion approximation. *Blätter der DGVFM*, **30** (1), 1–13.

318 Eisenberg, J. and Schmidli, H. (2011) Minimising expected discounted capital injections by reinsurance in a classical risk model. *Scand. Actuar. J.*, pp. 155–176.

319 Ekheden, E. and Hössjer, O. (2014) Pricing catastrophe risk in life (re)insurance. *Scand. Actuar. J.*, pp. 352–367.

320 El Methni, J., Gardes, L., Girard, S., and Guillou, A. (2012) Estimation of extreme quantiles from heavy and light tailed distribution functions. *J. Statist. Planning and Inference*, **142**, 2735–2747.

321 Eling, M. and Parnitzke, T. (2007) Dynamic financial analysis: Classification, conception, and implementation. *Risk Management and Insurance Review*, **10** (1), 33–50.

322 Eling, M. and Wirfs, J. (2016) *Cyber Risk: Too Big to Insure? Risk Transfer Options for a Mercurial Risk Class*, IVW HSG Schriftenreihe, Band 59.

323 Embrechts, P. (1983) A property of the generalized inverse Gaussian distribution with some applications. *J. Appl. Probab.*, **20**, 537–544.

324 Embrechts, P. (2000) Actuarial versus financial pricing of insurance. *J. Risk Finance*, **1** (4), 17–26.

325 Embrechts, P. and Frei, M. (2009) Panjer recursion versus FFT for compound distributions. *Math. Methods Oper. Res.*, **69** (3), 497–508.
326 Embrechts, P., Goldie, C., and Veraverbeke, N. (1982) Subexponentiality and infinite divisibility. *Z. Wahrscheinlichkeitsth.*, **49**, 335–347.
327 Embrechts, P., Grübel, R., and Pitts, S.M. (1993) Some applications of the fast Fourier transform algorithm in insurance mathematics. *Statist. Neerlandica*, **47** (1), 59–75.
328 Embrechts, P., Jensen, J., Maejima, M., and Teugels, J. (1985) Approximations for compound Poisson and Pólya processes. *Adv. Appl. Probab.*, **17**, 623–637.
329 Embrechts, P., Klüppelberg, C., and Mikosch, T. (1997) *Modelling Extremal Events*, Springer, Berlin.
330 Embrechts, P., Liu, H., and Wang, R. (2016) Quantile-based risk sharing. *Preprint, available at SSRN 2744142.*
331 Embrechts, P., Maejima, M., and Omey, E. (1984) A renewal theorem of Blackwell type. *Ann. Probab.*, **12**, 561–570.
332 Embrechts, P., Maejima, M., and Teugels, J. (1985) Asymptotic behaviour of compound distributions. *Astin Bull.*, **15**, 45–48.
333 Embrechts, P. and Meister, S. (1997) Pricing insurance derivatives, the case of cat-futures, in *Proc. 1995 Bowles Symposium on Securitization of Insurance Risk, Georgia State University Atlanta*, pp. 15–26.
334 Embrechts, P. and Puccetti, G. (2006) Bounds for functions of dependent risks. *Finance and Stochastics*, **10** (3), 341–352.
335 Embrechts, P. and Puccetti, G. (2010) Risk aggregation, in *Copula Theory and Its Applications*, Springer, pp. 111–126.
336 Embrechts, P., Puccetti, G., and Rüschendorf, L. (2013) Model uncertainty and VaR aggregation. *Journal of Banking and Finance*, **37** (8), 2750–2764.
337 Embrechts, P. and Veraverbeke, N. (1982) Estimates for the probability of ruin with special emphasis on the possibility of large claims. *Insurance Math. Econom.*, **1**, 55–72.
338 Embrechts, P., Wang, B., and Wang, R. (2015) Aggregation-robustness and model uncertainty of regulatory risk measures. *Finance and Stochastics*, **19** (4), 763–790.
339 Epstein, R. (1996) Catastrophic responses to catastrophic risks. *J. Risk Uncertainty*, **12**, 287–308.
340 Ericson, R. and Doyle, A. (2004) Catastrophe risk, insurance and terrorism. *Economy and Society*, **33** (2), 135–173.
341 Euler, L. (1748) *Introductio in Analysin Infinitorum*, vol. 1, Lausanne. Available at http://eulerarchive.maa.org//.
342 Fackler, M. (2011) Inflation and excess insurance, in *ASTIN Colloquium, Madrid.*
343 Fackler, M. (2013) Reinventing Pareto: Fits for both small and large losses, in *ASTIN Colloquium, Den Haag.*
344 Federal Office of Private Insurance (2006) Technisches Dokument zum Swiss Solvency Test.
345 Feller, W. (1971) *An Introduction to Probability Theory and its Applications*, vol. 2, John Wiley & Sons.
346 Feng, R. and Garrido, J. (2011) Actuarial applications of epidemiological models. *North American Actuarial Journal*, **15** (1), 112–136.
347 Ferrara, G. (1971) Distributions des sinistres incendie selon leur coût. *Astin Bull.*, **6**, 31–41.

348 Feyen, L., Dankers, R., Bódis, K., Salamon, P., and Barredo, J.I. (2012) Fluvial flood risk in Europe in present and future climates. *Climatic Change*, **112** (1), 47–62.
349 Filipović, D. (2009) Multi-level risk aggregation. *Astin Bull.*, **39** (2), 565–575.
350 Filipović, D. and Kupper, M. (2008) Optimal capital and risk transfers for group diversification. *Math. Finance*, **18** (1), 55–76.
351 Filipović, D. and Svindland, G. (2008) Optimal capital and risk allocations for law-and cash-invariant convex functions. *Finance and Stochastics*, **12** (3), 423–439.
352 Fischer, T. (2003) Risk capital allocation by coherent risk measures based on one-sided moments. *Insurance Math. Econom.*, **32**, 135–146.
353 Fisher, R.A. and Tippett, L.H.C. (1928) Limiting forms of the frequency distribution of the largest or smallest member of a sample, in *Mathematical Proceedings of the Cambridge Philosophical Society*, vol. 24, vol. 24, pp. 180–190.
354 Fitch Ratings (2015) Global reinsurance guide. Https://www.fitchratings.com.
355 Flåm, S.D. (2016) Borch's theorem, equal margins, and efficient allocation. *Insurance Math. Econom.*, **70**, 162–168.
356 Foss, S., Korshunov, D., and Zachary, S. (2013) *An Introduction to Heavy-Tailed and Subexponential Distributions*, Springer, New York, 2nd edn.
357 Fougeres, A.L., De Haan, L., and Mercadier, C. (2015) Bias correction in multivariate extremes. *The Annals of Statistics*, **43** (2), 903–934.
358 Franckx, E. (1963) Sur la fonction de distribution du sinistre le plus élevé. *Astin Bull.*, **2**, 415–424.
359 Frees, E. and Valdez, E. (1998) Understanding relationships using copulas. *North American Actuarial Journal*, **2** (1), 1–25.
360 Froot, K.A. (2001) The market for catastrophe risk: a clinical examination. *Journal of Financial Economics*, **60** (2), 529–571.
361 Froot, K.A. (2007) Risk management, capital budgeting, and capital structure policy for insurers and reinsurers. *J. Risk and Insurance*, **74** (2), 273–299.
362 Furman, E. and Zitikis, R. (2009) Weighted pricing functionals with applications to insurance: an overview. *North American Actuarial Journal*, **13** (4), 483–496.
363 Furrer, H., Michna, Z., and Weron, A. (1997) Stable Lévy motion approximation in collective risk theory. *Insurance Math. Econom.*, **20**, 97–114.
364 Gajek, L. and Zagrodny, D. (2000) Insurer's optimal reinsurance strategies. *Insurance Math. Econom.*, **27**, 105–112.
365 Gajek, L. and Zagrodny, D. (2004) Optimal reinsurance under general risk measures. *Insurance Math. Econom.*, **34** (2), 227–240.
366 Gajek, L. and Zagrodny, D. (2004) Reinsurance arrangements maximizing insurer's survival probability. *J. Risk and Insurance*, **71** (3), 421–435.
367 Galambos, J. (1978) *The Asymptotic Theory of Extreme Order Statistics*, J. Wiley and Sons, New York.
368 Gardes, L. and Girard, S. (2008) A moving window approach for nonparametric estimation of the conditional tail index. *Journal of Multivariate Analysis*, **99** (10), 2368–2388.
369 Gardes, L. and Girard, S. (2010) Conditional extremes from heavy-tailed distributions: An application to the estimation of extreme rainfall return levels. *Extremes*, **13** (2), 177–204.
370 Gardes, L. and Stupfler, G. (2014) Estimation of the conditional tail index using a smoothed local Hill estimator. *Extremes*, **17** (1), 45–75.

371 Garrido, J., Genest, C., and Schulz, J. (2016) Generalized linear models for dependent frequency and severity of insurance claims. *Insurance Math. Econom.*, **70**, 205–215.
372 Garven, J.R. (1987) On the application of finance theory to the insurance firm. *J. Financial Services Res.*, **1** (1), 57–76.
373 Garven, J.R., Hilliard, J.I., and Grace, M.F. (2014) Adverse selection in reinsurance markets. *The Geneva Risk and Insurance Review*, **39** (2), 222–253.
374 Garven, J.R. and Lamm-Tennant, J. (2003) The demand for reinsurance: Theory and empirical tests. *Insurance and Risk Management*, **7** (3), 217–237.
375 Gastel, R. (2004) *Reinsurance: Fundamentals and New Challenges*, 4th Edition, Insurance Information Institute.
376 Gatzert, N., Pokutta, S., and Vogl, N. (2017) Convergence of capital and insurance markets: Consistent pricing of index-linked catastrophic loss instruments. *J. Risk and Insurance*. Online.
377 Gatzert, N. and Schmeiser, H. (2011) On the risk situation of financial conglomerates: does diversification matter? *Financial Markets and Portfolio Management*, **25** (1), 3–26.
378 Geman, H. and Yor, M. (1997) Stochastic time changes in catastrophic option pricing. *Insurance Math. Econom.*, **21**, 185–193.
379 Gençay, R., Selçuk, F., and Ulugülyagci, A. (2003) High volatility, thick tails and extreme value theory in Value-at-Risk estimation. *Insurance Math. Econom.*, **33**, 337–356.
380 Gendron, M. and Crépeau, H. (1989) On the computation of the aggregate claim distribution when individual claims are inverse Gaussian. *Insurance Math. Econom.*, **8**, 251–258.
381 Genest, C. and Nešlehová, J. (2007) A primer on copulas for count data. *Astin Bull.*, **37** (2), 475–515.
382 Gerathewohl, K. (1976) *Rückversicherung - Grundlagen und Praxis*, Band I, Verlag Versicherungswirtschaft, Karlsruhe.
383 Gerber, H. (1979) *An Introduction to Mathematical Risk Theory*, Huebner Foundation Monograph 8, Homewood, Illinois.
384 Gerber, H. (1984) Chains of reinsurance. *Insurance Math. Econom.*, **3**, 43–48.
385 Gerber, H. (1984) Equilibria in a proportional reinsurance market. *Insurance Math. Econom.*, **3**, 97–100.
386 Gerber, H., Shiu, E., and Yang, H. (2016) A constraint-free approach to optimal reinsurance. *Preprint, University of Lausanne*.
387 Gerber, H.U. and Pafumi, G. (1998) Utility functions: from risk theory to finance. *North American Actuarial Journal*, **2** (3), 74–91.
388 Gerhold, S., Schmock, U., and Warnung, R. (2010) A generalization of Panjer's recursion and numerically stable risk aggregation. *Finance Stoch.*, **14** (1), 81–128.
389 Gibson, R., Habib, M.A., and Ziegler, A. (2014) Reinsurance or securitization: the case of natural catastrophe risk. *J. Math. Econom.*, **53**, 79–100.
390 Glasserman, P. (2004) *Monte Carlo Methods in Financial Engineering, Applications of Mathematics (New York)*, vol. 53, Springer-Verlag, New York.
391 Gnedenko, B. (1943) Sur la distribution limite du terme maximum d'une serie aléatoire. *Annals of Mathematics*, **4** (3), 423–453.
392 Göbel, D. (1993) Reinsurance worldwide - Opportunities, risks and perspectives. *Geneva Papers on Risk and Insurance*, **18**, 426–431.

393 Goegebeur, Y., Guillou, A., and Osmann, M. (2016) A local moment type estimator for an extreme quantile in regression with random covariates. *Communications in Statistics-Theory and Methods.* In press.
394 Gogol, D. (1994) An actuarial approach to property catastrophe cover rating. *Proc. Casualty Actuarial Soc.*, **81**, 1–35.
395 Gollier, C. (1987) Pareto-optimal risk sharing with fixed costs per claim. *Scand. Actuar. J.*, pp. 62–73.
396 Gollier, C. (1997) About the insurability of catastrophic risks. *Geneva Papers on Risk and Insurance*, **22**, 177–186.
397 Gollier, C. (2003) To insure or not to insure?: an insurance puzzle. *Geneva Papers on Risk and Insurance Theory*, **28**, 5–24.
398 Gomes, I., De Haan, L., and Rodrigues, L.H. (2008) Tail index estimation for heavy-tailed models: accommodation of bias in weighted log-excesses. *J. R. Stat. Soc. Ser. B Stat. Methodol.*, **70** (1), 31–52.
399 Gomesa, M.I. and Martins, M.J. (2002) Asymptotically unbiased estimators of the tail index based on external estimation of the second order parameter. *Extremes*, **5** (1), 5–31.
400 Goovaerts, M., de Vijlder, E., and Haezendonck, J. (1984) *Insurance Premiums: Theory and Applications*, North-Holland, Amsterdam.
401 Goovaerts, M., Kaas, R., Dhaene, J., and Tang, Q. (2003) A unified approach to generate risk measures. *Astin Bull.*, **33**, 173–191.
402 Goovaerts, M., Kaas, R., van Heerwaarden, A., and Bauwelinckx, T. (1990) *Effective Actuarial Methods*, Insurance Series, 3, North-Holland.
403 Gorvett, R. (1999) Insurance securitization: The development of a new asset class, in *Securitization of Risk*, Casualty Actuarial Society, pp. 133–173.
404 Grace, M., Klein, R., and Kleindorfer, P. (2000) The demand of homeowners insurance with bundled catastrophe coverages, in *Working Paper, Wharton Managing Catastrophic Risks Project*, University of Pennsylvania, Philadelphia.
405 Grandell, J. (1992) *Aspects of Risk Theory*, Springer, New York.
406 Grandell, J. (1997) *Mixed Poisson Processes*, vol. 77, Chapman & Hall, London.
407 Grandits, P., Kainhofer, R., and Temnov, G. (2010) On the impact of hidden trends for a compound Poisson model with Pareto-type claims. *Int. J. Theoret. Appl. Finance*, **13** (06), 959–978.
408 Green, P.J. and Silverman, B.W. (1993) *Nonparametric Regression and Generalized Linear Models: a Roughness Penalty Approach*, CRC Press.
409 Grossmann, M. (1977) *Rückversicherung - eine Einführung*, Peter Lang Verlag.
410 Grübel, R. (1987) On subordinated distributions and generalized renewal measures. *Ann. Probab.*, **15** (1), 394–415.
411 Grübel, R. and Hermesmeier, R. (1999) Computation of compound distributions I: Aliasing errors and exponential tilting. *Astin Bull.*, **29** (2), 197–214.
412 Gründl, H. and Schmeiser, H. (2002) Pricing double-trigger reinsurance contracts: financial versus actuarial approach. *J. Risk and Insurance*, **69** (4), 449–468.
413 Guerra, M. and Centeno, L. (2008) Optimal reinsurance policy: The adjustment coefficient and the expected utility criteria. *Insurance Math. Econom.*, **42** (2), 529–539.
414 Guerra, M. and Centeno, L. (2010) Optimal reinsurance for variance related premium calculation principles. *Astin Bull.*, **40** (1), 97–121.

415 Guillou, A., Naveau, P., and You, A. (2015) A folding methodology for multivariate extremes: estimation of the spectral probability measure and actuarial applications. *Scand. Actuar. J.*, pp. 549–572.

416 Gurenko, E. and Itigin, A. (2013) Reinsurance as capital optimization tool under solvency II. *World Bank Policy Research Working Paper 6306*.

417 Haas, S. (2012) *Optimal Reinsurance Forms and Solvency*, Ph.D. thesis, Université de Lausanne.

418 Hald, M. and Schmidli, H. (2004) On the maximisation of the adjustment coefficient under proportional reinsurance. *Astin Bull.*, **34** (1), 75–83.

419 Hall, P. (1982) On some simple estimates of an exponent of regular variation. *J. R. Stat. Soc. Ser. B Stat. Methodol.*, **44** (1), 37–42.

420 Hall, P. and Tajvidi, N. (2000) Nonparametric analysis of temporal trend when fitting parametric models to extreme-value data. *Statistical Science*, **15** (2), 153–167.

421 Harrington, S.E., Mann, S.V., and Niehaus, G. (1995) Insurer capital structure decisions and the viability of insurance derivatives. *J. Risk and Insurance*, **62** (3), 483–508.

422 Hartinger, J. and Kortschak, D. (2009) On the efficiency of the Asmussen–Kroese estimator and its application to stop-loss transforms. *Blätter der DGVFM*, **30** (2), 363–377.

423 Hashorva, E. (2003) On the number of near-maximum insurance claims under dependence. *Insurance Math. Econom.*, **32**, 37–49.

424 Hashorva, E. (2004) Bivariate maximum insurance claim and related point processes. *Statistics and Probability Letters*, **69** (2), 117–128.

425 Hashorva, E. (2007) On the asymptotic distribution of certain bivariate reinsurance treaties. *Insurance Math. Econom.*, **40** (2), 200–208.

426 Hashorva, E. and Li, J. (2013) ECOMOR and LCR reinsurance with gamma-like claims. *Insurance Math. Econom.*, **53** (1), 206–215.

427 Haslip, G.G. and Kaishev, V.K. (2010) Pricing of reinsurance contracts in the presence of catastrophe bonds. *Astin Bull.*, **40** (01), 307–329.

428 Hastie, T.J. and Tibshirani, R.J. (1990) *Generalized Additive Models*, vol. 43, CRC Press.

429 Haueter, N.V. and Jones, G. (2016) *Managing Risk in Reinsurance: From City Fires to Global Warming*, Oxford University Press.

430 Heijnen, B. (1989) Perturbation calculus in risk theory: Applications to chains and trees of reinsurance. *Insurance Math. Econom.*, **8**, 97–104.

431 Heijnen, B. (1990) Best upper and lower bounds on modified stop-loss premiums in case of known range, mode, mean and variance of the original risk. *Insurance Math. Econom.*, **9**, 207–220.

432 Heijnen, B. and Gerber, H. (1986) On the small risk approximation. *Insurance Math. Econom.*, **5**, 151–157.

433 Heilmann, W.R. (1982) An approach to the retention problem based on claim experience. *Blätter der DGVFM*, **15**, 397–404.

434 Heilmann, W.R. (1985) Transformations of claim distributions. *Mitt. Ver. Schweiz. Versich. Math.*, pp. 57–69.

435 Heilmann, W.R. (1988) *Fundamentals of Risk Theory*, Verlag Versicherungswirtschaft e.V., Karlsruhe.

436 Heneka, P. and Ruck, B. (2008) A damage model for the assessment of storm damage to buildings. *Engineering Structures*, **30** (12), 3603–3609.

437 Hess, A. (1998) *Financial Reinsurance*, Verlag Versicherungswirtschaft e.V., Karlsruhe.

438 Hess, K., Liewald, A., and Schmidt, K. (2002) An extension of Panjer's recursion. *Astin Bull.*, **32**, 283–297.

439 Hess, K.T. and Schmidt, K.D. (2004) Optimal premium plans for reinsurance with reinstatements. *Astin Bull.*, **34** (2), 299–313.

440 Hesselager, O. (1990) Some results on optimal reinsurance in terms of the adjustment coefficient. *Scand. Actuar. J.*, pp. 80–95.

441 Hesselager, O. (1993) A class of conjugate priors with applications to excess-of-loss reinsurance. *Astin Bull.*, **23**, 77–93.

442 Hill, B.M. (1975) A simple general approach to inference about the tail of a distribution. *The Annals of Statistics*, **3** (5), 1163–1174.

443 Hipp, C. (1985) Approximation of aggregate claims distributions by compound Poisson distributions. *Insurance Math. Econom.*, **4** (4), 227–232.

444 Hipp, C. (2004) Stochastic control with application in insurance, in *Stochastic Methods in Finance*, Springer, pp. 127–164.

445 Hipp, C. (2006) Speedy convolution algorithms and Panjer recursions for phase-type distributions. *Insurance Math. Econom.*, **38** (1), 176–188.

446 Hipp, C. (2016) Stochastic control for insurance: Models, strategies and numerics. *Preprint.*

447 Hipp, C. and Taksar, M. (2010) Optimal non-proportional reinsurance control. *Insurance Math. Econom.*, **47** (2), 246–254.

448 Hipp, C. and Vogt, M. (2003) Optimal dynamic XL reinsurance. *Astin Bull.*, **33** (02), 193–207.

449 Hogg, R. and Klugman, S. (1983) On the estimation of long tailed distributions with actuarial applications. *J. Econometrics*, **23**, 91–102.

450 Højgaard, B. and Taksar, M. (2001) Optimal risk control for a large corporation in the presence of returns on investments. *Finance and Stochastics*, **5** (4), 527–547.

451 Holland, D.M. (2009) A brief history of reinsurance. *Reinsurance News, Special Edition*, pp. 4–29.

452 Hu, X., Yang, H., and Zhang, L. (2015) Optimal retention for a stop-loss reinsurance with incomplete information. *Insurance Math. Econom.*, **65**, 15–21.

453 Hürlimann, W. (1994) A note on experience rating, reinsurance and premium principles. *Insurance Math. Econom.*, **14**, 197–204.

454 Hürlimann, W. (1997) On quasi-mean value principles. *Blätter der DGVFM*, **23**, 1–16.

455 Hürlimann, W. (2005) Excess of loss reinsurance with reinstatements revisited. *Astin Bull.*, **35** (1), 211–238.

456 IAA (2009) *Measurement of liabilities for insurance contracts: current estimates and risk margins*, Risk Margin Working Group, International Actuarial Association.

457 Ignatov, Z.G., Kaishev, V.K., and Krachunov, R.S. (2004) Optimal retention levels, given the joint survival of cedent and reinsurer. *Scand. Actuar. J.*, pp. 401–430.

458 Irgens, C. and Paulsen, J. (2004) Optimal control of risk exposure, reinsurance and investments for insurance portfolios. *Insurance Math. Econom.*, **35** (1), 21–51.

459 Ivanovs, J. and Boxma, O. (2015) A bivariate risk model with mutual deficit coverage. *Insurance Math. Econom.*, **64**, 126–134.

460 Jacque, L. and Tapiero, C. (1987) Premium valuation in international insurance. *Scand. Actuar. J.*, pp. 50–61.

461 Jaimungal, S. and Wang, T. (2006) Catastrophe options with stochastic interest rates and compound Poisson losses. *Insurance Math. Econom.*, **38** (3), 469–483.

462 Jang, J.W. and Krvavych, Y. (2004) Arbitrage-free premium calculation for extreme losses using the shot noise process and the Esscher transform. *Insurance Math. Econom.*, **35** (1), 97–111.

463 Jensen, J. (1991) Saddlepoint approximations to the distribution of the total claim amount in some recent risk models. *Scand. Actuar. J.*, pp. 154–168.

464 Jensen, J. (1995) *Saddlepoint Approximations*, Oxford University Press, Oxford.

465 Jiang, J. and Tang, Q. (2008) Reinsurance under the LCR and ECOMOR treaties with emphasis on light-tailed claims. *Insurance Math. Econom.*, **43** (3), 431–436.

466 Jiang, W., Ren, J., and Zitikis, R. (2017) Optimal reinsurance policies when the interests of both the cedent and the reinsurer are taken into account. *Risks*, **5** (1), 11.

467 Joe, H. (1997) *Multivariate Models and Dependence Concepts*, Chapman & Hall, London.

468 Johnson, N.L. and Kotz, S. (1969) *Distributions in Statistics: Discrete Distributions*, Houghton Mifflin Co., Boston, Massachusetts.

469 Jones, R. and Casti, J. (2000) Insurance world: a process for understanding risk flows due to catastrophes in the insurance/reinsurance industry. *Euroconference IIASA, Laxenburg.*

470 Jongman, B., Kreibich, H., Apel, H., Barredo, J., Bates, P., Feyen, L., Gericke, A., Neal, J., Aerts, J., and Ward, P. (2012) Comparative flood damage model assessment: towards a European approach. *Natural Hazards and Earth System Sciences*, **12** (12), 3733–3752.

471 Jouini, E., Schachermayer, W., and Touzi, N. (2008) Optimal risk sharing for law invariant monetary utility functions. *Mathematical Finance*, **18** (2), 269–292.

472 Juneja, S. and Shahabuddin, P. (2002) Simulating heavy-tailed processes using delayed hazard rate twisting. *ACM Tomacs*, **12**, 94–118.

473 Jung, J. (1964) On the use of extreme values to estimate the premium for an excess-of-loss reinsurance. *Astin Bull.*, **3**, 178–184.

474 Kaas, R. (1993) How to (and how not to) compute stop-loss premiums in practice. *Insurance Math. Econom.*, **13**, 214–254.

475 Kaas, R., Goovaerts, M., Dhaene, J., and Denuit, M. (2001) *Modern Actuarial Risk Theory*, Kluwer Academic Publishers, Boston.

476 Kaas, R., Goovaerts, M., Dhaene, J., and Denuit, M. (2008) *Modern Actuarial Risk Theory, using R*, Springer Science and Business Media.

477 Kaishev, V.K. and Dimitrova, D.S. (2006) Excess of loss reinsurance under joint survival optimality. *Insurance Math. Econom.*, **39** (3), 376–389.

478 Kaluszka, M. (2001) Optimal reinsurance under mean-variance premium principles. *Insurance Math. Econom.*, **28**, 61–67.

479 Kaluszka, M. (2004) An extension of Arrow's result on optimality of a stop loss contract. *Insurance Math. Econom.*, **35** (3), 527–536.

480 Kaluszka, M. (2004) An extension of the Gerber-Bühlmann-Jewell conditions for optimal risk sharing. *Astin Bull.*, **34** (1), 27–48.

481 Kaluszka, M. (2004) Mean-variance optimal reinsurance arrangements. *Scand. Actuar. J.*, pp. 28–41.

482 Kaplan, E.L. and Meier, P. (1958) Nonparametric estimation from incomplete observations. *Journal of the American Statistical Association*, **53** (282), 457–481.

483 Kaufmann, R., Gadmer, A., and Klett, R. (2001) Introduction to dynamic financial analysis. *Astin Bull.*, **31**, 213–249.

484 Kchouk, B. and Mailhot, M. (2016) Reciprocal reinsurance treaties under an optimal and fair joint survival probability. *Variance,* (forthcoming).

485 Keller, P. (2007) Group diversification. *The Geneva Papers on Risk and Insurance: Issues and Practice,* **32** (3), 382–392.

486 Kestemont, R. and Paris, J. (1985) Sur l'ájustement du nombre des sinistres. *Mitt. Ver. Schweiz. Versich. Math.*, pp. 157–164.

487 Kijko, A. and Singh, M. (2011) Statistical tools for maximum possible earthquake magnitude estimation. *Acta Geophysica,* **59** (4), 674–700.

488 Kiriliouk, A., Rootzén, H., Segers, J. and Wadsworth, J.L. (2016) Peaks over Thresholds modelling with multivariate generalized Pareto distributions. arXiv:1612.01773.

489 Klaassen, P. and van Eeghen, I. (2009) *Economic Capital: How It Works, and What Every Manager Needs to Know,* Elsevier.

490 Klawa, M. and Ulbrich, U. (2003) A model for the estimation of storm losses and the identification of severe winter storms in Germany. *Natural Hazards and Earth System Science,* **3** (6), 725–732.

491 Klugman, S.A., Panjer, H.H., and Willmot, G.E. (2012) *Loss Models: From Data to Decisions,* Wiley Series in Probability and Statistics, John Wiley & Sons, Hoboken, NJ, 4th edn.

492 Klugman, S.A., Panjer, H.H., and Willmot, G.E. (2013) *Loss Models: Further Topics,* John Wiley & Sons.

493 Klüppelberg, C. (1988) Subexponential distributions and integrated tails. *J. Appl. Probabability,* **25**, 132–141.

494 Knispel, T., Laeven, R.J., and Svindland, G. (2016) Robust optimal risk sharing and risk premia in expanding pools. *Insurance Math. Econom.* In press.

495 Knuth, D. (1969) *The Art of Computer Programming: Vol 2: Seminumerical Algorithms,* Addison-Wesley.

496 Kopf, E. (1929) Notes on the origin and development of reinsurance. *Proceedings of the Casualty Actuarial Society,* pp. 22–91.

497 Korn, E., Korn, R., and Kroisandt, G. (2009) *Monte Carlo Methods in Financial and Actuarial Models,* Chapman and Hall/CRC, London.

498 Korn, R., Menkens, O., and Steffensen, M. (2012) Worst-case-optimal dynamic reinsurance for large claims. *Eur. Actuar. J.*, **2** (1), 21–48.

499 Kozlowski, R. and Mathewson, S. (1995) Measuring and managing catastrophic risk. *Journal of Actuarial Practice,* **3**, 211–241.

500 Kozubowski, T.J. and Podgórski, K. (2009) Distributional properties of the negative binomial Lévy process. *Prob. Math. Stat.*, **29** (1), 43–71.

501 Kremer, E. (1983) Distribution-free upper bounds on the premiums of the LCR and ECOMOR treaties. *Insurance Math. Econom.*, **2**, 209–213.

502 Kremer, E. (1990) The asymptotic efficiency of largest claims reinsurance treaties. *Astin Bull.*, **20**, 11–22.

503 Kremer, E. (1990) An elementary upper bound for the loading of a stop-loss cover. *Scand. Actuar. J.*, pp. 105–108.

504 Kremer, E. (1991) Large claims in credibility. *Blätter der DGVFM,* **20**, 123–150.

505 Kremer, E. (1992) The total claim amount of largest claims reinsurance treaties revisited. *Blätter der DGVFM,* **20**, 431–439.

506 Kremer, E. (2006) Net premium of the drop down excess of loss cover. *Nonlinear Analysis: Real World Applications,* **7** (3), 478–485.

507 Kreps, R. (1990) Reinsurer risk loads from marginal surplus requirements. *Proc. Casualty Actuarial Soc.*, **77**, 196–203.

508 Krieter, F., Bloch, A., Meyer, P., Schmid, E., and Bernegger, S. (1995) Catastrophe portfolio and capital needs of a reinsurance company. *Trans. 25th Intern. Congress Actuaries*, pp. 199–210.

509 Kroll, Y. and Nye, D. (1991) Reinsurance retention levels for property/liability firms: a managerial portfolio selection framework. *Insurance Math. Econom.*, **10**, 109–123.

510 Krvavych, Y. (2013) Making use of internal capital models, in *Proceedings of the Int. ASTIN Colloquium*, The Hague.

511 Krvavych, Y. (2014) Uncertainty in catastrophe modelling, in *Int. Congress of Actuaries, Washington.*

512 Krvavych, Y. (2016) Insurance ERM and advanced uses of internal models in a post Solvency II world, in *Int. ASTIN Colloquium, Lisbon.*

513 Krvavych, Y. and Sherris, M. (2006) Enhancing insurer value through reinsurance optimization. *Insurance Math. Econom.*, **38** (3), 495–517.

514 Kull, A. (2009) Sharing risk—an economic perspective. *Astin Bull.*, **39** (2), 591–613.

515 Kunreuther, H. (2002) The role of insurance in managing extreme events: implications for terrorism coverage. *Business Economics*, **37** (2), 6–16.

516 Kupper, J. (1962) Wahrscheinlichkeitstheoretische Modelle in der Schadenversicherung. Teil I: Die Schadenzahl. *Blätter der DGVFM*, **5**, 451–503.

517 Kupper, J. (1962) Wahrscheinlichkeitstheoretische Modelle in der Schadenversicherung. Teil II: Schadenhöhe und Totalschaden. *Blätter der DGVFM*, **6**, 95–130.

518 Kupper, J. (1963) Some aspects of cumulative risk. *Astin Bull.*, **3**, 85–103.

519 Kupper, J. (1971) Contributions to the theory of the largest claim cover. *Astin Bull.*, **6**, 134–146.

520 Kupper, J. (1972) Kapazität und Höchstschadenversicherung. *Mitt. Ver. Schweiz. Versich. Math.*, pp. 249–258.

521 Kyprianou, A.E. (2014) *Fluctuations of Lévy Processes with Applications. Introductory Lectures*, Springer, Berlin, 2nd edn.

522 Ladoucette, S.A. and Teugels, J.L. (2006) Analysis of risk measures for reinsurance layers. *Insurance Math. Econom.*, **38** (3), 630–639.

523 Ladoucette, S.A. and Teugels, J.L. (2006) The largest claims treaty ECOMOR, in *Progress in industrial mathematics at ECMI 2004, Math. Ind.*, vol. 8, Springer, Berlin, pp. 422–426.

524 Ladoucette, S.A. and Teugels, J.L. (2006) Reinsurance of large claims. *J. Comput. Appl. Math.*, **186** (1), 163–190.

525 Ladoucette, S.A. and Teugels, J.L. (2007) Asymptotics for ratios with applications to reinsurance. *Meth. Comput. Appl. Prob.*, **9** (2), 225–242.

526 Lampaert, I. and Walhin, J.F. (2005) On the optimality of proportional reinsurance. *Scand. Actuar. J.*, pp. 225–239.

527 Landriault, D., Willmot, G.E., and Xu, D. (2014) On the analysis of time dependent claims in a class of birth process claim count models. *Insurance Math. Econom.*, **58**, 168–173.

528 Lane, M. (2012) *Alternative (Re)insurance Strategies*, Risk Books, London.

529 Leadbetter, R. (1995) On high level exceedance modeling and tail inference. *J. Statistical Planning & Inference*, **45**, 247–260.

530 Ledford, A. and Tawn, J. (1997) Modelling dependence within joint tail regions. *J. Roy. Statist. Soc.: Series B*, **59**, 475–499.

531 Lee, D., Li, W.K., and Wong, T.S.T. (2012) Modeling insurance claims via a mixture exponential model combined with peaks-over-threshold approach. *Insurance Math. Econom.*, **51** (3), 538–550.

532 Lee, J.P. and Yu, M.T. (2007) Valuation of catastrophe reinsurance with catastrophe bonds. *Insurance Math. Econom.*, **41** (2), 264–278.

533 Lee, S. and Lin, X.S. (2012) Modeling dependent risks with multivariate Erlang mixtures. *Astin Bull.*, **42** (1), 153–180.

534 Lee, S.C. and Lin, X.S. (2010) Modeling and evaluating insurance losses via mixtures of Erlang distributions. *North American Actuarial Journal*, **14** (1), 107–130.

535 Lemaire, J. (1973) Sur la détermination d'un contrat optimal de réassurance. *Astin Bull.*, **7**, 165–180.

536 Lemaire, J. (1977) Exchange de risques entre assureurs et théorie des jeux. *Astin Bull.*, **9** (1-2), 155–180.

537 Lemaire, J. and Quairière, J.P. (1986) Chains of reinsurance revisited. *Astin Bull.*, **16**, 77–88.

538 Lemieux, C. (2009) *Monte Carlo and Quasi-Monte Carlo Sampling*, Springer, New York.

539 Lescourret, L. and Robert, C.Y. (2006) Extreme dependence of multivariate catastrophic losses. *Scand. Actuar. J.*, pp. 203–225.

540 Li, Y. and Pakes, A. (2001) On the number of near-maximum insurance claims. *Insurance Math. Econom.*, **28**, 309–323.

541 Liang, Z. and Bayraktar, E. (2014) Optimal reinsurance and investment with unobservable claim size and intensity. *Insurance Math. Econom.*, **55**, 156–166.

542 Liang, Z. and Yuen, K.C. (2016) Optimal dynamic reinsurance with dependent risks: variance premium principle. *Scand. Actuar. J.*, pp. 18–36.

543 Liebwein, P. (2009) *Klassische und moderne Formen der Rückversicherung*, Verlag Versicherungswirtschaft, Karlsruhe.

544 Lin, S.K., Chang, C.C., and Powers, M.R. (2009) The valuation of contingent capital with catastrophe risks. *Insurance: Math. Econom.*, **45** (1), 65–73.

545 Lindskog, F. and McNeil, A.J. (2003) Common Poisson shock models: applications to insurance and credit risk modelling. *Astin Bull.*, **33** (2), 209–238.

546 Lippe, S., Sedlmair, H., and Witting, T. (1991) Praxisrelevante Aspekte der Burning-Cost-Kalkulation. *Blätter der DGVFM*, **20**, 97–121.

547 Lorson, J. and Wagner, J. (2014) The pricing of hedging longevity risk with the help of annuity securitizations: An application to the German market. *J. Risk Finance*, **15** (4), 385–416.

548 Luciano, E. and Kast, R. (2001) A value at risk approach to background risk. *Geneva Papers on Risk and Insurance*, **26**, 91–115.

549 Luciano, E. and Schoutens, W. (2006) A multivariate jump-driven financial asset model. *Quantitative Finance*, **6** (5), 385–402.

550 Ludkovski, M. and Rüschendorf, L. (2008) On comonotonicity of Pareto optimal risk sharing. *Statistics & Probability Letters*, **78** (10), 1181–1188.

551 Lundberg, O. (1940) *On Random Processes and their Application to Sickness and Accident Statistics*, University of Stockholm thesis, Uppsala.

552 Lüthy, H. (1989) Die Entwicklung von Rentnerbeständen als stochastischer Prozess. *Blätter der DGVFM*, **19**, 29–45.

553 L'Ecuyer, P. and Lemieux, C. (2005) Recent advances in randomized quasi-Monte Carlo methods, in *Modeling Uncertainty*, Springer, pp. 419–474.

554 Maccaferri, S., Carboni, J., and Campolongo, F. (2012) Natural catastrophes: risk relevance and insurance coverage in the EU. *JRC67329. Joint Research Centre, Ispra, Italy*.

555 Mack, T. (1997) *Schadenversicherungsmathematik*, Verlag Versicherungswirtschaft e.V., Karlsruhe.

556 Mack, T. and Fackler, M. (2003) Exposure rating in liability reinsurance. *Blätter der DGVFM*, **26**, 229–238.

557 Mahul, O. and Wright, B.D. (2004) Implications of incomplete performance for optimal insurance. *Economica*, **71** (284), 661–670.

558 Mai, J.F. and Scherer, M. (2012) *Simulating Copulas: Stochastic Models, Sampling Algorithms and Applications*, vol. 4, World Scientific.

559 Major, J.A. (2009) The firm-value risk model. *Available at SSRN 2610675*.

560 Malamud, S., Rui, H., and Whinston, A. (2016) Optimal reinsurance with multiple tranches. *J. Mathematical Econ.*, **65**, 71–82.

561 Mango, D., Major, J., Adler, A., and Bunick, C. (2013) Capital tranching: A raroc approach to assessing reinsurance cost effectiveness. *Variance*, 7 (1), 82–91.

562 Mata, A. (2000) Pricing excess of loss reinsurance with reinstatements. *Astin Bull.*, **30**, 349–368.

563 Mata, A. (2004) Burning cost, in *Encyclopedia of Actuarial Science*, John Wiley, New York.

564 Matulla, C., Schöner, W., Alexandersson, H., Von Storch, H., and Wang, X. (2008) European storminess: late nineteenth century to present. *Climate Dynamics*, **31** (2-3), 125–130.

565 Mayuzumi, T. (1995) How to cope with wind storms. Simulation to search for the most efficient way to stabilize the insurance company management using a mix of reinsured and reserve for catastrophe loss in Japan. *Trans. 25th Intern. Congress Actuaries*, pp. 225–262.

566 McCullagh, P. and Nelder, J.A. (1989) *Generalized Linear Models*, Monographs on Statistics and Applied Probability, Chapman & Hall, London.

567 McNeil, A. (1997) Estimating the tails of loss severity distributions using extreme value theory. *Astin Bull.*, **27**, 117–137.

568 McNeil, A.J., Frey, R., and Embrechts, P. (2015) *Quantitative Risk Management*, Princeton Series in Finance, Princeton University Press, Princeton, NJ, revised edn.

569 Meier, U.B. and Outreville, J. (2006) Business cycles in insurance and reinsurance: the case of France, Germany and Switzerland. *J. Risk Finance*, 7 (2), 160–176.

570 Meier, U.B. and Outreville, J. (2010) Business cycles in insurance and reinsurance: international diversification effects. *Applied Financial Economics*, **20** (8), 659–668.

571 Meng, H., Siu, T.K., and Yang, H. (2016) Optimal insurance risk control with multiple reinsurers. *J. Computational Appl. Math.*, **306**, 40–52.

572 Merz, B., Kreibich, H., Schwarze, R., and Thieken, A. (2010) Assessment of economic flood damage. *Natural Hazards and Earth System Sciences*, **10** (8), 1697–1724.

573 Merz, B., Kreibich, H., Thieken, A., and Schmidtke, R. (2004) Estimation uncertainty of direct monetary flood damage to buildings. *Natural Hazards and Earth System Science*, **4** (1), 153–163.

574 Meyers, G. (1995) Managing the catastrophic risk. *CAS Discussion Papers*, pp. 111–150.

575 Meyers, G. and Koller, J. (1999) Catastrophe risk securitization - Insurer and investor perspectives. *CAS Discussion Papers*, pp. 1–48.

576 Meyers, G.G. (2007) The common shock model for correlated insurance losses. *Variance*, **1** (2007), 40–52.

577 Mikosch, T. (2009) *Non-Life Insurance Mathematics*, Universitext, Springer-Verlag, Berlin, 2nd edn.

578 Miljkovic, T. and Grün, B. (2016) Modeling loss data using mixtures of distributions. *Insurance Math. Econom.*, **70**, 387–396.

579 Mohr, S., Kunz, M., and Geyer, B. (2015) Hail potential in Europe based on a regional climate model hindcast. *Geophysical Research Letters*, **42** (24), 904–912.

580 Møller, T. (2004) Stochastic orders in dynamic reinsurance markets. *Finance and Stochastics*, **8** (4), 479–499.

581 Monahan, J. (2012) The individual risk assessment of terrorism. *Psychology, Public Policy, and Law*, **18** (2), 167.

582 Muermann, A. (2003) Actuarially consistent valuation of catastrophe derivatives. *The Wharton Financial Institution Center Working Paper Series*, pp. 3–18.

583 Muermann, A. (2008) Market price of insurance risk implied by catastrophe derivatives. *North American Actuarial Journal*, **12** (3), 221–227.

584 MunichRe (2001) *Risk Transfer to the Capital Markets: Using the Capital Markets in Insurance Risk Management*, Munich Re ART Solutions, Munich.

585 MunichRe (2010) *Reinsurance: a Basic Guide to Facultative and Treaty Reinsurance*, Munich Reinsurance America, Princeton, NJ.

586 Mürmann, A. (2004) Catastrophe derivatives, in *Encyclopedia of Actuarial Science*, John Wiley and Sons, Chichester.

587 Naveau, P., Huser, R., Ribereau, P., and Hannart, A. (2016) Modeling jointly low, moderate, and heavy rainfall intensities without a threshold selection. *Water Resources Research*, **52**, 2753–2769.

588 Neftci, S. (2000) Value at risk calculation, extreme events, and tail estimation. *J. Derivatives*, **7**, 23–37.

589 Nelsen, R. (1999) *An Introduction to Copulas*, Lecture Notes in Statistics 139, Springer, Berlin.

590 Niederreiter, H. (1992) *Random Number Generation and Quasi-Monte Carlo Methods*, Society for Industrial and Applied Mathematics, Philadelphia, Pennsylvania.

591 Niedrig, T. and Gründl, H. (2015) The effects of contingent convertible (coco) bonds on insurers' capital requirements under solvency ii. *The Geneva Papers on Risk and Insurance Issues and Practice*, **40** (3), 416–443.

592 Niehaus, G. (2002) The allocation of catastrophe risk. *J. Banking & Finance*, **26** (2), 585–596.

593 O'Brien, T. (1997) Hedging strategies using catastrophe insurance options. *Insurance Math. Econom.*, **21**, 153–162.

594 Ohlin, J. (1969) On a class of measures of dispersion with application to optimal reinsurance. *Astin Bull.*, **5** (2), 249–266.

595 Omey, E. and Teugels, J. (2002) Weighted renewal functions: a hierarchical approach. *Adv. Appl. Probab.*, **34**, 394–415.

596 Omey, E. and Willekens, E. (1986) Second order behavior of the tail of a subordinated probability distribution. *Stoch. Proc. Appl.*, **21**, 339–353.

597 Omey, E. and Willekens, E. (1987) Second order behaviour of distributions subordinate to a distribution with finite mean. *Comm. Statist., Stoch. Models*, **3**, 311–342.

598 Outreville, J. (1998) Size and concentration patterns of the world's largest reinsurance companies. *Geneva Papers on Risk and Insurance*, **23**, 68–80.

599 Outreville, J.F. (2012) The world's largest reinsurance groups: A look at names, numbers and countries from 1980 to 2010. *Insurance and Risk Management*, **80**, 137–156.

600 Panjer, H. (1980) The aggregate claims distribution and stop-loss reinsurance. *Trans. 21th Intern. Congress Actuaries*, pp. 523–545.

601 Panjer, H. (1981) Recursive evaluation of a family of compound distributions. *Astin Bull.*, **12**, 22–26.

602 Panjer, H. and Willmot, G. (1984) Models for the distribution of aggregate claims in risk theory. *Trans. Society Actuaries*, **36**, 399–446.

603 Panning, W.H. (2013) Managing the invisible: Identifying value-maximizing combinations of risk and capital. *North American Actuarial Journal*, **17** (1), 13–28.

604 Papush, D., Patrik, G., and Podgaits, F. (2001) Approximations of the aggregate loss distribution, in *CAS Forum*, Winter, pp. 175–186.

605 Park, S.C. and Xie, X. (2014) Reinsurance and systemic risk: The impact of reinsurer downgrading on property–casualty insurers. *J. Risk and Insurance*, **81** (3), 587–622.

606 Parodi, P. (2014) *Pricing in General Insurance*, CRC Press.

607 Patrat, C. and Huygues-Beaufond, C. (1993) Rate making for natural events coverage in the USA. *SCOR Notes*.

608 Paudel, Y. (2012) A comparative study of public-private catastrophe insurance systems: Lessons from current practices. *The Geneva Papers on Risk and Insurance Issues and Practice*, **37** (2), 257–285.

609 Pechlivanides, P. (1978) Optimal reinsurance and dividend payment strategies. *Astin Bull.*, **10**, 34–46.

610 Peng, L. (2014) Joint tail of ECOMOR and LCR reinsurance treaties. *Insurance Math. Econom.*, **58**, 116–120.

611 Pentikäinen, T. (1987) Approximate evaluation of the distribution function of aggregate claims. *Astin Bull.*, **17**, 15–39.

612 Pentikäinen, T. (1996) Credit insurance, risk of catastrophic periods. *XXVII Astin Colloquium*, pp. 95–106.

613 Pesenti, S.M., Millossovich, P., and Tsanakas, A. (2016) Robustness regions for measures of risk aggregation. *Dependence Modelling*, **4**, 338–367.

614 Pesonen, M. (1984) Optimal reinsurances. *Scand. Actuar. J.*, pp. 65–90.

615 Pflug, G.C. and Römisch, W. (2007) *Modeling, Measuring and Managing Risk*, vol. 190, World Scientific.

616 Phillips, R.D., Cummins, J.D., and Allen, F. (1998) Financial pricing of insurance in the multiple-line insurance company. *J. Risk and Insurance*, **65** (4), 597–636.

617 Picard, P. (2000) On the design of optimal insurance policies under manipulation of audit cost. *International Economic Review*, **41** (4), 1049–1071.

618 Pigeon, M. and Denuit, M. (2011) Composite lognormal–Pareto model with random threshold. *Scand. Actuar. J.*, pp. 177–192.

619 Pinelis, I. (1985) Asymptotic equivalence of the probabilities of large deviations for sums and maximum of independent random variables (Russian), in *Limit Theorems of Probability Theory, Trudy Inst. Mat.*, vol. 5, "Nauka" Sibirsk. Otdel., Novosibirsk, pp. 44–173, 176.

620 Pita, G.L., Pinelli, J.P., Gurley, K.R., and Hamid, S. (2013) Hurricane vulnerability modeling: Development and future trends. *Journal of Wind Engineering and Industrial Aerodynamics*, **114**, 96–105.

621 Pitera, M. and Schmidt, T. (2016) Unbiased estimation of risk. *Preprint, available as arXiv:1603.02615*.

622 Pitman, E. (1980) Subexponential distribution functions. *J. Austral. Math. Soc.*, **29**, 337–347.

623 Planchet, F. (2013) Modélisation du risque de pandémie dans solvabilité 2. *Insurance and Risk Management*, **81**.

624 Plantin, G. (2006) Does reinsurance need reinsurers? *J. Risk and Insurance*, **73** (1), 153–168.

625 Powers, M. (2003) Leapfrogging the variance: The financial management of extreme event risk. *J. Risk Finance*, **4**, 26–39.

626 Powers, M. and Shubik, M. (2001) Toward a theory of reinsurance and retrocession. *Insurance Math. Econom.*, **29**, 271–290.

627 Powers, M.R. and Shubik, M. (2006) A "square-root rule" for reinsurance. *Revista Contabilidade & Finanças*, **17** (2), 101–107.

628 Preischl, M. (2016) Bounds on integrals with respect to multivariate copulas. *Dependence Modelling*. In press.

629 Preischl, M., Thonhauser, S., and Tichy, R. (2016) Integral equations, quasi-Monte Carlo methods and risk modeling. *Preprint, Graz University of Technology*.

630 Pressacco, F. (1979) Values and prices in a reinsurance market. *Astin Bull.*, **10**, 263–273.

631 Prettenthaler, F. and Albrecher, H. (2009) *Hochwasser und dessen Versicherung in Oesterreich*, Studien zum Klimawandel, Verlag der Oesterreichischen Akademie der Wissenschaften, Wien.

632 Prettenthaler, F., Albrecher, H., Asadi, P., and Köberl, J. (2017) On flood risk pooling in Europe. *Natural Hazards*, **88**(1), 1–20.

633 Prettenthaler, F., Albrecher, H., Köberl, J., and Kortschak, D. (2012) Risk and insurability of storm damages to residential buildings in Austria. *The Geneva Papers on Risk and Insurance - Issues and Practice*, **37**, 340–364.

634 Promislow, S. and Young, V. (2005) Unifying framework for optimal insurance. *Insurance Math. Econom.*, **36** (3), 347–364.

635 Puccetti, G. and Rüschendorf, L. (2012) Computation of sharp bounds on the distribution of a function of dependent risks. *J. Computational Appl. Math.*, **236** (7), 1833–1840.

636 Puccetti, G., Rüschendorf, L., and Manko, D. (2016) Var bounds for joint portfolios with dependence constraints. *Dependence Modelling*, **4**, 368–381.

637 Punter, A. (2002) Reinventing re/insurance for the twenty-first century. *Geneva Papers on Risk and Insurance*, **27**, 102–112.

638 Radtke, M., Schmidt, K.D., and Schnaus, A. (2016) *Handbook of Loss Reserving*, EAA Series, Springer, Heidelberg.

639 Ramachandran, G. (1974) Extreme value theory and large fire losses. *Astin Bull.*, **7**, 293–310.

640 Ramlau-Hansen, H. (1988) A solvency study in non-life insurance. Part I. Analyses of fire, windstorm and glass claims. *Scand. Actuar. J.*, pp. 3–34.

641 Raviv, A. (1979) The design of an optimal insurance policy. *The American Economic Review*, **69** (1), 84–96.

642 Reich, A. (1985) Eine Charakterisierung des Standardabweichungsprinzips. *Blätter der DGVFM*, **17**, 93–102.

643 Reijnen, R., Albers, W., and Kallenberg, W.C. (2005) Approximations for stop-loss reinsurance premiums. *Insurance Math. Econom.*, **36** (3), 237–250.

644 Reiss, R.D. and Thomas, M. (1999) A new class of Bayesian estimators in Paretian excess-of-loss reinsurance. *Astin Bull.*, **29**, 339–349.

645 Reiss, R.D. and Thomas, M. (2001) *Statistical Analysis of Extreme Values. From Insurance, Finance, Hydrology and Other Fields*, Birkhäuser, Basel, 2nd edn.

646 Resnick, S. (1997) Discussion of the Danish data on large fire insurance losses. *Astin Bull.*, **27**, 139–151.

647 Reynkens, T., Verbelen, R., Beirlant, J., and Antonio, K. (2016) Global fits using splicing of mixtures of Erlangs and extreme value distributions. *Preprint, K.U. Leuven.*

648 Riebesell, P. (1936) *Einführung in die Sachversicherungsmathematik*, Berlin.

649 Riegel, U. (2008) Generalizations of common ILF models. *Blätter der DGVFM*, **29** (1), 45–71.

650 Riegel, U. (2010) On fire exposure rating and the impact of the risk profile type. *Astin Bull.*, **40** (2), 727–777.

651 Robert, C.Y. and Segers, J. (2008) Tails of random sums of a heavy-tailed number of light-tailed terms. *Insurance Math. Econom.*, **43** (1), 85–92.

652 Rolski, T., Schmidli, H., Schmidt, V., and Teugels, J. (1999) *Stochastic Processes for Insurance and Finance*, Wiley Series in Probability and Statistics, John Wiley & Sons Ltd., Chichester.

653 Romera, R. and Runggaldier, W. (2012) Ruin probabilities in a finite-horizon risk model with investment and reinsurance. *J. Appl. Probability*, **49** (4), 954–966.

654 Rootzén, H. and Tajvidi, N. (2006) Multivariate generalized Pareto distributions. *Bernoulli*, **12** (5), 917–930.

655 Rose, A. and Huyck, C.K. (2016) Improving catastrophe modeling for business interruption insurance needs. *Risk Analysis*, **36** (10), 1896–1915.

656 Rose, A. and Lim, D. (2002) Business interruption losses from natural hazards: conceptual and methodological issues in the case of the Northridge earthquake. *Global Environmental Change Part B: Environmental Hazards*, **4** (1), 1–14.

657 Rosiński, J. and Samorodnitsky, G. (1993) Distributions of subadditive functionals of sample paths of infinitely divisible processes. *Ann. Probab.*, **21**, 996–1014.

658 Rudolph, C. (2014) A generalization of Panjer's recursion for dependent claim numbers and an approximation of Poisson mixture models, PhD Thesis, Vienna University of Technology.

659 Runnenberg, J. and Goovaerts, M. (1985) Bounds on compound distributions and stop-loss premiums. *Insurance Math. Econom.*, **4**, 287–293.

660 Ruohonen, M. (1988) On a model for the claim number process. *Astin Bull.*, **18**, 57–68.

661 Ryder, J. and Kimber, D. (1977) Earthquakes and windstorm - Natural disasters. *Astin Bull.*, **9**, 317–324.

662 Rymaszewski, P., Schmeiser, H., and Wagner, J. (2012) Under what conditions is an insurance guaranty fund beneficial for policyholders? *J. Risk and Insurance*, **79** (3), 785–815.

663 Rytgaard, M. (1996) On calculating the risk premium for an excess-of-loss cover with an annual aggregate deductible and a limited number of reinstatements. *Colloq. XXVII Astin*, pp. 82–94.

664 Samson, D. and Thomas, H. (1985) Decision analysis models in reinsurance. *European J. Operations Res.*, **19**, 201–211.

665 Sanders, D. (1995) When the wind blows: An introduction to catastrophe excess-of-loss reinsurance, in *CAS Forum*, vol. Fall, pp. 157–228.

666 Sato, K. (1999) *Lévy Processes and Infinite Divisibility*, Cambridge University Press Cambridge.

667 Scarrott, C. and MacDonald, A. (2012) A review of extreme value threshold estimation and uncertainty quantification. *REVSTAT–Statistical Journal*, **10** (1), 33–60.

668 Schäl, M. (1998) On piecewise deterministic Markov control processes: control of jumps and of risk processes in insurance. *Insurance Math. Econom.*, **22** (1), 75–91.

669 Schäl, M. (2004) On discrete-time dynamic programming in insurance: exponential utility and minimizing the ruin probability. *Scand. Actuar. J.*, pp. 189–210.

670 Scherer, M., Schmid, L., and Schmidt, T. (2012) Shot-noise driven multivariate default models. *Eur. Actuar. J.*, **2** (2), 161–186.

671 Schlütter, S. and Gründl, H. (2012) Who benefits from building insurance group? A welfare analysis of optimal group capital management. *The Geneva Papers on Risk and Insurance-Issues and Practice*, **37** (3), 571–593.

672 Schmeiser, H., Wagner, J., and Zemp, A. (2014) Proposal for a capital market-based guaranty scheme for the financial industry. *Eur. J. Finance*, **20** (12), 1133–1160.

673 Schmidli, H. (1996) Lundberg inequalities for a Cox model with a piecewise constant intensity. *J. Appl. Prob.*, **33** (1), 196–120.

674 Schmidli, H. (2001) Optimal proportional reinsurance policies in a dynamic setting. *Scand. Actuar. J.*, pp. 55–68.

675 Schmidli, H. (2002) On minimizing the ruin probability by investment and reinsurance. *Annals of Applied Probability*, **12** (3), 890–907.

676 Schmidli, H. (2008) *Stochastic Control in Insurance*, Probability and its Applications, Springer, London.

677 Schmidt, K. (1995) *Lectures on Risk Theory*, Teubner, Berlin.

678 Schmidt, T. (2014) Catastrophe insurance modeled by shot-noise processes. *Risks*, **2** (1), 3–24.

679 Schmitter, H. (1987) Eine optimale Kombination von proportionalem und nichtproportionalem Selbstbehalt. *Mitt. Ver. Schweiz. Versich. Math.*, pp. 229–236.

680 Schmock, U. (1999) Estimating the value of the WINCAT coupons of the Winterthur insurance convertible bond: a study of the model risk. *Astin Bull.*, **29**, 101–163.

681 Schnieper, R. (1993) The insurance of catastrophe risks. *SCOR Notes.*

682 Schnieper, R. (1993) Praktische Erfahrungen mit Grossschadenverteilungen. *Mitt. Ver. Schweiz. Versich. Math.*, pp. 149–165.

683 Schoutens, W. (2003) *Lévy Processes in Finance*, Wiley, New York.

684 Schröter, K. (1990) On a family of counting distributions and recursions for related distributions. *Scand. Actuar. J.*, pp. 161–175.

685 Schweizer, M. (2001) From actuarial to financial valuation principles. *Insurance Math. Econom.*, **28** (1), 31–47.

686 Schwepcke, A. (2004) *Rückversicherung*, Verlag Versicherungswirtschaft, Karlsruhe.

687 Scollnik, D.P. (2007) On composite lognormal-Pareto models. *Scand. Actuar. J.*, pp. 20–33.

688 Scollnik, D.P. and Sun, C. (2012) Modeling with Weibull-Pareto models. *North American Actuarial Journal*, **16** (2), 260–272.

689 SCOR (2008) From principle-based risk management to solvency requirements. *Swiss solvency test documentation, Zurich.*
690 Seal, H. (1969) *Stochastic Theory of a Risk Business*, John Wiley, New York.
691 Seal, H. (1974) The numerical calculation of $u(w, t)$ the probability of non-ruin in an interval $(0, t)$. *Scand. Actuar. J.*, pp. 121–139.
692 Seal, H. (1978) From aggregate claims distribution to the probability of ruin. *Astin Bull.*, **10**, 47–53.
693 Seal, H. (1978) Survival probabilities; the goal of risk theory, *Tech. Rep.*.
694 Seal, H. (1980) Survival probabilities based on Pareto claim distributions. *Astin Bull.*, **11**, 61–71.
695 Selch, D.A. (2016) *A Multivariate Cox Process with Simultaneous Jump Arrivals and its Application in Insurance Modelling*, Ph.D. thesis, Technische Universität München.
696 Shaked, M. and Shanthikumar, J. (1994) *Stochastic Orders and Their Applications*, Academic Press, Boston, MA.
697 Sherris, M. (2003) Economic valuation: Something old, something new. *Austr. Actuar. J.*, **9** (4), 625–665.
698 Shi, P., Feng, X., and Ivantsova, A. (2015) Dependent frequency–severity modeling of insurance claims. *Insurance Math. Econom.*, **64**, 417–428.
699 Shumway, R.H. (1988) *Applied Time Series Analysis*, Prentice-Hall, Englewood Cliffs, NJ.
700 Sichel, H. (1971) On a family of discrete distributions particularly suited to represent long tailed frequency data, in *Proceedings of the 3rd Symposium on Mathematics and Statistics, Pretoria, CSIR*, pp. 51–97.
701 Silvestrov, D. and Teugels, J. (1999) Limit theorems for extremes with random sample size. *Adv. Appl. Probab.*, **30**, 777–806.
702 Siu, T. and Yang, H. (1999) Subjective risk measures: Bayesian predictive scenarios analysis. *Insurance Math. Econom.*, **25**, 157–169.
703 Smith, R., Canelo, E., and Di Dio, A. (1997) Reinventing reinsurance using the capital markets. *Geneva Papers on Risk and Insurance*, **22**, 26–37.
704 Smith, W. (1972) *On the tails of queueing-time distributions*, Mimeo Series No. 830, Department of Statistics, University North-Carolina, Chapel Hill.
705 Sobol', I.M. (1967) On the distribution of points in a cube and the approximate evaluation of integrals. *USSR Comput. Math. Math. Phys.*, **7**, 86–112.
706 Sondermann, D. (1991) Reinsurance in arbitrage-free markets. *Insurance Math. Econom.*, **10**, 191–202.
707 Sparre Andersen, E. (1957) On the collective theory of risk in case of contagion between claims. *Trans. 15th Intern. Congress Actuaries*, **II**, 219–229.
708 Stanard, J. and John, R. (1990) Evaluating the effect of reinsurance contract terms. *Proc. Casualty Actuarial Soc.*, **77** (1), 1–41.
709 Steenackers, A. and Goovaerts, M. (1992) Optimal reinsurance from the viewpoint of the cedent, in *Proc. Intern. Congress Actuaries*, vol. 2, pp. 271–300.
710 Strain, R. (1983) *Reinsurance*, The College of Insurance, New York.
711 Straub, E. (1988) *Non-Life Insurance Mathematics*, Springer-Verlag.
712 Strickler, P. (1960) Rückversicherung des Kumulrisikos in der Lebensversicherung, in *Transactions of the Sixteenth International Congress of Actuaries*, vol. 1, Cambridge University Press.

713 Stupfler, G. *et al.* (2013) A moment estimator for the conditional extreme-value index. *Electronic Journal of Statistics*, **7**, 2298–2343.
714 Subramanian, A. and Wang, J. (2015) Catastrophe risk transfer. *Preprint, available at SSRN 2321415.*
715 Suijs, J., De Waegenaere, A., and Borm, P. (1998) Stochastic cooperative games in insurance. *Insurance Math. Econom.*, **22**, 209–228.
716 Sundt, B. (1991) *An Introduction to Non-Life Insurance Mathematics*, Verlag Versicherungswirtschaft e.V., Karlsruhe, Second edn.
717 Sundt, B. (1991) On asymptotic rates on line in excess of loss reinsurance. *Insurance Math. Econom.*, **10**, 61–67.
718 Sundt, B. (1991) On excess of loss reinsurance with reinstatements. *Schweiz. Verein. Versicherungsmath. Mitt.*, (1), 51–66.
719 Sundt, B. (1992) On allocation of excess-of-loss premiums. *Astin Bull.*, **22**, 167–176.
720 Sundt, B. (2002) Recursive evaluation of aggregate claims distributions. *Insurance Math. Econom.*, **30**, 297–322.
721 Sundt, B. and Jewell, W. (1981) Further results of recursive evaluation of compound distributions. *Astin Bull.*, **12**, 27–39.
722 Sundt, B. and Vernic, R. (2009) *Recursions for Convolutions and Compound Distributions with Insurance Applications*, Springer Science & Business Media.
723 Swiss Re (1999) Alternative risk transfer (ART) for corporations: a passing fashion or risk management for the 21st century? *Sigma*, **2**.
724 Swiss Re (2007) *Pandemic influenza: A 21st century model for mortality shocks*, Technical publishing, life and health. SwissRe.
725 Swiss Re (2013) The essential guide to reinsurance.
726 Tan, K.S. and Weng, C. (2014) Empirical approach for optimal reinsurance design. *North American Actuarial Journal*, **18** (2), 315–342.
727 Tan, K.S., Weng, C., and Zhang, Y. (2009) VaR and CTE criteria for optimal quota-share and stop-loss reinsurance. *North American Actuarial Journal*, **13** (4), 459–482.
728 Tan, K.S., Weng, C., and Zhang, Y. (2011) Optimality of general reinsurance contracts under CTE risk measure. *Insurance Math. Econom.*, **49** (2), 175–187.
729 Taylor, G. (1978) An investigation of the use of weighted averages in the estimation of the mean of a long-tailed size distribution. *Astin Bull.*, **10**, 87–98.
730 Taylor, G. (1982) Estimation of outstanding reinsurance recoveries on the basis of incomplete information. *Insurance Math. Econom.*, **1**, 3–11.
731 Taylor, G. (1985) A heuristic review of some ruin theory results. *Astin Bull.*, **15**, 73–88.
732 Taylor, G. (1992) Risk exchange I: A unification of some existing results. *Scand. Actuar. J.*, pp. 15–39.
733 Taylor, G. (1992) Risk exchange II: Optimal reinsurance contracts. *Scand. Actuar. J.*, pp. 40–59.
734 Taylor, G. (1997) Reserving consecutive layers of inwards excess-of-loss reinsurance. *Insurance Math. Econom.*, **20**, 225–242.
735 Ter Berg, P. (1984) An outline of the power-ratio-gamma distribution. *Proc. 4 countries Astin Coll.*, pp. 243–250.
736 Ter Berg, P. (1994) Deductibles and the inverse Gaussian distribution. *Astin Bull.*, **24**, 319–323.

737 Teugels, J. (1975) The class of sub-exponential distributions. *Ann. Probability*, **3**, 1000–1011.
738 Teugels, J. (1981) Remarks on large claims. *Bull. Inst. Internat. Statist.*, **49**, 1490–1500.
739 Teugels, J. (1985) Approximation and estimation of some compound distributions. *Insurance Math. Econom.*, **4**, 143–153.
740 Teugels, J. and Willmot, G. (1987) Approximations for stop-loss premiums. *Insurance Math. Econom.*, **6**, 195–202.
741 Teugels, J.L. and Sundt, B. (eds) (2004) *Encyclopedia of Actuarial Science*, 3 vols, John Wiley & Sons, Hoboken, NJ.
742 Thépaut, A. (1950) Une nouvelle forme de réassurance. le traité d'excédent du coût moyen relatif (ECOMOR). *Bull. Trimestriel Inst. Actuair. Français*, **49**, 273.
743 Thomas, R.L. (2012) Underwriting terrorism risk. *J. Civil Rights and Economic Development*, **18** (2), 8.
744 Thyrion, P. (1969) Extension of the collective risk theory. *Scand. Actuar. J.*, pp. 84–98.
745 Tijms, H.C. (1994) *Stochastic Models: An Algorithmic Approach*, vol. 303, John Wiley & Sons Inc.
746 Trufin, J., Albrecher, H., and Denuit, M.M. (2011) Properties of a risk measure derived from ruin theory. *The Geneva Risk and Insurance Review*, **36** (2), 174–188.
747 Turnbull, B.W. (1976) The empirical distribution function with arbitrarily grouped, censored and truncated data. *J. Roy. J. R. Stat. Soc. Ser. B Stat. Methodol.*, **38** (3), 290–295.
748 Upreti, V. and Adams, M. (2015) The strategic role of reinsurance in the United Kingdom's (UK) non-life insurance market. *J. Banking & Finance*, **61**, 206–219.
749 Vajda, S. (1962) Minimum variance reinsurance. *Astin Bull.*, **2** (2), 257–260.
750 Van Broekhoven, H., Alm, E., Tuominen, T., Hellman, A., and Dziworski (2006) Actuarial reflections on pandemic risk and its consequences. *Groupe Consultatif Actuariel Report.*
751 Van der Hoek, J. and Sherris, M. (2001) A class of non-expected utility risk measures and implications for asset allocations. *Insurance Math. Econom.*, **28**, 69–82.
752 Van Heerwaarden, A., Kaas, R., and Goovaerts, M. (1989) Optimal reinsurance in relation to ordering of risks. *Insurance Math. Econom.*, **8**, 11–17.
753 Vaugirard, V.E. (2003) Valuing catastrophe bonds by Monte Carlo simulations. *Applied Math. Fin.*, **10** (1), 75–90.
754 Venter, G. (1991) Premium calculation implications of reinsurance without arbitrage. *Astin Bull.*, **21**, 223–230.
755 Venter, G. (2001) Measuring value in reinsurance, in *CAS Forum*, Summer, pp. 179–199.
756 Venter, G., Barnett, J., and Owen, M. (2004) Market value of risk transfer: Catastrophe reinsurance case, in *International AFIR Colloquium.*
757 Verbelen, R., Antonio, K., and Claeskens, G. (2015) Multivariate mixtures of Erlangs for density estimation under censoring. *Lifetime Data Analysis*, **22** (3), 1–27.
758 Verbelen, R., Gong, L., Antonio, K., Badescu, A., and Lin, S. (2015) Fitting mixtures of Erlangs to censored and truncated data using the EM algorithm. *Astin Bull.*, **45** (3), 729–758.
759 Verlaak, R. (2001) *Optimal Reinsurance Programs.* Master's thesis, Katholieke Universiteit Leuven, Belgium.

760 Verlaak, R. and Beirlant, J. (2003) Optimal reinsurance programs: an optimal combination of several reinsurance protections on a heterogeneous insurance portfolio. *Insurance Math. Econom.*, **33** (2), 381–403.

761 Verlaak, R., Hürlimann, W., and Beirlant, J. (2009) Quasi-likelihood estimation of benchmark rates for excess of loss reinsurance programs. *Astin Bull.*, **39** (2), 429–452.

762 Von Dahlen, S. (2007) Finite reinsurance: How does it concern supervisors? Some efficiency considerations in the light of prevailing regulatory aims. *The Geneva Papers on Risk and Insurance-Issues and Practice*, **32** (3), 283–300.

763 Wacek, M. (2013) Discussion on "capital tranching: A raroc approach to assessing reinsurance cost effectiveness". *Variance*, **7** (2), 101–109.

764 Wakuri, M. and Yasuhara, Y. (1977) Earthquake insurance in Japan. *Astin Bull.*, **9**, 329–364.

765 Walhin, J. (2012) *La Reassurance*, Larcier, 2nd edn.

766 Walhin, J., Herfurth, L., and De Longueville, P. (2001) The practical pricing of excess of loss treaties: actuarial, financial, economic and commercial apects. *Belgian Actuarial Bulletin*, **1** (1), 1–17.

767 Walhin, J. and Paris, J. (2000) The effect of excess of loss reinsurance with reinstatements on the cedent's portfolio. *Blätter der DGVFM*, **24**, 616–627.

768 Walhin, J. and Paris, J. (2001) Excess of loss reinsurance with reinstatements: Premium calculation and ruin probability for the cedent. *Blätter der DGVFM*, **25**, 257–270.

769 Wang, S. (1995) Insurance pricing and increasing limits ratemaking by proportional hazards transforms. *Insurance Math. Econom.*, **17**, 43–54.

770 Wang, S. (1996) Premium calculation by transforming the layer premium density. *Astin Bull.*, **26**, 71–92.

771 Wang, S. (1997) Implementation of PH-transforms in ratemaking, in *CAS Forum*, Winter, pp. 291–316.

772 Wang, S. (1998) Aggregation of correlated risk portfolios: models and algorithms, in *Proceedings of the Casualty Actuarial Society*, vol. 85, pp. 848–939.

773 Wang, S., Young, V., and Panjer, H. (1997) Axiomatic characterization of insurance prices. *Insurance Math. Econom.*, **21** (2), 173–183.

774 Wang, X. and Fang, K.T. (2003) The effective dimension and quasi-Monte Carlo integration. *Journal of Complexity*, **19** (2), 101–124.

775 Waters, H. (1983) Some mathematical aspects of reinsurance. *Insurance Math. Econom.*, **2**, 17–26.

776 Watson Jr, C.C. and Johnson, M.E. (2004) Hurricane loss estimation models. *Bulletin of the American Meteorological Society*, **85** (11), 1713.

777 Weissman, I. (1978) Estimation of parameters and large quantiles based on the k largest observations. *Journal of the American Statistical Association*, **73** (364), 812–815.

778 Weng, C. and Zhuang, S.C. (2017) CDF formulation for solving an optimal reinsurance problem. *Scand. Actuar. J.*, pp. 395–418.

779 Willekens, E. (1988) The structure of the class of sub-exponential distributions. *Probab. Theory Relat. Fields*, **77**, 567–581.

780 Willekens, E. and Teugels, J.L. (1992) Asymptotic expansions for waiting time probabilities in an $M/G/1$ queue with long-tailed service time. *Queueing Systems Theory Appl.*, **10** (4), 295–311.

781 Williams, R. (1979) Computing the probability density function for aggregate claims. *Canadian Institute of Actuaries*, pp. 1–9.

782 Willmot, G. (1986) Mixed compound Poisson distributions. *Astin Bull.*, **16**, S59–S79.

783 Willmot, G. (1987) The Poisson-inverse Gaussian as an alternative to the negative binomial. *Scand. Actuar. J.*, pp. 113–127.

784 Willmot, G. (1988) Sundt and Jewell's family of discrete distributions. *Astin Bull.*, **18**, 17–29.

785 Willmot, G. (1989) The total claims distributions under inflationary conditions. *Scand. Actuar. J.*, pp. 1–12.

786 Willmot, G. (1990) Asymptotic tail behaviour of Poisson mixtures with applications. *Adv. Appl. Probability*, **22**, 147–159.

787 Willmot, G. and Panjer, H. (1987) Difference equation approaches in evaluation on compound distributions. *Insurance Math. Econom.*, **6**, 43–56.

788 Willmot, G. and Sprott, D. (1994) A note on parameters orthogonal to the mean. *Biometrika*, **81** (2), 409–412.

789 Willmot, G. and Sundt, B. (1989) On evaluation of the Delaporte distribution and related distributions. *Scand. Actuar. J.*, pp. 101–113.

790 Willmot, G.E. and Woo, J.K. (2007) On the class of Erlang mixtures with risk theoretic applications. *North American Actuarial Journal*, **11** (2), 99–115.

791 Wilson, R. (1968) The theory of syndicates. *Econometrica*, **36** (1), 119–132.

792 Wilson, T. (2015) *Value and Capital Management*, John Wiley & Sons.

793 Wirch, J. and Hardy, M. (1999) A synthesis of risk measures for capital adequacy. *Insurance Math. Econom.*, **25**, 337–347.

794 Woo, G. (2011) *Calculating Catastrophe*, World Scientific.

795 Wood, S. (2006) *Generalized Additive Models: An Introduction with R*, CRC Press.

796 Worms, J. and Worms, R. (2014) New estimators of the extreme value index under random right censoring, for heavy-tailed distributions. *Extremes*, **17** (2), 337–358.

797 Wüthrich, M. and Merz, M. (2008) *Stochastic Claims Reserving Methods in Insurance*, Wiley, Chichester.

798 Xu, L., Bricker, D., and Kortanek, K. (1998) Bounds for stop-loss premium under restrictions on I-divergence. *Insurance Math. Econom.*, **23**, 119–139.

799 Yaari, M.E. (1987) The dual theory of choice under risk. *Econometrica*, **55** (1), 95–115.

800 Young, V. (1999) Optimal insurance under Wang's premium principle. *Insurance Math. Econom.*, **25**, 109–122.

801 Young, V. (2004) Premium calculation principles, in *Encyclopedia of Actuarial Science*, John Wiley, New York.

802 Zajdenweber, D. (1996) Extreme values in business interruption insurance. *J. Risk Insurance*, **63**, 95–110.

803 Zanjani, G. (2002) Pricing and capital allocation in catastrophe insurance. *J. Financial Economics*, **65** (2), 283–305.

804 Zecchin, M. (1987) Calculation of the maximum retentions in XL reinsurance. *Insurance Math. Econom.*, **6**, 169–178.

805 Zeckhauser, R. (1996) 19th Annual Lecture of the Geneva Association Insurance and Catastrophes. *Geneva Papers on Risk and Insurance*, **21**, 3–21.

806 Zhang, T. and Kou, S. (2010) Nonparametric inference of doubly stochastic Poisson process data via the kernel method. *Ann. Appl. Stat.*, **4** (4), 1913.

807 Zhang, Y. (2013) *Reinsurance Arrangements Minimizing the Total Required Capital*, Casualty Actuarial Society, E-Forum, Spring.

808 Zhou, C. and Wu, C. (2008) Optimal insurance under the insurer's risk constraint. *Insurance Math. Econom.*, **42** (3), 992–999.
809 Zhu, Y., Chi, Y., and Weng, C. (2014) Multivariate reinsurance designs for minimizing an insurer's capital requirement. *Insurance Math. Econom.*, **59**, 144–155.
810 Zhuang, S.C., Weng, C., Tan, K.S., and Assa, H. (2016) Marginal indemnification function formulation for optimal reinsurance. *Insurance Math. Econom.*, **67**, 65–76.
811 Ziegel, J.F. (2014) Coherence and elicitability. *Mathematical Finance*, **26**, 901–918.
812 Zweifel, P. and Eisen, R. (2012) *Insurance Economics*, Springer Science & Business Media.

Index

Reinsurance: Actuarial and Statistical Aspects, First Edition.
Hansjörg Albrecher, Jan Beirlant and Jozef L. Teugels.

e

f

i

k

l

m

r

s

Printed and bound by CPI Group (UK) Ltd, Croydon, CR0 4YY

27/10/2024

14580219-0003